GGE
Biplot Analysis
A Graphical Tool for Breeders, Geneticists, and Agronomists

GGE Biplot Analysis

A Graphical Tool for Breeders, Geneticists, and Agronomists

Weikai Yan • Manjit S. Kang

CRC PRESS

Boca Raton   London   New York   Washington, D.C.

Library of Congress Cataloging-in-Publication Data

Yan, Weikai.
  GGE biplot analysis : a graphical tool for breeders, geneticists, and agronomists / Weikai Yan and Manjit S. Kang.
      p.   cm.
    Includes bibliographical references (p.).
    ISBN 0-8493-1338-4 (alk. paper)
    1. Plant breeding--Statistical methods. 2. Crops--Genetics--Statistical methods. 3. Genotype-environmental interaction. I. Kang, Manjit S. II. Title.

SB123 .Y364 2002
631.5′2′072--dc21   2002067091

This book contains information obtained from authentic and highly regarded sources. Reprinted material is quoted with permission, and sources are indicated. A wide variety of references are listed. Reasonable efforts have been made to publish reliable data and information, but the author and the publisher cannot assume responsibility for the validity of all materials or for the consequences of their use.

Neither this book nor any part may be reproduced or transmitted in any form or by any means, electronic or mechanical, including photocopying, microfilming, and recording, or by any information storage or retrieval system, without prior permission in writing from the publisher.

The consent of CRC Press LLC does not extend to copying for general distribution, for promotion, for creating new works, or for resale. Specific permission must be obtained in writing from CRC Press LLC for such copying.

Direct all inquiries to CRC Press LLC, 2000 N.W. Corporate Blvd., Boca Raton, Florida 33431.

Trademark Notice: Product or corporate names may be trademarks or registered trademarks, and are used only for identification and explanation, without intent to infringe.

Visit the CRC Press Web site at www.crcpress.com

© 2003 by CRC Press LLC

No claim to original U.S. Government works
International Standard Book Number 0-8493-1338-4
Library of Congress Card Number 2002067091
Printed in the United States of America 1 2 3 4 5 6 7 8 9 0
Printed on acid-free paper

Foreword

"One picture is worth ten thousand words."

One of the most well-developed and acute faculties of the human brain is the ability to perceive, analyze, and interpret complex visual information. Patterns, or visual relationships, are the easiest to understand quickly. Comprehension of visual information is much better and faster than linear or tabular numerical information.

Extensive databases of complex information are being meticulously accumulated and maintained in many research programs, but they are not being utilized to their full potential because interpretation of them is too complicated. The principal component methodology and the GGE biplot software described in this book can change that by making interpretation of complex data sets as easy as a few keystrokes and a few minutes of viewing easy-to-understand graphic displays.

Using GGE biplot software requires minimal preparation of data — simple two-way data spreadsheets are the basic input format. The initial graphic output can be easily modified from drop-down menus. The data set being analyzed can be manipulated by removing various parameters and entries during the real-time massaging of the data set and different statistical analyses can be performed. Several alternative ways of viewing the output are possible to assist in interpretation of the relationships among the parameters, among the entries, and their interactions, and it is all done "live" in real time. Information can be analyzed faster, and can be understood more completely, using the GGE biplot software visual display than the data tables and graphic displays generated by most spreadsheets and database systems. The graphic output can be printed to produce a hardcopy version, and/or saved as an image and used in presentations or publications. A log file of the relevant statistics and correlation matrices is also produced and automatically saved for later detailed reference.

The GGE biplot can help anyone learn more about the interactions among a multitude of traits/parameters *in their own data* than ever before. GGE software has tremendous potential value to plant breeders, agronomists, pathologists, physiologists, nutritionists, and anyone working in an applied science field. It is currently being used in the cereal breeding program at the University of Guelph to evaluate overall agronomic merit, quality, genotype X environment interaction for numerous traits, genotype X trait interactions, and trait X environment interactions in breeding lines being advanced through the testing system, to select parents and parental combinations for crossing, to evaluate relationships among traits (especially quality), to identify determinants of yield and quality factors in the populations, and to assess the discriminating value and stability of various testing locations.

GGE biplots have been utilized in chromosome mapping of morphological and molecular markers, and to identify and locate QTL markers in doubled haploid barley mapping populations. The biplot concept is also used to determine the physiological associations of morphological traits and kernel characteristics with quality factors. This is especially valuable when both negative and positive correlations exist among several desired traits. Biplots can also be used to identify and quantify the effects of environment on one or more traits across a range of genotypes, and the effects the environment may be having on the relationships among those traits. This is particularly useful when selecting among elite lines to advance to more extensive and intensive trials, and in choosing parents for the next cycle of improvement in complex traits with interactions among the components.

Biplots have been used in evaluating students' performance in a course to determine which assignments and exercises are most likely to contribute to their understanding of concepts, based on association with performance on later examinations.

Essentially, any two-way data set with multiple entries and multiple parameters can be analyzed using GGE biplot; the result will be a better understanding of the interactions among entries, among parameters, and interactions between entries and parameters. The GGE biplot software offers convenience, the resolving power of principal component analysis, speed, and simplicity of interpretation. The potential applications of the principal component analysis technique using this amazingly friendly and informative software are limited only by the imagination of the researcher. GGE biplot is simple and fun to use; it is much like a useful and productive video game for scientists.

Duane E. Falk, Ph.D.
Associate Professor and Cereal Breeder
Department of Plant Agriculture
University of Guelph
Guelph, Ontario, Canada

and

Adjunct Associate Professor
Department of Plant Biology
Faculty of Natural and Agricultural Sciences
University of Western Australia
Crawley, Western Australia

Preface

The human population of the world is currently growing at the rate of 1 billion people per 10 to 12 years. Projections are that, by 2050, world population will increase from the current 6 billion to about 10 billion. During the past 50 years, agricultural research and technology transfer have helped increase the output of world crops two and a half-fold. Ruttan (1998), while summarizing the world's future food situation, referred to the "2–4–6–8" scenario, which means a doubling of population, a quadrupling of agricultural production, a sextupling of energy production, and an octupling of the size of the global economy by 2050. Currently, more than a billion people can be categorized as the world's absolute poor, subsisting on less than $1 of income per day, and 800 million of these do not have secure access to food (McCalla, 2001). The challenge for agricultural researchers to meet the food demand is astounding.

From the perspective of food security, the stability of agricultural production is as important as, if not more important than, the magnitude of output (Wittwer, 1998). Food production is very much a function of climate, which in itself is unpredictable; the principal characteristic of climate is variability (Wittwer, 1998). The consultative group on international agricultural research (CGIAR) warns, "Agricultural growth has to be achieved with methods that preserve the productivity of natural resources, without further damage to the Earth's previous life support systems — land, water, flora, and fauna — that are already under stress" (Harsch, 2001).

Agricultural production may be increased through increased efficiency in utilization of resources such as increased productivity per unit of land and of money, and through a better understanding and utilization of genotype-by-environment interaction (GEI). Stuber et al. (1999) considered GEI as one of the factors that gave impetus to research in the application of molecular-marker technology and genomics to plant breeding. The GEI and stability of crop performance across environments are expected to become more relevant issues in the 21st century as greater emphasis is placed on sustainable agricultural systems.

GEI is a major concern among breeders, geneticists, production agronomists, and farmers because of its universal presence and consequences. The occurrence of GEI necessitates multi-environment trials (MET) and has resulted in the development and use of the numerous measures of stability. Understanding and management of GEI has gone through several phases. The first phase was the realization of GEI as a confounding factor in cultivar selection and plant breeding early in the 20th century. The second phase was the concentrated study of GEI, which led to the development of numerous measures of stability (reviewed in Lin and Binns, 1994; Kang and Gauch, 1996; Kang, 1998). The third phase was the integration of the genotype main effect (G) and GEI. An integration of the two was needed because in practical breeding, selection only for stability is inconsequential if production level is ignored. Concepts such as crossover interaction (Baker, 1988a), stability-modified yield (Kang, 1993), rank-based statistics (Hühn, 1996), statistics to differentiate crossover from noncrossover interactions (Crossa and Cornelius, 1997), and methods of identifying the "which-won-where" pattern of MET data (Gauch and Zobel, 1997) emerged; all reflect the contemporary understanding and use of G and GEI in selection. The most recent development along this line is the development of the genotype and the genotype-by-environment (GGE) biplot methodology (Yan et al., 2000).

Since 2000, Yan has received tremendous support for his GGE biplot methodology from colleagues worldwide. The GGE biplot methodology drew the attention of many plant breeders and other researchers for two reasons. First, it explicitly and necessarily requires that genotype (G) and (GE) interaction, i.e., GGE, be regarded as integral parts in cultivar evaluation and plant breeding. Second, it presents GGE using the biplot technique (Gabriel, 1971) in a way that many important questions, such as the "which-won-where" pattern, mean performance and stability of genotypes, discriminating ability and representativeness of environments, etc., can be addressed graphically.

Although GGE biplot analysis was initially used only for dissecting GGE and visualizing MET data, its application has been extended to any data set that has a two-way structure. In the area of plant breeding in particular, the GGE biplot has been used to address important questions a breeder or researcher is likely to ask. Thus far, it has been applied to genotype-by-trait data, genotype-by-marker data, quantitive trait loci (QTL)-mapping data, diallel-cross data, and host genotype-by-pathogen strain data. Undoubtedly, with the fertile imagination of researchers engaged in crop breeding and production, additional applications to other types of two-way data will be found in time.

To facilitate the use of the GGE biplot methodology by researchers with only limited familiarity with computer applications and statistics, a Windows application called GGEbiplot has been developed (Yan, 2001). The GGEbiplot software has evolved into a comprehensive and convenient tool in quantitative genetics and plant breeding.

This book was envisioned during a meeting between Manjit Kang and Weikai Yan at the annual American Society of Agronomy meeting in Minneapolis, MN in November 2000, to make this useful technology available on a wider scale to plant breeders, geneticists, college teachers, and graduate students. The book is organized into three sections: Section I. GEI and stability analysis (Chapters 1 and 2); Section II. GGE biplot and MET data analysis (Chapters 3 to 5); and Section III. GGEbiplot software and applications in analyzing other types of two-way data (Chapters 6 through 11).

In preparing the book, we have been cognizant of the needs of researchers, teachers, and students of plant breeding, quantitative genetics, and genomics. We trust that readers will find the book stimulating and useful, as we do. The book is extensively illustrated and a person with a few courses in genetics and statistics should be able to comprehend easily the concepts and applications. It should also be useful to all researchers in other areas who must deal with large two-way data sets with complex patterns. We trust that this book will provide a powerful tool to breeders and production agronomists, and make a significant contribution toward helping meet the challenges of food production and food security that the world faces today.

Weikai Yan acknowledges that he benefited from his association and interactions with Professor L. A. Hunt at the University of Guelph and Professor D. H. Wallace at Cornell University, and from stimulating communications with Drs. Hugh Gauch and Rich Zobel (both at Cornell University at the time) — two of the major advocates of the additive main effects and multiplicative interaction effects (AMMI) model for analyzing MET data. Professor Paul L. Cornelius at the University of Kentucky and Dr. Jose Crossa at CIMMYT provided valuable comments and suggestions during the pre-publication phase of Yan et al. (2000). We thank John Sulzycki of CRC Press for his role in making this project a reality.

Weikai Yan
Guelph, Ontario

Manjit S. Kang
Baton Rouge, Louisiana

Contents

SECTION I Genotype-by-Environment Interaction and Stability Analysis 1

Chapter 1 Genotype-by-Environment Interaction 3
1.1 Heredity and Environment 3
    1.1.1 Qualitative Traits 4
    1.1.2 Quantitative Traits 4
1.2 Genotype-by-Environment Interaction 5
1.3 Implications of GEI in Crop Breeding 7
    1.3.1 The Breeding Phase 7
        1.3.1.1 Environmental (E), Genetic (G), and GE Interaction Contributions 7
        1.3.1.2 Additional Breeding Programs 8
        1.3.1.3 Genetic Gain Reduction 8
        1.3.1.4 Increased Field Evaluation Cost 8
        1.3.1.5 Early- and Advanced-Performance Testing 8
    1.3.2 The Performance Evaluation Phase 8
        1.3.2.1 Problem in Identifying Superior Cultivars 8
        1.3.2.2 Increased Cost of Variety Testing 9
1.4 Causes of Genotype-by-Environment Interaction 9

Chapter 2 Stability Analyses in Plant Breeding and Performance Trials 11
2.1 Stability Concepts and Statistics 11
    2.1.1 Static vs. Dynamic Concept 11
    2.1.2 Stability Statistics 12
    2.1.3 Simultaneous Selection for Yield and Stability 14
    2.1.4 Contributions of Environmental Variables to Stability 14
    2.1.5 Stability Variance for Unbalanced Data 15
2.2 Dealing with Genotype-by-Environment Interaction 16
    2.2.1 Correct Genetic Cause(s) of GE Interaction 17
    2.2.2 Characterize Genotypes and Environments 17
    2.2.3 QTL-by-Environment Interaction 17
    2.2.4 Breeding for Stability/Reliability of Performance 18
    2.2.5 Early Multi-Environment Testing 18
    2.2.6 Resource Allocation 19
2.3 GGE Biplot: Genotype + GE Interaction 19

SECTION II GGE Biplot and Multi-Environment Trial Data Analysis 21

Chapter 3 Theory of Biplot 23

3.1 The Concept of Biplot 23
   3.1.1 Matrix Multiplication 23
   3.1.2 Presenting the Matrix Multiplication Process in a Biplot 24
3.2 The Inner-Product Property of a Biplot 26
3.3 Visualizing the Biplot 29
   3.3.1 Visual Comparison of the Row Elements within a Single Column 29
   3.3.2 Visual Comparison of the Column Elements within a Single Row 29
   3.3.3 Visual Comparison of Two Row Factors for Each of the Columns 30
   3.3.4 Visual Comparison of Two Column Factors for Each of the Rows 30
   3.3.5 Visual Identification of the Largest Column Factor within a Row 32
   3.3.6 Visual Identification of the Largest Row Factor within a Column 32
3.4 Relationships among Columns and among Rows 34
   3.4.1 Relationships among Columns 34
   3.4.2 Relationships among Rows 35
3.5 Biplot Analysis of Two-Way Data 36

Chapter 4 Introduction to GGE Biplot 39

4.1 The Concept of GGE and GGE Biplot 39
4.2 The Basic Model for a GGE Biplot 41
4.3 Methods of Singular Value Partitioning 42
   4.3.1 Genotype-Focused Scaling 42
   4.3.2 Environment-Focused Scaling 44
   4.3.3 Symmetric Scaling 44
   4.3.4 Equal-Space Scaling 44
   4.3.5 Merits of Different Scaling Methods 47
4.4 An Alternative Model for GGE Biplot 49
4.5 Three Types of Data Transformation 56
4.6 Generating a GGE Biplot Using Conventional Methods 57

Chapter 5 Biplot Analysis of Multi-Environment Trial Data 63

5.1 Objectives of Multi-Environment Trial Data Analysis 63
5.2 Simple Comparisons Using GGE Biplot 65
   5.2.1 Performance of Different Cultivars in a Given Environment 65
   5.2.2 Relative Adaptation of a Given Cultivar in Different Environments 67
   5.2.3 Comparison of Two Cultivars 67
5.3 Mega-Environment Investigation 73
   5.3.1 "Which-Won-Where" Pattern of an MET Dataset 73
   5.3.2 Mega-Environment Investigation 77
5.4 Cultivar Evaluation for a Given Mega-Environment 83
   5.4.1 Cultivar Evaluation Based on Mean Performance and Stability 83
   5.4.2 The Average Environment Coordinate 85
   5.4.3 How Important is Stability? 88

5.5 Evaluation of Test Environments 89
    5.5.1 Interrelationships among Environments..89
    5.5.2 Discriminating Ability of the Test Environments 91
    5.5.3 Environment Ranking Based on Both Discriminating Ability
          and Representativeness...91
    5.5.4 Environments for Positive and Negative Selection.......................................92
5.6 Comparison with the AMMI Biplot..93
5.7 Interpreting Genotype-by-Environment Interaction..94
    5.7.1 The General Idea 94
    5.7.2 Causes of GE Interaction Represented by PC1...96
    5.7.3 Causes of GE Interaction Represented by PC2...98
    5.7.4 Some Comments on the Approach 98

SECTION III GGE Biplot Software and Applications to Other Types of Two-Way Data 101

Chapter 6 GGE Biplot Software — The Solution for GGE Biplot Analyses......................103

6.1 The Need for GGE Biplot Software 103
6.2 The Terminology of Entries and Testers...104
6.3 Preparing Data File for GGEbiplot...104
    6.3.1 The Observation Data Format...104
    6.3.2 The Matrix Data Format 106
6.4 Organization of GGEbiplot Software 107
    6.4.1 File..110
    6.4.2 View..111
    6.4.3 Visualization 111
    6.4.4 Find QTL..113
    6.4.5 Format...113
    6.4.6 Models 114
    6.4.7 Data Manipulation 114
    6.4.8 Biplots...115
    6.4.9 Scaling (Singular Value Partitioning)..115
    6.4.10 Accessories 116
    6.4.11 Help..117
    6.4.12 Log File 117
6.5 Functions for a Genotype-by-Environment Dataset 117
6.6 Functions for a Genotype-by-Trait Dataset...118
6.7 Functions for a QTL-Mapping Dataset...118
6.8 Functions for a Diallel-Cross Dataset...119
6.9 Functions for a Genotype-by-Strain Dataset 119
6.10 Application of GGEbiplot to Other Types of Two-Way Data..................................119
6.11 GGEbiplot Continues to Evolve 119
    6.11.1 Interactive Stepwise Regression..120
    6.11.2 Interactive QQE Biplot..120
    6.11.3 Three-Way Data Input and Visualization...120
    6.11.4 Interactive Statistics...120

Chapter 7 Cultivar Evaluation Based on Multiple Traits 121
7.1 Why Multiple Traits? 121
7.2 Cultivar Evaluation Based on Multiple Traits 122
    7.2.1 Which is Good at What 122
    7.2.2 Which is Bad at What 122
    7.2.3 Comparison between Two Cultivars 127
    7.2.4 Interrelationship among Traits 127
    7.2.5 The Triangle of Grain Yield, Loaf Volume, and Flour Extraction 127
    7.2.6 Cultivar Evaluation Based on Individual Traits 127
    7.2.7 Cultivar Evaluation Based on Two Traits 135
7.3 Identifying Traits for Indirect Selection for Loaf Volume 135
7.4 Identification of Redundant Traits 142
7.5 Comparing Cultivars as Packages of Traits 142
    7.5.1 Comparing New Genotypes with the Standard Cultivar 144
    7.5.2 What is Good with the Standard Cultivar? 144
    7.5.3 Traits for Indirect Selection for Cookie-Making Quality 144
7.6 Investigation of Different Selection Strategies 150
7.7 Systems Understanding of Crop Improvement 150
    7.7.1 Systems Understanding, Independent Culling, and the Breeder's Eye 150
    7.7.2 Enlarging the System Capacity by Reinforcing Factors outside the System 158
7.8 Three-Mode Principal Component Analysis and Visualization 158

Chapter 8 QTL Identification Using GGEbiplot 159
8.1 Why Biplot? 159
8.2 Data Source and Model 160
    8.2.1 The Data 160
    8.2.2 The Model 160
8.3 Grouping of Linked Markers 160
8.4 Gene Mapping Using Biplot 162
8.5 QTL Identification via GGEbiplot 175
    8.5.1 Strategy for QTL Identification 175
    8.5.2 QTL Mapping of Selected Traits 175
        8.5.2.1 Yield 175
        8.5.2.2 Heading Date 177
        8.5.2.3 Maturity 182
        8.5.2.4 Plant Height 182
        8.5.2.5 Lodging 182
        8.5.2.6 Test Weight 182
        8.5.2.7 Kernel Weight 182
8.6 Interconnectedness among Traits and Pleiotropic Effects of a Given Locus 182
8.7 Understanding DH Lines through the Biplot Pattern 192
    8.7.1 Visualizing Marker and Trait Values of the DH Lines 192
    8.7.2 Marker Nearest to the QTL 192
    8.7.3 Estimating Missing Values Based on the Biplot Pattern 198
8.8 QTL and GE Interaction 202

Chapter 9 Biplot Analysis of Diallel Data 207

9.1 Model for Biplot Analysis of Diallel Data 207
9.2 General Combining Ability of Parents 208
9.3 Specific Combining Ability of Parents 210
9.4 Heterotic Groups 210
9.5 The Best Testers for Assessing General Combining Ability of Parents....................210
9.6 The Best Crosses 215
9.7 Hypothesis on the Genetic Constitution of Parents..215
9.8 Targeting a Large Dataset 217
    9.8.1 Shrinking the Dataset by Removing Similar Parents 221
    9.8.2 The Best Crosses 221
    9.8.3 GCA and SCA 221
    9.8.4 Best Tester 221
    9.8.5 Heterotic Groups 221
    9.8.6 Genetic Constitutions of Parents with Regard to PSB Resistance.................221
9.9 Advantages and Disadvantages of the Biplot Approach 225

Chapter 10 Biplot Analysis of Host Genotype-by-Pathogen Strain Interactions 229

10.1 Vertical vs. Horizontal Resistance 229
10.2 Genotype-by-Strain Interaction for Barley Net Blotch 230
    10.2.1 Model for Studying Genotype-by-Strain Interaction....................................230
    10.2.2 Biplots with Barley Lines as Entries 231
    10.2.3 Biplots with Net Blotch Isolates as Entries..233
10.3 Genotype-by-Strain Interaction for Wheat *Fusarium* Head Blight 239

Chapter 11 Biplot Analysis to Detect Synergism between Genotypes of Different Species 247

11.1 Genotype-by-Strain Interaction for Nitrogen-Fixation 247
    11.1.1 Biplot with *Frankia* Strains as Entries..248
    11.1.2 Biplot with *Casuarina* Species as Entries 248
11.2 Wheat–Maize Interaction for Wheat Haploid Embryo Formation...........................251
    11.2.1 Biplot with Maize Genotypes as Entries 251
    11.2.2 Biplot with Wheat Genotypes as Entries...251

References 255

Index 263

Section I

Genotype-by-Environment Interaction and Stability Analysis

1 Genotype-by-Environment Interaction

SUMMARY

This chapter introduces the reader to concepts of genotype, environment, phenotype, and genotype-by-environment interaction (GEI). Contributions and implications of GEI in crop breeding and causes of GEI are briefly discussed. Examples provided show how even the highly heritable qualitative traits can be influenced by the environment. The concept of norm of reaction or plasticity is illustrated. The entire chapter is designed to help understand and appreciate GEI.

GEI is an extremely important issue in crop breeding and production (Kang, 1990; Kang and Gauch, 1996; Cooper and Hammer, 1996). To understand the relationship between crop performance and environment, we begin with the fundamentals — definitions of genotype (G), environment (E), phenotype (P), and GEI. We show how environment has an influence on qualitative and quantitative traits. We also discuss the implications and causes of GEI relative to crop breeding.

1.1 HEREDITY AND ENVIRONMENT

Heredity may be defined as the transmission of traits or characteristics from parents to offspring through genes. G refers to an individual's genetic makeup — the nucleotide sequences of DNA, a gene or genes, that are transmitted from parents to offspring. A gene is a short segment of chromosome, DNA, that can function as a unit of selection. P refers to physical appearance or discernible trait of an individual, which is dependent on expression of a genotype in E. E may be defined as the sum total of circumstances surrounding or affecting an organism or a group of organisms. Gene expression is the realization of the genetic blueprint encoded in the nucleic acids (Rédei, 1998). Gene expression may be modified, enhanced, silenced, or timed by the regulatory mechanisms of the cell in response to internal and external factors (Rédei, 1998). The genotype may specify a range of phenotypic expressions that are called the norm of reaction, or plasticity, which is simply the expression of variability in the phenotype of individuals of identical genotype (Bradshaw, 1965). Genotypes developed under relatively constant environments are expected to display much less plasticity. The norm of reaction for one genotype may differ from that for another genotype. Norm of reaction is illustrated in Figure 1.1, which shows there is a correlation between the environmental factor or the amount of water and the phenotype. This should help us understand how the same genotype, when subjected to different environments, can produce a range of phenotypes.

The genetic constitution of an individual generally remains constant from one environment to another; occasionally, spontaneous mutations do occur. Therefore, any phenotypic variation or norm of reaction for a specific genotype is attributable to the environment. In general, genes do not absolutely determine the phenotype, but they permit a range of expressions, depending on the genetic background, developmental and tissue-specificity conditions, and the environment (Rédei, 1998).

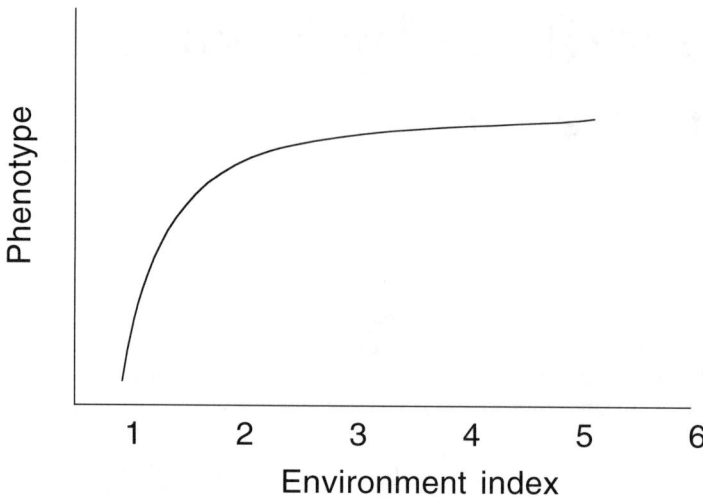

FIGURE 1.1 Norm or range of reaction for a phenotype (e.g., plant height) of a genotype to increasing levels of an environmental factor (e.g., water).

1.1.1 Qualitative Traits

Qualitative traits, which are controlled by one or two genes, are generally highly heritable; i.e., phenotype is a highly reliable indicator/predictor of genotype. Nevertheless, qualitative traits also, occasionally, may be subject to environmental modification. For example:

- The sun-red gene in maize (*Zea mays* L.) produces red kernels when they are exposed to light; in the absence of light, kernels are white (Burns, 1969).
- Gardeners manipulate the soil pH to grow hydrangeas (Singleton, 1967). When hydrangeas are grown on acidic soil, they produce beautiful blue flowers, but on alkaline soil, they produce off-white, faintly pink, less attractive flowers.
- Reversal of dominance is another excellent example of an environmental influence on heredity. When squash plants with green fruits and vine-type stems are crossed with those with yellow fruits and bushy growth (Shifriss, 1947), F_2 segregation ratio observed at flowering is 9 green, bushy: 3 green, vine: 3 yellow, bushy: 1 yellow vine; whereas, at maturity, a reversal of dominance occurs (1 green, bushy: 3 green, vine: 3 yellow, bushy: 9 yellow vine). This example clearly points out that dominance of a gene is not always absolute. Environment can and does modify the phenotypic expression of a genotype.

1.1.2 Quantitative Traits

Quantitative traits exhibit continuous variation patterns that are attributable to polygenic control and environmental or nongenetic factors. Because of the considerable influence of environment, quantitative traits generally have relatively low heritability.

In 1903, Wilhelm Johannsen provided a clear distinction between G and P for traits exhibiting continuous variation, i.e., quantitative traits. He recognized that continuous variation was partly heritable, and that it was attributable to two components — heredity and environment (Rédei, 1982). In 1909, Johannsen coined the following crucial terms: gene for what Mendel called factor, genotype, and phenotype.

Quantitative traits are generally influenced much more by the environment than qualitative traits are. As environmental influence increases, the reliability of phenotype as a predictor of genotype decreases.

Genotype and environmental conditions are the two main components that determine a phenotype. Gene expression is, in part, environmentally induced and regulated. The better the characterization of the environment, the better our understanding of the relationship between phenotype and environment. In general, a phenotype can be expressed as follows if interaction between G and E is not important or is ignored:

$$P = G + E.$$

1.2 GENOTYPE-BY-ENVIRONMENT INTERACTION

Understanding the relationship between crop performance and environment has long been a key issue for plant breeders and geneticists. Crop performance, the observed phenotype, is a function of genotype — variety or cultivar, environment, and GEI. GEI is said to occur when different cultivars or genotypes respond differently to diverse environments. Researchers agree that GEI is important only when it is significant and causes significant change in genotype ranks in different environments, i.e., different genotypes are superior in different environments (Haldane, 1946). For GEI to be detected via statistical procedures, there must be at least two distinct genotypes or cultivars evaluated in at least two different environments. Thus, a basic model that includes GEI is:

$$P = G + E + GE.$$

This model can be written from a statistical standpoint as: $P_{ij} = \mu + G_i + E_j + (GE)_{ij}$. Here, μ is the overall mean. It follows from this model that, for a given genotype, there can be many phenotypes depending upon the environment and GEI. The minimum number of genotypes and environments for such a model is two. The various components of this model can be explained as follows (Simmonds, 1981).

Genotype\Environment→ ↓	E1	E2	Difference (E effect)
G1	a	c	$\Delta 1 = c - a$
G2	b	d	$\Delta 2 = d - b$
Difference (G effect)	$\Delta 3 = b - a$	$\Delta 4 = d - c$	

GE interaction:

$$(\Delta 2 - \Delta 1) = (\Delta 4 - \Delta 3) \text{ or } (d - b) - (c - a) = (d - c) - (b - a)$$

or

$$(\Delta 1 + \Delta 4) = (\Delta 2 + \Delta 3) \text{ or } (c - a) + (d - c) = (d - b) + (b - a).$$

The genotype effect, $\Delta 3$, represents the difference between genotypes in environment E1 and $\Delta 4$ represents the difference between genotypes in environment E2. The environmental effect, $\Delta 1$, represents change attributable to environments for genotype G1 and $\Delta 2$ is the change attributable to environments for genotype G2.

$$\text{Total effect (T)} = G + E + GE = (d - a)$$

or

$$GE = T - G - E.$$

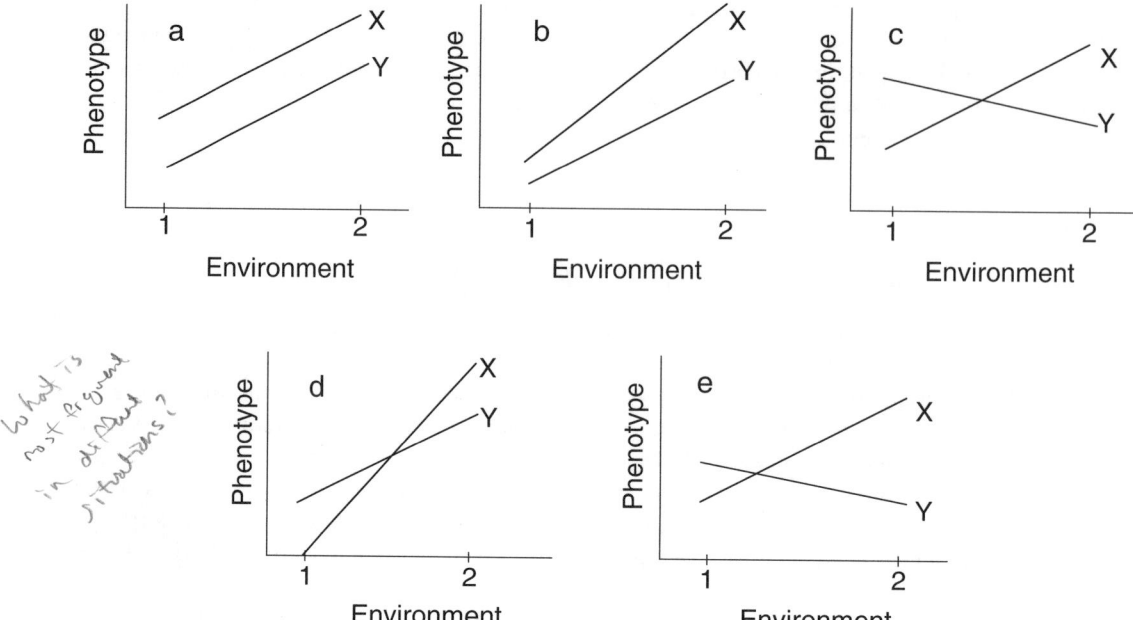

FIGURE 1.2 Graphical representation of types of GEI: (a) no interaction — X and Y responses parallel in the two environments; (b) noncrossover type of interaction — both X and Y increase but unequal intergenotypic difference in the two environments; (c) crossover interaction — genotypic modification by environment in opposite direction but intergenotypic difference remains the same; (d) crossover interaction — unequal intergenotypic difference but both X and Y increase; and (e) crossover interaction — unequal intergenotypic difference in the two environments: X shows an increase whereas Y shows a decrease in environment 2.

Hall (2001) uses the above relationships and provides examples in Tables 2.1 and 2.2 about interaction and the main effects for genotypes and environments. In Table 2.1, he presents cultivar × location interaction, whereas in Table 2.2, he reports cultivar × environment interaction.

GEI can be visualized by comparing the relative performance of X and Y genotypes in two environments. The basic types and magnitudes of interaction are illustrated in the diagram (Figure 1.2a to 1.2e).

GEI is graphically represented as genotypes' norms of reaction that are nonparallel or nonadditive. The first graph (Figure 1.2a) illustrates that genotypes X and Y respond to or perform similarly in the two environments, because their responses are parallel. Genotype X performs better than genotype Y in both environments. The norms of reaction for the two genotypes are additive. The intergenotypic difference remains unchanged in the two environments, and the direction of environmental modification of genotypes is the same. Performance of both X and Y increases. This case represents the absence of interaction.

The second graph (Figure 1.2b) illustrates a noncrossover type of interaction. The genotypes X and Y respond differently to the two environments, but their ranks remain unchanged. The response of the two genotypes to environments is not additive. The magnitude of intergenotypic difference increases and the environmental modification of the two genotypes is in the same direction. Performance of both X and Y increases.

The next graph (Figure 1.2c) represents a crossover, rank change, type of interaction. The direction of environmental modification of genotypes X and Y is opposite. The performance of X increases and that of Y decreases. The genotypic ranks change between the two environments, but the magnitude of intergenotypic difference remains unchanged. Crossover types of interaction are important to breeders and production agronomists.

Genotype-by-Environment Interaction

Figure 1.2d also represents a crossover interaction as genotypes switch ranks between the two environments. In this case, the magnitude of intergenotypic difference changes. In environment 1, the difference between genotypes X and Y is smaller than that in environment 2 and the direction of environmental modification of the two genotypes is the same. Performance of both X and Y increases. X and Y are not specialized genotypes because there is crossover type of interaction.

The last graph (Figure 1.2e) also illustrates a crossover interaction. The environmental modification here is in the opposite direction — performance of X increases but that of Y decreases. This situation is different from the one illustrated in Figure 1.2c in that here the magnitude of intergenotypic difference increases between environments.

In practice, the situation is much more complex than that illustrated in Figures 1.2a through 1.2e because breeders generally must deal with many genotypes and environments. With large numbers of intermediary effects, the number of types of interaction can become extremely large (Allard and Bradshaw, 1964). With only two genotypes and two environments, and with only a single criterion, at least four different types of interaction are possible; with 10 different genotypes and 10 different environments, 400 types of interaction are possible, which would certainly make their consequences, implications, and interpretation more difficult to comprehend (Allard, 1999). The reader may refer to Yan and Hunt (1998) who reviewed complexities of GEI under the following subheadings: (1) detection of GEI, (2) quantification of GEI, (3) interpretation of GEI, (4) GEI and mega-environments, and (5) GEI and plant breeding.

1.3 IMPLICATIONS OF GEI IN CROP BREEDING

Agricultural researchers have long been cognizant of the various implications of GEI in breeding programs (Mooers, 1921; Yates and Cochran, 1938). GEI has a negative impact on heritability. The lower the heritability of a trait, the greater the difficulty in improving that trait via selection. Understanding the structure and nature of GEI is important in plant breeding programs because a significant GEI can seriously impair efforts in selecting superior genotypes relative to new crop introductions and cultivar development programs (Shafii and Price, 1998). Information on the structure and nature of GEI is particularly useful to breeders because it can help determine if they need to develop cultivars for all environments of interest or if they should develop specific cultivars for specific target environments (Bridges, 1989). Gauch and Zobel (1996) explained the importance of GEI as: "Were there no interaction, a single variety of wheat (*Triticum aestivum* L.) or corn (*Zea mays* L.) or any other crop would yield the most the world over, and furthermore the variety trial need be conducted at only one location to provide universal results. And were there no noise, experimental results would be exact, identifying the best variety without error, and there would be no need for replication. So, one replicate at one location would identify that one best wheat variety that flourishes worldwide."

A crop cultivar development program has two major components — the breeding phase and the performance evaluation phase. Importance and implications of GEI during these phases are discussed next.

1.3.1 THE BREEDING PHASE

1.3.1.1 Environmental (E), Genetic (G), and GE Interaction Contributions

GEI confounds precise partitioning of the contributions of improved cultivars and improved environment or technology to yield (Silvey, 1981). In a barley trial, estimates of environmental contribution to yield were 10 to 30% and those of genetic contribution were 30 to 60%; GEI accounted for the remaining 25 to 45% of yield gain (Simmonds, 1981). Comparable estimates from wheat for the same period were: 40 to 60% attributable to E, 25 to 40% to G, and 15 to 25% to GEI (Simmonds, 1981).

1.3.1.2 Additional Breeding Programs

In the presence of a significant GEI, it may be necessary to partition a cultivar development program with different objectives for different regions and for different weather conditions (Busey, 1983). The GEI necessitates the development of separate populations for each region where genotypic rankings drastically change (McKeand et al., 1990). The establishment of separate programs is, no doubt, expensive, but it should yield greater genetic gains. The selection of genotypes that generally perform well across many locations is less expensive, but it also produces reduced gain.

1.3.1.3 Genetic Gain Reduction

Losses of potential genetic gain as a result of GEI may be determined as follows (Matheson and Cotterill, 1990):

$$C = 1 - [(\sigma_G^2 + \sigma^2)^{1/2}/(\sigma_G^2 + \sigma_{GE}^2 + \sigma^2)^{1/2}],$$

where C = loss of potential gain, σ_G^2 = genetic variance component, σ^2 = error variance, and σ_{GE}^2 = GE variance component. There are several ways in which effects of GEI on breeding programs can be estimated (Matheson and Raymond, 1984).

Heritability of a trait is a key component in determining genetic advance from selection (Nyquist, 1991). As a component of the total phenotypic variance (the denominator in any heritability equation), GEI affects heritability negatively. The larger the GEI component, the smaller the heritability estimate; thus, progress from selection would be reduced as well.

1.3.1.4 Increased Field Evaluation Cost

Cost of field evaluation of breeding lines goes up when a large GEI necessitates testing of cultivars in numerous environments, such as locations and years, to obtain reliable results. If the weather patterns or management practices differ in the target areas, testing has to be done at several sites representative of each target area.

1.3.1.5 Early- and Advanced-Performance Testing

If cultivars are to be used in marginal areas or fringes of a crop's adaptation or specific adaptation, testing must begin in those areas as early as possible. In the early stages of a testing program (LeClerg, 1966) as well as in advanced yield trials (Sprague and Federer, 1951), it is advisable to have as many locations as possible but with one replication per location vis-à-vis multiple replications at fewer locations. Advanced corn-belt single crosses of maize are generally evaluated at more than 100 single-replicate locations, and additional locations help identify widely adapted single crosses and more adaptedness-related genes (Troyer, 1997). Kang (1998) outlined strategies for early testing in both clonally and nonclonally propagated crops.

1.3.2 THE PERFORMANCE EVALUATION PHASE

1.3.2.1 Problem in Identifying Superior Cultivars

As the magnitude of a significant interaction between two factors increases, the usefulness and reliability of the main effects correspondingly decrease. GEI reduces the correlation between phenotypic and genotypic values, increasing the difficulty in identifying truly superior genotypes across environments, especially in the presence of crossover GEI.

1.3.2.2 Increased Cost of Variety Testing

Multi-environment testing (MET) makes it possible to identify cultivars that perform consistently from year to year, displaying a small temporal variability and those that perform consistently from location to location, displaying a small spatial variability. Temporal stability is desired by and beneficial to growers, whereas spatial stability is desired by and is beneficial to seed companies and breeders. The cost of variety testing goes up when there is a significant interaction between genotypes and environments. The extensive testing, however, allows application of stability analyses to identify genotypes that are broadly adapted. Such analyses also make it possible to identify those genotypes that are specifically adapted to a particular environment.

1.4 CAUSES OF GENOTYPE-BY-ENVIRONMENT INTERACTION

When a significant GEI is present, researchers usually are interested to know the causes of the interaction in order to make accurate predictions of genotype performance under a variety of environments. Understanding genotypic responses to individual factors aids in interpreting and exploiting GEI. At a level other than optimal, an environmental factor represents a stress. Differences in the rate of increase in genotypic response at sub-optimal levels reflect differences in efficiency, and differences in the rate of decrease in genotypic response at super-optimal levels reflect differences in tolerance (Baker, 1988b). For example, when drought occurs, i.e., water is at sub-optimal levels, water-use efficient genotypes can be identified. At super-optimal levels of flooding, one can identify plants that are flood tolerant.

Plants respond to a host of environmental signals: nutrients, toxic elements, salts in the soil solution, gases in the atmosphere, light of different wavelengths, mechanical stimuli, gravity, wounding, pests, pathogens, and symbionts (Crispeels, 1994). The extent of an individual's adaptability to environmental conditions reflects the extent and sophistication of the controls over the synthesis and action of specific proteins (Smith, 1990). Adaptability is a quantitative estimation of the range of environmental conditions to which a particular genotype can adapt and is determined by the extent and sophistication of its plastic traits (Smith, 1990). Plants that have incorporated a variety of environmental signals into their developmental pathways possess a wide range of adaptive capacities (Scandalios, 1990). The diversity of pathways among plants to communicate information both within and between cells suggests a large number of possible ways in which plants can perceive and transduce environmental stresses and changes (Leigh, 1993).

It is now known that many abiotic stresses are related to a group of chemicals called superoxides. Activated oxygen species — endogenous, byproducts of normal metabolism and exogenous, triggered by environmental factors — are highly reactive molecules capable of causing extensive damage to plant cells (Scandalios, 1990). The effects can range from simple inhibition of enzyme function to the production of random lesions in proteins and nucleic acids, and peroxidation of membrane lipids. Loss of membrane integrity caused by peroxidation, along with direct damage to enzymatic and structural proteins and their respective genes, can decrease metabolic efficiency resulting in decreased mitochondrial and chloroplast functions. This can lower a plant's ability to fix carbon and to properly use the resulting products, hence reducing yield.

Biotic stresses are a major constraint to crop productivity. Differences in insect and disease resistance among genotypes can be associated with stable or unstable performance across environments. For example, Baker (1990) concluded that many of the observed crossover interactions in wheat were manifestations of differences in disease resistance or some other highly heritable trait. Gravois et al. (1990) also implicated disease resistance or susceptibility as a factor that contributed to GEI.

Nutritional deficiencies or disorders may be related to the absence of certain genes in the plant. These nutritional disorders or deficiencies can be responsible for differential plant performance. Two nutritional disorders are controlled by single gene loci, susceptibility to magnesium and boron deficiency, in celery (Pope and Munger, 1953a,b). A single recessive gene locus controls iron deficiency symptoms in a soybean introduction (Devine, 1982). Copper deficiency in *triticale* was associated with hairy peduncle (*Hp*) gene located on chromosome 5R (Graham, 1982). The presence of GE interaction for potassium-use efficiency in wheat has been reported (Woodend and Glass, 1993). Additional single gene controls for nutrient uptake or efficiency are given in Blum (1988).

Differential survival rates of genotypes also could be responsible for GEI. Genetic and environmental factors and their interactions affect the number of seeds each genotype produces and the proportion of seeds of each genotype that reaches maturity (Allard, 1960). To survive in competition or stress, a plant may produce a reduced number of viable seed. Such reproductive adjustment may be differential among genotypes.

Differential responses among genotypes to herbicides and allelochemicals could result in GEI. Crop plants are often stressed by carryover herbicide or post-emergence herbicide applications (Einhellig, 1996). Along with herbicide stress, weed–crop or crop–crop allelopathy can occur (Einhellig, 1996). Plant breeders should be able to improve plant resistance to allelochemicals (Miller, 1996).

Other major abiotic stresses include atmospheric pollutants; soil stresses, such as salinity, acidity, and mineral toxicity and deficiency; temperature, heat and cold; water, drought and flooding; and tillage operations (Steiner et al., 1984; Unsworth and Fuhrer, 1993; Specht and Laing, 1993). Precipitation and altitude of trial locations, when used as covariates in tree species, explained interaction for survival traits (Alia et al., 1997). Also, altitude was responsible for interaction for wood volume (Alia et al., 1997). Genetic variation exists for plant responses to many stress factors. There is considerable potential for breeding for tolerance to air pollutants (Unsworth and Fuhrer, 1993). Differential response of genotypes to these stresses could be a cause for GEI. Genotypic difference in disease resistance can be a common cause of noncrossover GEI, and genes related to plant maturity and height may be causes of crossover GEI (Yan and Hunt, 1998, 2002).

The detection of GEI in trials and breeders' desire to handle these interactions appropriately has led to the development of procedures that are generically called stability analyses. The numerous stability statistics available to the plant breeder and to the production agronomist provide different strategies and approaches of dealing with GEI. These issues will be discussed in detail in Chapter 2.

2 Stability Analyses in Plant Breeding and Performance Trials

SUMMARY

This chapter introduces the reader to the important concepts of stability and stability-related statistics. The myriads of stability statistics that have been developed are briefly mentioned. The point is emphasized that genotype-by-environment interaction (GEI) is a breeding issue and is appropriately dealt with in breeding programs. Strategies are outlined for handling GEI in plant breeding. The genotype and genotype-by-environment (GGE) biplot method is introduced and its versatility and merits are briefly discussed.

GEI has been a continuing challenge for breeders and has significant implications in both applied plant and animal breeding programs (Kang, 2002). Breeders have found the GEI issue to be intriguing and perplexing because there is no uniform way to handle GEI and usefully integrate it into breeding programs (Kang, 1990; Kang and Gauch, 1996; Cooper and Hammer, 1996).

GEI has led to the development of concepts of stability or consistency of performance. In the next section, the various concepts of stability and stability statistics are discussed.

2.1 STABILITY CONCEPTS AND STATISTICS

Stability is an important concept for plant breeders interested in analyzing GEI data (Denis et al., 1996). Some researchers prefer to use the term sensitivity analysis instead of stability analysis (Dyke et al., 1995). Breeders seek performance stability or a lack of GEI against uncontrollable forces, such as weather, whereas a positive GEI may be sought for controllable factors, such as fertilizer and other inputs. For certain genotypes to do well, additional agronomic inputs may be necessary, which tends to create a GE correlation. For a detailed discussion of the GE correlation concept, refer to Kang (1998).

2.1.1 STATIC VS. DYNAMIC CONCEPT

Stability has many concepts. The static concept of stability implies that a genotype has a stable performance across environments, with no among-environment variance, i.e., a genotype is nonresponsive to increased levels of inputs. This type of stability, also referred to as biological concept of stability (Becker, 1981), is not desirable in production agriculture.

The dynamic concept implies that a genotype's performance is stable, but for each environment, its performance corresponds to the estimated or predicted level. The estimated or predicted level and the level of actual performance should agree (Becker and Leon, 1988). This concept has been referred to as the agronomic concept (Becker, 1981).

Lin et al. (1986) classified stability statistics into four groups:

- Group A — based on deviation from average genotype effect (DG); represents sums of squares
- Group B — based on GEI; represents sums of squares
- Groups C and D — based on either DG or GEI; represent regression coefficient or deviations from regression

Lin et al. (1986) further assigned the four groups to three classes of stability:

- Group A — Type 1 stability
- Groups B and C — Type 2 stability
- Group D — Type 3 stability

In Type 1 stability, which is equivalent to biological stability, a genotype is regarded as stable if its among-environment variance is small. In Type 2 stability, which is equivalent to agronomic stability, a genotype is regarded as stable if its response to environments is parallel to the mean response of all genotypes in a test. In Type 3 stability, a genotype is regarded as stable if the residual mean square following regression of genotype performance or yield on environmental index is small (Lin et al., 1986). Lin and Binns (1988a) proposed a Type 4 stability concept on the basis of predictable and unpredictable nongenetic variation; the predictable component related to locations and the unpredictable component related to years. They suggested the use of a regression approach for the predictable portion. The mean square for years-within-locations for each genotype as a measure of the unpredictable variation was referred to as Type 4 stability.

2.1.2 Stability Statistics

Kang and Miller (1984) discussed merits and demerits of several methods. Shukla's stability variance (Shukla, 1972) and Wricke's ecovalence (Wricke, 1962) are identical statistics for determining stability — rank correlation coefficient = 1.00 (Kang et al., 1987). These types of measures are useful to breeders and agronomists because they can pinpoint contributions of individual genotypes in a test to total GEI. They also can be used to identify locations with a similar GEI pattern (Glaz et al., 1985).

Stable performance relative to crop yield and quality traits across a wide range of growing conditions is desirable for management, marketing, and profit (Gutierrez et al., 1994). To breed for improved stability of performance and broad adaptation, reliable selection criteria must be used. Various stability statistics and concepts have been developed for this purpose. Several methods have been developed to analyze and interpret GEI (Lin et al., 1986; Becker and Leon, 1988; Kang and Gauch, 1996; Piepho, 1998). The earliest approach was the linear regression analysis (Mooers, 1921; Yates and Cochran, 1938). The regression approach was popularized in the 1960s and 1970s (Finlay and Wilkinson, 1963; Eberhart and Russell, 1966, 1969; Tai, 1971). In this approach, regression graphs are used to predict adaptability of genotypes. According to Robbertse (1989), GEI variance is the only important factor in the regression ratio that is the basis for the regression graphs. The graphs are based on the following ratio of the phenotypic performance of a specific clone or family (S) from several environments to the mean phenotypic performance of all clones or families (M) from the same environments:

$$\frac{\sigma^2_{p(S)}}{\frac{1}{N}\sum_i \sigma^2_{P(M_i)}}$$

where $\sigma^2_{P(S)}$ is the phenotypic variance of a specific clone or family, S, in an environment, and $1/N \sum \sigma^2_{P(M)} = \sigma^2_{P(M\_1)} + \sigma^2_{P(M\_2)} \ldots + \sigma^2_{P(M\_N)}$, N = total number of clones or families and M_1 = mean for clone or family number 1, and so on.

In multi-environment trials (MET), $\sigma_P^2 = \sigma_G^2 + \sigma_E^2 + \sigma_{GEI}^2$. For a homogeneous genotype (e.g., a clone), σ_G^2 is zero (Robbertse, 1989). For nonhomogeneous families, σ_G^2 may be minimized statistically. σ_E^2 also can be minimized. Thus, the regression ratio, instead of being:

$$\frac{\sigma_{G(S)}^2 + \sigma_{E(S)}^2 + \sigma_{GE(S)}^2}{\frac{1}{N}\sum_i (\sigma_{G(M_i)}^2 + \sigma_{E(M_i)}^2 + \sigma_{GE(M_i)}^2)}$$

will be:

$$\frac{\sigma_{GE(S)}^2}{\frac{1}{N}\sum_i \sigma_{GE(M_i)}^2}$$

Statistical methods that have recently received attention are pattern analysis (DeLacy et al., 1996b) and the additive main effects and multiplicative interaction (AMMI) model (Gauch and Zobel, 1996). The AMMI model incorporates both additive and multiplicative components of the two-way structure that can, more effectively, account for the underlying interaction (Shafii and Price, 1998). Integrating results obtained from biplot graphics with those of the genotypic stability analyses allows clustering of genotypes based on similarity of response and the degree of stability of performance across environments (Shafii and Price, 1998). The AMMI model or variations thereof (e.g., factor regression or FANOVA) have been used for interpreting GEI in various situations (Baril, 1992; Gutierrez et al., 1994; Yan and Hunt, 1998). Biadditive factorial regression models, which encompass both factorial regression and biadditive AMMI models, also have been evaluated (Brancourt-Hulmel et al., 2000). Gimelfarb (1994) used an additive–multiplicative approximation of GEI, demonstrating that GEI tends to cause a substantial nonlinearity in offspring–parent regression and a reversed response to directional selection.

The shifted multiplicative model (SHMM) (Crossa et al., 1996; Cornelius et al., 1996), the nonparametric methods of Hühn (1996) that are based on cultivar ranks, analyses based upon the probability of outperforming a check (Eskridge, 1996), and Kang's rank-sum method (Kang, 1988, 1993) are some of the other methods that have been considered by plant breeders. The methods of Hühn (1996) and Kang (1988, 1993) integrate yield and stability into one statistic that can be used as a selection criterion.

Flores et al. (1998) compared 22 univariate and multivariate methods to analyze GEI. These 22 methods were classified into three main groups (Flores et al., 1998):

- **Group 1** statistics are mostly associated with yield level and show little or no correlation with stability parameters. Group 1 includes: YIELD — mean yield for comparison or reference; PI — Lin and Binns (1988b) superiority measure; UPGMA — unweighted pair–group method fusion strategy (Sokal and Michener, 1958); FOXRANK — stratified ranking technique of Fox et al. (1990); and FOXROS — cluster analysis of Fox and Rosielle (1982).
- **Group 2** both yield and stability of performance are considered simultaneously to reduce the effect of GEI. Group 2 includes: S6O — a nonparametric method based on ranks (Hühn, 1979; Nassar and Hühn, 1987); PPCC — principal coordinate analysis of Westcott (1987); STAR — a star is drawn for every genotype and the length of each star arm represents mean yield in an environment (Flores, 1993); AMMI — additive main effects and mutiplicative interaction model (Kempton, 1984; Zobel et al., 1988);

and KANG — assigned yield ranks and stability variance ranks are summed, rank-sum method of Kang (1988).
- **Group 3** emphasizes only stability. Group 3 includes: TAI — linear response of a genotype to the environmental effects and deviation from the linear response (Tai, 1971); LIN — cluster analysis of Lin (1982); CA — correspondence analysis (Hill, 1974 and Lopez, 1990); SHUKLA — an unbiased estimate using Shukla's stability variance (Shukla, 1972); and EBRAS — yield stability measured via the regression approach (Eberhart and Russell, 1966).

Tai and Coleman (1999) effectively used path coefficient analysis to investigate GEI in potato. The path analysis method has not found much favor with most researchers. Nevertheless, Tai has expounded on the merits of this method (Tai, 1990).

Hussein et al. (2000) provided a comprehensive SAS program for computing univariate and multivariate stability statistics for balanced data. Their program provides estimates of more than 15 stability-related statistics.

Piepho (2000) proposed a mixed-model method to detect quantitative trait loci (QTLs) with significant effects and to characterize the stability of effects across environments. He treated environment main effects as random, which meant that both environmental main effects and QTL × E interaction effects could be regarded as random.

2.1.3 SIMULTANEOUS SELECTION FOR YIELD AND STABILITY

Simultaneous selection for yield and stability of performance is an important consideration in breeding programs. Researchers have attempted to jointly capture the effects of genotype and GEI. No methods developed so far have been universally adopted.

Kang and Pham (1991) discussed several methods of simultaneous selection for yield and stability and their interrelationships. An emphasis on stability in the selection process helps reduce Type II errors and benefits and protects growers (Kang, 1993). The development and use of yield-stability statistic (YS_i) has enabled incorporation of stability in the selection process (Kang, 1993). A computer program (STABLE) for calculating this statistic is freely available (Kang and Magari, 1995). Researchers have found Kang's yield-stability statistic (Kang, 1993) to be useful for recommending varieties (Pazdernik et al., 1997; Rao et al., 2002). Dashiell et al. (1994) and Fernandez (1991) evaluated the usefulness of several stability statistics for simultaneously selecting for high yield and stability of performance in soybean.

Annicchiarico (1997) developed an SAS-based computer program (STABSAS) that computes different measures of genotype stability. Recently, a comprehensive SAS program called SASG × ESTAB has become available (Hussein et al., 2000). This program computes univariate and multivariate stability statistics for balanced data. Specifically, the program calculates Tai's α and λ statistics, Wricke's ecovalence, Shukla's stability variance, Hanson's genotypic stability, Plaisted and Peterson's θ_i, Plaisted's $\theta_{i\text{-}bar}$, Francis and Kannenberg's environmental variance and coefficient of variance, and the rank-based nonparametric stability statistics of Hühn, Kang's rank sum, and the stratified rank analysis of genotypes. The program also calculates Type 4 stability, superiority measure, the desirability index of genotype performance, and pairwise GEI of genotypes with checks.

2.1.4 CONTRIBUTIONS OF ENVIRONMENTAL VARIABLES TO STABILITY

Yield components and plant characteristics — resistance or susceptibility to biotic and abiotic stress factors, such as resistance to pests and tolerance to environmental-stress factors, affect yield stability. By determining factors responsible for GEI or stability or instability, breeders can attempt to improve cultivar stability. If instability is caused by susceptibility to a disease, breeding for resistance to the disease should increase genotypic stability by assuring a relatively high performance in disease-prone environments.

Methods of assessing contributions of weather variables and other factors or covariates that contribute to GEI are available (Shukla, 1972; Denis, 1988; Magari et al., 1997; Van Eeuwijk et al., 1996). Contributions of different environmental variables to GEI have been estimated (e.g., Saeed and Francis, 1984; Kang and Gorman, 1989; Rameau and Denis, 1992; Charmet et al., 1993). Using data from multi-environment maize hybrid yield trials, Kang and Gorman (1989) removed from GEI heterogeneity caused by maximum and minimum temperatures, rainfall, and relative humidity. The percent heterogeneity removed via environmental index, which is the difference between mean of all genotypes in ith environment and grand mean, was 9.61% of total GEI. Heterogeneity removed via other covariates was lower. Similar results were reported for sorghum (Gorman et al., 1989), soybean (Kang et al., 1989), and other maize trials (Magari and Kang, 1993).

Because of the complexity of GEI, the relative contribution of a single environmental variable is very small in comparison with the total number of variables affecting GEI. A combination of two or more environmental variables helps remove more heterogeneity from GEI than individual variables do. Methods developed by Van Eeuwijk et al. (1996) may be useful for this purpose. Recently, Magari et al. (1997) determined contributions of individual environmental factors and combinations thereof to total GEI. They identified precipitation as the most important environmental factor that contributed to GEI for ear moisture loss rate in maize. Precipitation + growing-degree days from planting to black-layer maturity (GDD-BL) and relative humidity + GDD-BL were the two-factor combinations that explained a larger amount of GEI interaction than other combinations did.

Biadditive factorial regression models also can involve environmental covariates related to each deviation (Brancourt-Hulmel et al., 2000). The covariates included environmental main effect, sum of water deficits, an indicator of nitrogen stress, sum of daily radiation, high temperature, pressure of powdery mildew, and lodging. The models explained about 75% of the interaction sum of squares. The biadditive factorial biplot provided relevant information about the interaction of the genotypes with respect to environmental covariates.

2.1.5 STABILITY VARIANCE FOR UNBALANCED DATA

When a set of genotypes is grown in a specific set of environments, oftentimes a balanced data set is not possible, especially when a wide range of environments is used, or long-term trials are conducted. Hybrids/varieties are continually replaced year after year. Also, the number of replications may not be equal for all genotypes because experimental plots may be discarded for one reason or another. In such cases, plant breeders must deal with unbalanced data.

Researchers have used different approaches for studying GEI from unbalanced data (Freeman, 1975; Zhang and Geng, 1986; Piepho, 1994; Kang and Magari, 1996). Usually environmental effects are regarded as random and cultivar effects as fixed. Mixed model equations (MME) are useful in such situations (Henderson, 1975).

Kang and Magari (1996) used an unbalanced dataset from maize hybrid yield trials involving 11 hybrids grown at 4 locations for 4 years (1985 to 1988) to demonstrate a new procedure. Not all hybrids were included in all years and locations. Experiments were conducted using a randomized complete-block design with four replications per location. A few experimental plots were discarded during the course of the experiments for various reasons.

For calculating phenotypic stability variances, restricted maximum likelihood (REML) using the EM-type algorithm was employed (Patterson and Thompson, 1971). All variances were computed by iterating on MME and tested for H_0: $\sigma^2_{g(k)e} = 0$. MME gave best linear unbiased estimators (BLUE) for hybrids or fixed effects and best linear unbiased predictors (BLUP) for random effects (Henderson, 1975). Hybrid yields, BLUE, were the solutions for hybrid means from the MME equations.

For stability interpretation, $\sigma^2_{g(k)e}$ has the same properties as σ^2_i. It is a Type 2 stability statistic (Lin et al., 1986) or a statistic related to agronomic concept of stability (Becker, 1981). High values of $\sigma^2_{g(k)e}$ indicate that genotypes are not stable or that they interacted differentially with the environments.

REML was used to calculate stability variances because it can be effectively adapted to unbalanced data (Searle, 1987). The method is translation invariant and REML estimators of $\sigma^2_{g(k)e}$ will always be positive if positive starting values are used (Harville, 1977). Calculation of REML stability variances for unbalanced data allows one to obtain a reliable estimate of stability parameters as well as overcome the difficulties of manipulating unbalanced data. SAS_STABLE, a computer program for calculating stability variances, is available (Magari and Kang, 1997).

Balzarini et al. (2002) have recently illustrated the use of the BLUP methodology, mixed-model approach, to improve predictions of genotype performance in MET. The use of mixed models to analyze advanced variety trials offers the potential to improve predictive precision. It also allows researchers to objectively integrate GEI stability measures with mean performance (Balzarini et al., 2002).

2.2 DEALING WITH GENOTYPE-BY-ENVIRONMENT INTERACTION

Inconsistencies of genotype performance in different test environments make the job of a breeder difficult because no genotype is consistently superior in all test environments. In such situations, breeders may look for genotypes that perform relatively consistently across test environments, stable or broadly adapted genotypes, or choose different specifically adapted genotypes for production in different environments.

Three obvious ways of dealing with GEI in a breeding program (Eisemann et al., 1990) are: (1) ignore it, (2) avoid it, or (3) exploit it. Additional discussion on the three possible strategies is given in Chapter 5. Most breeders agree that GEI should not be ignored when it is significant and of crossover rank change type and indeed can be useful. Lack of consistency in genotype performance across environments provides additional information for the breeder (Busey, 1983). In addition to justifying the need for additional broad-based testing in different environments, the degree of inconsistency can help predict the variability expected among different farms (Busey, 1983). Another approach involves grouping similar environments or mega-environments via a cluster analysis. Within a cluster, environments would be more or less homogeneous, and genotypes evaluated in them would not be expected to show crossover interactions. To identify genotypes with broad adaptation (i.e., stable performance) across many sites, clustering of test environments or genotypes, however, is not advisable.

One may determine stability of performance across diverse environments and analyze and interpret genotypic and environmental differences. This approach allows researchers to identify genotypes that exhibit stable performance across diverse test environments, ascertain the causes of GEI, and devise strategies to correct the problems, or exploit them. For a known cause of unstable performance, either the genotype could be improved by genetic means or the proper environment of inputs and management could be provided to maximize productivity.

Broad adaptation or stability of performance reliability across environments helps conserve limited resources. P. Annicchiarico (personal communication, June 2000) uses the term adaptation in relation to locations, areas, regions, farming systems, or other aspects whose effects can be known in advance, prior to planting. He advocates breeding for wide or specific adaptation to a given region but with emphasis always on high stability. To achieve greater success, crop environments must be characterized as fully as possible for developing cultivars with wide adaptability or judiciously targeting appropriate cultivars to production environments. The assessment of the potential for genotype-by-location interaction from multi-location trials is important in crop improvement because these effects can be exploited for raising yields in a target region (Annicchiarico, 1999).

Some of the important strategies for minimizing undesirable aspects of and exploiting the beneficial potential of GEI through appropriate breeding, genetic, and statistical methodologies are outlined in the next section.

2.2.1 Correct Genetic Cause(s) of GE Interaction

Resistance or tolerance to any type of stress, biotic or abiotic, is essential for stable performance (Khush, 1993; Duvick, 1996). Sources of increased crop productivity include enhanced yield potential, heterosis, modified plant types, improved yield stability, gene pyramiding, and exotic and transgenic germplasm (Khush, 1993). The first important step toward developing resistance is to identify the factor(s) that are responsible for GEI.

If the causes (e.g., disease susceptibility) of interaction are traits with monogenic or digenic inheritance, solutions are relatively easy. Resistance or tolerance to stresses can be incorporated into commercial products. For traits with more complex inheritance (e.g., quantitative traits), population improvement via a cyclic selection procedure may need to be practiced before parental lines for hybrids or cultivars are developed.

Whether GEI should be avoided or exploited depends on whether or not environmental differences are predictable. If GEI is attributable to unpredictable environmental factors, such as year-to-year variation in weather variables, GEI should be avoided by selecting varieties with stable performance under a range of conditions. If GEI is caused by variations in predictable factors, such as soil type and cultural practices, GEI can be exploited by selecting varieties specifically adapted to an environment.

2.2.2 Characterize Genotypes and Environments

To address GEI concerns caused by various stresses, breeders need as much information about genotypes as possible. They also need to characterize environments (micro and macro) as fully as possible. Information on soil characteristics and ranges of weather variables and stresses that plant materials will be exposed to is a prerequisite to targeting appropriate cultivars to specific environments and maximizing productivity.

2.2.3 QTL-by-Environment Interaction

The GEI must be well understood to optimize its use (Alia et al., 1997). It is important to gain insights into the nature of GEI via molecular markers (Magari and Kang, 1997; Kang, 2000; Stuber and LeDeaux, 2000). Recent advances in molecular genetics have provided some of the best tools for obtaining deeper insights into the molecular mechanisms associated with GEI. Molecular markers, such as restriction-fragment length polymorphisms (RFLPs), microsatellites, single nucleotide polymorphisms (SNPs), and amplified fragment length polymorphisms (AFLPs), can be used to determine their association with QTL. By using genetic markers, it is possible to detect and map QTL (Beavis and Keim, 1996; Stuber and LeDeaux, 2000).

For quantitative traits, methodologies, such as marker-assisted selection, should be beneficial in dissecting and using GEI. Marker-assisted selection for desirable QTLs should be more reliable than phenotypic selection. It is highly desirable to identify QTL for a complex trait, such as high yield, expressed in a number of environments.

The QTL mapping has led to studies focusing on QTL-by-environment interaction (Paterson et al., 1991; Beavis and Keim, 1996; Sari-Gorla et al., 1997). Data must be obtained in multiple environments to investigate QTL × E interaction. Linkage between QTL and molecular markers can help identify genomic regions that show stable responses across diverse environments. This should enable breeders to manipulate QTL in the same fashion as single genes. This could substantially reduce breeding and evaluation time. Stuber et al. (1999) cautioned, however, that little was known about the stability of QTL alleles when transferred to different genetic backgrounds and when evaluated in varying environments.

The usual method, combined analyses, for evaluating QTL × E interaction has been problematic by increasing the number of parameters in the mapping model. A method of QTL × E analysis based on classical GEI analysis (Korol et al., 1995; Ronin et al., 1995) allows detection of QTL × E interaction across a large number of environments without causing an increased number of parameters in the mapping model. Jensen et al. (1995) proposed a mapping model that included environmental factors, which reduced somewhat the parameter problem.

2.2.4 Breeding for Stability/Reliability of Performance

GEI effects are directly involved in the determination of adaptability (Robbertse, 1989). Evans (1993) pointed to the need for developing new cultivars with broad adaptation to a number of diverse environments — selection for adaptability and the need of farmers to use new cultivars with reliable or consistent performance from year to year — reliability. Genetic improvement for low-input conditions requires capitalizing on GEI. Slower or limited gains in low-input or stress environments suggest that conventional high-input management of breeding nurseries and evaluation trials do not effectively select genotypes with improved performance under low-input levels (Smith et al., 1990). Selection for tolerance to stress generally reduces mean yield in nonstress environments and selection for mean productivity generally increases mean yields in both stress and nonstress environments (Rosielle and Hamblin, 1981).

From the standpoint of individual growers, stability across years, temporal stability, is most important. A breeder could test cultivars or lines for 10 to 15 years and identify those that have a high temporal stability. Then crosses may be made among the most stable cultivars to create source material or germplasm for subsequent development of inbred lines or pure lines. Therefore, extensive cultivar testing across years can be useful for cultivar development. Progress from selection would depend on heritability or repeatability of the stability statistic used.

Stability of cultivars would be enhanced if multiple resistance/tolerance to stress factors were incorporated into the germplasm used for cultivar development. If every cultivar — different genotypes, possessed equal resistance or tolerance to every major stress encountered in diverse target environments, GEI would be reduced to nonsignificant levels. On the other hand, if genotypes possessed differential levels of resistance — a heterogeneous group and, if somehow we could make all target environments as homogeneous as possible, GEI would again be reduced to nonsignificant levels. Since we do not have any control over unpredictable environments, such as year-to-year variation, the practical approach would be to equip genotypes with essential attributes, such as stress resistance or tolerance.

Stability analyses can be used to identify durable resistance to pathogens (Jenns et al., 1982). If a cultivar-by-isolate pathogen interaction exists, it would be necessary to identify a cultivar that has general resistance instead of specific resistance.

Another strategy might be to use indirect selection for the trait of interest. For indirect selection to be effective, the following requirements should be met (Sherrard et al., 1985; Jones, 1992):

- Yield component or biochemical or physiological trait must be easier to measure than yield or the complex trait itself
- Causal correlation must exist between the trait and yield
- Trait must have heritable variation
- Screening test should be simple, accurate, economical, and rapid; preferably capable of being used at seedling stage at any time of the year

2.2.5 Early Multi-Environment Testing

Because seed is usually in short supply in the early stages of a breeding program, extensive testing cannot be done. In clonally propagated crops, such as sugarcane or potato, one stalk of sugarcane or one tuber of potato can, however, be divided into at least two pieces and planted in more than one environment. Similarly, in cereal crops, if one has 20 kernels, one could plant 10 seeds each in two diverse environments. In the absence of a GEI, one would simply obtain a better evaluation of the genotypes, but if GEI were present, one would obtain precious information about consistency or inconsistency of genotype performance early in the program. This strategy would prevent genetic erosion resulting from testing done only in one environment.

2.2.6 Resource Allocation

GEI offers an opportunity to judiciously allocate resources in a breeding program (Pandey and Gardner, 1992; Magari et al., 1996). Carter et al. (1983) estimated that, at a low level of treatment × environment interaction such as 10% of error variance, testing in at least two environments was necessary to detect treatment differences of 20%, and it required at least seven environments to detect smaller (10%) treatment differences for growth analysis experiments in soybean. With a larger magnitude of interaction, a larger number of environments would be needed for a given level of precision in treatment differences.

Magari et al. (1996) used multi-environment, different planting dates, data for ear moisture loss rate in maize that exhibited a significant planting date × genotype interaction. Relative efficiency (RE) for the benchmark protocol was based on 11 plants/replication, three replications, and three planting dates (RE = 100%). The RE for five plants/plot in four replications and three planting dates was equivalent to that for the benchmark protocol. An RE of 100% also could be achieved with a sample of four planting dates, three replications, and three to four plants/plot.

2.3 GGE BIPLOT: GENOTYPE + GE INTERACTION

The biplot method originated with Gabriel (1971), and its use was subsequently expanded by Kempton (1984) and Zobel et al. (1988). The extensive usefulness of GGE biplot, where G = genotype effect and GE = genotype-by-environment effect, has only recently been elucidated (Yan et al., 2000). The GGE Biplot approach has strongly captured the imagination of plant breeders and production agronomists.

The GGE biplot is a multi-faceted tool in quantitative genetic analyses and plant breeding. In addition to dissecting GEI, GGE Biplot helps analyze genotype-by-trait data, genotype-by-marker data, and diallel cross data (Yan et al., 2000, 2001; Yan, 2001; Yan and Hunt, 2001, 2002; Yan and Rajcan, 2002). These aspects make GGE biplot a most comprehensive tool in quantitative genetics and plant breeding. The versatility of the GGE biplot analyses and software is elucidated in the next several chapters.

Section II

GGE Biplot and Multi-Environment Trial Data Analysis

3 Theory of Biplot

SUMMARY

This chapter introduces the theory of biplot through examining the process of matrix multiplication. A rank-two matrix results from multiplying one matrix with two columns with another matrix with two rows. A biplot displays a rank-two matrix through plotting its two component matrices in a scatter diagram. A biplot not only displays the two component matrices but also the product matrix, though implicitly. The elements of the product matrix are recovered through the inner-product property of the biplot. Moreover, a biplot allows visualization of the product matrix from all perspectives. These include:

- Ranking of the rows in each column
- Ranking of the columns in each row
- Comparison of two rows for all columns
- Comparison of two columns for all rows
- Identification of the largest value (row) within each column
- Identification of the largest value (column) within each row
- Visualization of the interrelationships among rows
- Visualization of the interrelationships among columns

These form the basis for visualizing two-way datasets. If the mathematics looks complicated, one may just skip this chapter. Nevertheless, it will not hinder one's ability to understand and apply the genotype and genotype-by-interaction (GGE) biplot methodology to one's data.

3.1 THE CONCEPT OF BIPLOT

3.1.1 MATRIX MULTIPLICATION

Two matrices can only be multiplied if the number of columns in the first matrix located on the left side of the multiplication sign equals the number of rows in the second matrix located on the right side of the multiplication sign. The product matrix assumes the number of rows of the left matrix and the number of columns of the right matrix. Thus the left matrix may be conveniently referred to as the row matrix and the right matrix the column matrix.

Assume that we have a row matrix:

$$G = \begin{bmatrix} 4 & 3 \\ -3 & 3 \\ 1 & -3 \\ 4 & 0 \end{bmatrix}$$

and a column matrix:

$$E = \begin{bmatrix} 2 & -3 & 3 \\ 4 & 1 & -2 \end{bmatrix}.$$

Since G has two columns and E has two rows, they can be multiplied; and since G has four rows and E has three columns, the product matrix Y will have four rows and three columns:

$$Y = G \times E = \begin{bmatrix} 4 & 3 \\ -3 & 3 \\ 1 & -3 \\ 4 & 0 \end{bmatrix} \begin{bmatrix} 2 & -3 & 3 \\ 4 & 1 & -2 \end{bmatrix} = \begin{bmatrix} 20 & -9 & 6 \\ 6 & 12 & -15 \\ -10 & -6 & 9 \\ 8 & -12 & 12 \end{bmatrix}.$$

The rule of matrix multiplication is that an element in Y is the sum of the product between each element of a row in the row matrix (G) and each element of a column in the column matrix (E). Thus, the $4 \times 3 = 12$ elements in Y are calculated as follows:

$Y11 = G11 \times E11 + G12 \times E21 = (4)(2) + (3)(4) = 20$
$Y12 = G11 \times E12 + G12 \times E22 = (4)(-3) + (3)(1) = -9$
$Y13 = G11 \times E13 + G12 \times E23 = (4)(3) + (3)(-2) = 6$

$Y21 = G21 \times E11 + G22 \times E21 = (-3)(2) + (3)(4) = 6$
$Y22 = G21 \times E12 + G22 \times E22 = (-3)(-3) + (3)(1) = 12$
$Y23 = G21 \times E13 + G22 \times E23 = (-3)(3) + (3)(-2) = -15$

$Y31 = G31 \times E11 + G32 \times E21 = (1)(2) + (-3)(4) = -10$
$Y32 = G31 \times E12 + G32 \times E22 = (1)(-3) + (-3)(1) = -6$
$Y33 = G31 \times E13 + G32 \times E23 = (1)(3) + (-3)(-2) = 9$

$Y41 = G41 \times E11 + G42 \times E21 = (4)(2) + (0)(4) = 8$
$Y42 = G41 \times E12 + G42 \times E22 = (4)(-3) + (0)(1) = -12$
$Y43 = G41 \times E13 + G42 \times E23 = (4)(3) + (0)(-2) = 12$

Y is called a rank-two matrix, since it results from multiplication of G, which has two columns, and E, which has two rows. A rank-two matrix can be displayed by a two-dimensional biplot (Gabriel, 1971).

3.1.2 Presenting the Matrix Multiplication Process in a Biplot

Now, we present matrices G and E in the form of a table (Table 3.1), in which the two columns of G are named as x and y, and the two rows of E are also named x and y. Likewise, we present matrix Y in the form of a table (Table 3.2), which indicates the row and column origin of each element. For example, Y11 is summation of the product between the first row of matrix G (G1) and the first column of matrix E (E1), and Y32 is the product between the third row of matrix G (G3) and the second column of matrix E (E2), etc.

If the x values are plotted against the y values in Table 3.1, a two-dimensional scatter diagram is generated, in which each row of G is represented by a point. Similarly, each column of E is represented by a point (Figure 3.1). This scatter diagram is called a biplot because it displays both the row and column matrices. The biplot not only presents the two original matrices, G and E, but it also displays their product, matrix Y. We show how Y is displayed next.

Theory of Biplot

TABLE 3.1
The Matrices G and E to be Multiplied

Names	x	y
Row matrix (G)		
G1	4	3
G2	−3	3
G3	1	−3
G4	4	0
Column matrix (E)		
E1	2	4
E2	−3	1
E3	3	−2

TABLE 3.2
Elements of the Product Matrix Y

	Columns		
Rows	E1	E2	E3
G1	20	−9	6
G2	6	12	−15
G3	−10	−6	9
G4	8	−12	12

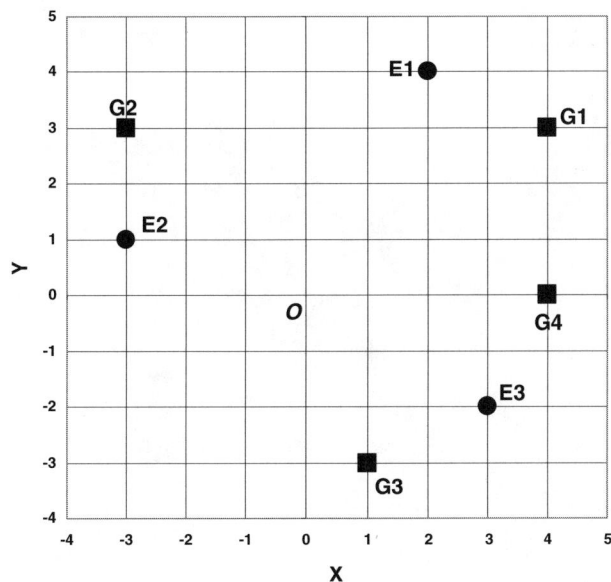

FIGURE 3.1 A biplot that displays both matrix G, which has four rows and two columns, and matrix E, which has two rows and three columns. G and E are multiplied to produce matrix Y with four rows and three columns.

3.2 THE INNER-PRODUCT PROPERTY OF A BIPLOT

As indicated earlier, the element Y11 in matrix Y is calculated as:

$$Y11 = G11 \times E11 + G12 \times E21 = (4)(2) + (3)(4) = 20.$$

In Figure 3.2, G11 and G12 are denoted as $x1$ and $y1$, respectively, and E11 and E21 are denoted as $x2$ and $y2$, respectively. Therefore:

$$Y11 = x1 x2 + y1 y2 .$$

The distance between the biplot origin O and the marker of E1, i.e., $\overline{OE1}$, is called the vector of E1, and the distance between the origin O and the marker of G1, i.e., $\overline{OG1}$, is called the vector of G1.

Figure 3.2 gives the following relationships:

$$x_1 = \overline{OG1} \cos \beta$$

$$y_1 = \overline{OG1} \sin \beta$$

and

$$x_2 = \overline{OE1} \cos(\alpha 11 + \beta)$$

$$y_2 = \overline{OE1} \sin(\alpha 11 + \beta).$$

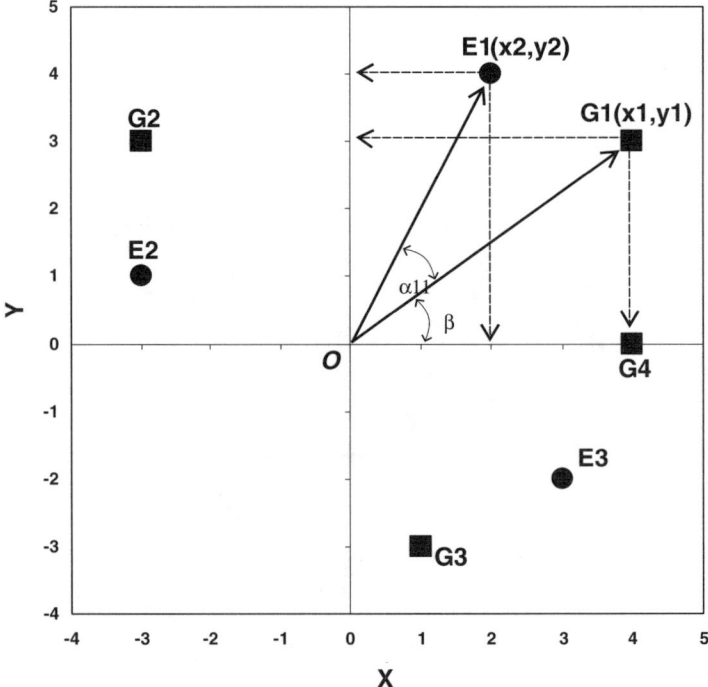

FIGURE 3.2 Derivation of the inner product from which elements of a rank-two matrix can be graphically recovered from its row vectors and column vectors. The row markers are defined by the row matrix and the column markers are defined by the column matrix.

Therefore:

$$Y11 = \overline{OG1}\,\overline{OE1}[(\cos\beta\cos(\alpha11+\beta)) + \sin\beta\sin(\alpha11+\beta)],$$

which can be simplified to:

$$Y11 = \overline{OG1}\,\overline{OE1}\cos[\beta - (\alpha11+\beta)]$$

$$= \overline{OG1}\,\overline{OE1}\cos(-\alpha11)$$

or

$$Y11 = \overline{OG1}\cos\alpha11\,\overline{OE1}. \qquad (3.1)$$

Thus, the element of the first row and first column in Y is the product of the vector of the first row ($\overline{OG1}$), the vector of the first column ($\overline{OE1}$), and the cosine of the angle between them ($\alpha11$) (Figure 3.2). To verify, the vectors can be calculated as follows:

$$\overline{OG1} = \sqrt{x1^2 + y1^2} = \sqrt{4^2 + 3^2} = 5$$

and

$$\overline{OE1} = \sqrt{x2^2 + y2^2} = \sqrt{2^2 + 4^2} = 4.471.$$

For the angle between the two vectors, since:

$$\alpha11 + \beta = arctg(y2/x2) = arctg(4/2) = 63.4°$$

and

$$\beta = arctg(y1/x1) = arctg(3/4) = 36.9°,$$

we have

$$\alpha11 = 63.4° - 36.9° = 26.5°$$

and

$$\cos\alpha11 = \cos 26.5° = 0.894.$$

Thus,

$$Y11 = \overline{OG1}\cos\alpha11\,\overline{OE1} = (5)(0.8944)(4.471) = 20.$$

Hence, Equation 3.1 is verified. To visualize Y11 from the biplot directly, Equation 3.1 can be written as follows:

$$Y11 = \left(\overline{OG1}\cos\alpha 11\right)\overline{OE1} = \overline{OP_{ge}}\,\overline{OE1} \tag{3.2}$$

where $\overline{OP_{ge}} = \overline{OG1}\cos\alpha 11 = (5)(0.8944) = 4.472$ is the graphic projection of vector G1 (i.e., $\overline{OG1}$) onto vector E1 ($\overline{OE1}$) (Figure 3.3a).

Alternatively, Equation 3.1 can be written as:

$$Y11 = \overline{OG1}\left(\cos\alpha 11\overline{OE1}\right) = \overline{OG1}\,\overline{OP_{eg}} \tag{3.3}$$

where $\overline{OP_{eg}} = \cos\alpha 11 \overline{OE1} = (0.894)(4.472) = 4$ is the graphic projection of vector E1 onto vector G1 (Figure 3.3b). Based on Equation 3.2, we have Y11 = (4.472)(4.472) = 20, and based on Equation 3.3, we have Y11 = (4)(5) = 20.

Equation 3.1 can be generalized as:

$$Y_{ij} = \overline{OG_i}\cos\alpha_{ij}\overline{OE_j} \tag{3.4}$$

which is referred to as the inner-product property of a biplot, where Y_{ij} is the element of Y in row i and column j, $\overline{OG_i}$ is vector of G_i, which is the distance between the biplot origin and the marker of G_i, $\overline{OE_j}$ is vector of E_j, which is the distance between the biplot origin and the marker of E_j, and α_{ij} is the angle between the vectors $\overline{OG_i}$ and $\overline{OE_j}$.

Note that $\overline{OG_i}$ and $\overline{OE_j}$ will never be negative, but cos α_{ij} can be positive or negative, depending on α_{ij}. Consequently, the sign of Y_{ij} is solely determined by α_{ij}. Y_{ij} will be zero if α_{ij} = 90°; Y_{ij} will be positive if α_{ij} < 90°; and Y_{ij} will be negative if α_{ij} > 90°.

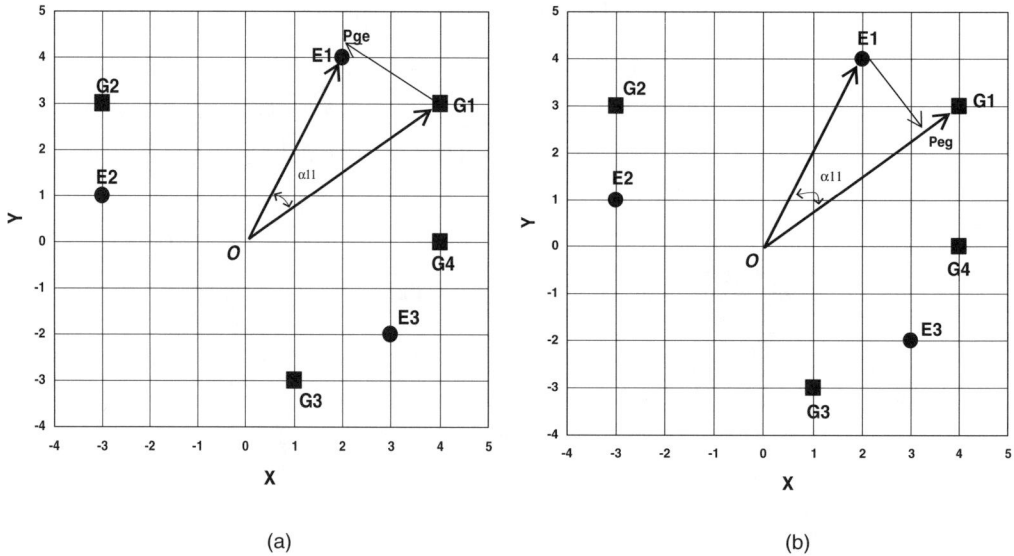

FIGURE 3.3 Inner-product of a biplot. The element (1,1) in the product matrix can be recovered either by (a) the vector of G1 multiplied by its projection onto the vector of E1 or (b) by the vector of E1 multiplied by its projection onto the vector of G1.

Theory of Biplot

3.3 VISUALIZING THE BIPLOT

As has been demonstrated, the biplot (Figure 3.1) displays not only the rows in matrix G and the columns in matrix E but also the product matrix Y between G and E. Moreover, in the light of Equation 3.4, the biplot also allows visualization of the relationships between the rows and the columns in different perspectives. A direct application of the inner-product property of a biplot is visual comparison of the row elements within a single column and of the column elements within a single row. An extended application is visual comparison of two rows for each of the columns and of two columns for each of the rows. A further extension is visual identification of the row that has the largest value within each column, and visual identification of the column that has the largest value within each row. The last application can be used to group the columns and the rows. These applications are detailed in this section.

3.3.1 VISUAL COMPARISON OF THE ROW ELEMENTS WITHIN A SINGLE COLUMN

Equation 3.4 can be rewritten as:

$$Y_{ij} = \left(\overline{OG_i}\cos\alpha_{ij}\right)\overline{OE_j} = \overline{OP_{ij}}\,\overline{OE_j} \tag{3.5}$$

where $\overline{OP_{ij}}$ is the projection of the vector of row i, $\overline{OG_i}$, onto the vector of column j, $\overline{OE_j}$. Since $\overline{OE_j}$ is common for all the elements in column E_j, Equation 3.5 can be further rewritten as:

$$Y_{ij}/\overline{OE_j} = \overline{OP_{ij}}. \tag{3.6}$$

In other words, the relative magnitude of the row elements in column E_j, i.e., $Y_{ij}/\overline{OE_j}$, can be visualized by simply comparing their projections onto the column vector $\overline{OP_{ij}}$. This is demonstrated in Figure 3.4 for column E1, where the projections of the four rows, G1 to G4, onto the vector of E1 are denoted as P11 to P41, respectively. Their projections are in the rank order of P11 > P41 > P21 > P31, among which P31 was negative; it was opposite to marker E1. Thus, Figure 3.4 indicates Y11 > Y41 > Y21 > 0 > Y31. This is exactly the case in matrix Y, where Y11, Y41, Y21, and Y31 are 20, 8, 6, and –10, respectively (Table 3.2).

3.3.2 VISUAL COMPARISON OF THE COLUMN ELEMENTS WITHIN A SINGLE ROW

Analogous to the previous section, the element at row i and column j in matrix Y (Y_{ij}) can also be expressed as the product of the vector of row i, $\overline{OG_i}$, and the projection of the vector of column j ($\overline{OE_j}$) onto $\overline{OG_i}$, which is $\overline{OP_{ji}}$:

$$Y_{ij} = \overline{OG_i}\left(\cos\alpha_{ij}\overline{OE_j}\right) = \overline{OG_i}\,\overline{OP_{ji}}. \tag{3.7}$$

Also, since all elements in row i share the same row vector, $\overline{OG_i}$, the relative magnitude of the elements within row i can be evaluated by simply visualizing the projections $\overline{OP_{ij}}$:

$$Y_{ij}/\overline{OG_i} = \overline{OP_{ji}}. \tag{3.8}$$

Figure 3.5 illustrates graphical comparison among the elements in row 1 of matrix Y. Since P11 > P13 > P12, the following relationship also holds: Y11 > Y13 > Y12. Since $\alpha 12 > 90°$ and, therefore, P12 is negative, Y12 also must be negative. This is exactly the case in Table 3.2, where Y11, Y13, and Y12 are 20, 6, and –9, respectively.

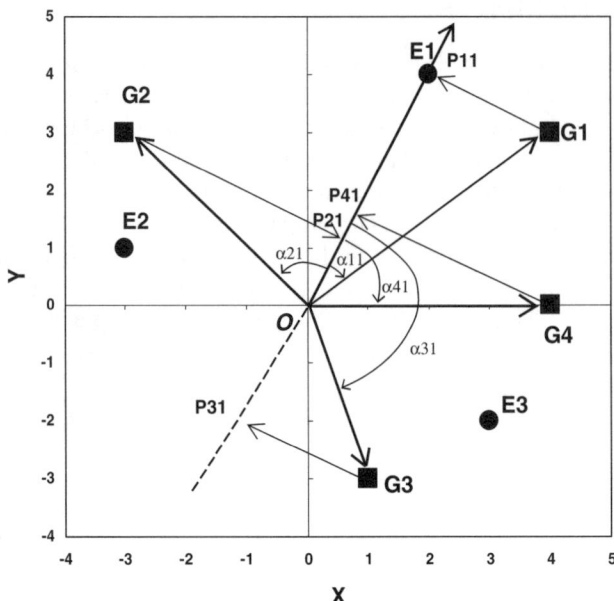

FIGURE 3.4 Comparison among the elements in the first column of the product matrix by visualizing their projections onto the vector of E1. Since P11 > P41 > P21 > 0 > P31; Y11 > Y41 > Y21 > 0 > Y31.

3.3.3 Visual Comparison of Two Row Factors for Each of the Columns

Figure 3.6a illustrates that two rows can be graphically compared for their values in combination with each of the columns. To compare rows G2 and G4, we draw a line, referred to as the connector line, that connects the markers of the two rows, and then draw another line, referred to as the perpendicular line, that passes through the biplot origin and is perpendicular to the connector line. Based on the inner-product property discussed earlier, G2 and G4 should have the same values in combination with any imaginary column factors that happen to be located on the perpendicular line, because they share the same projection length, OP, onto this line. Moreover, G2 and G4 will be equal only in combination with column factors that are located on the perpendicular line; they will differ for all other possible column factors. E2 is located on the same side of the perpendicular line as G2, suggesting that G2 should have a greater value than G4 relative to E2. Indeed, Y22 = 12 and Y42 = –12 (Table 3.2). On the contrary, E1 and E3 are located on the other side of the perpendicular line, i.e., on the same side as G4, suggesting that G4 has greater values than G2 with regard to these two column factors, i.e., E1 and E2. Again these can be verified from Table 3.2, where Y21 (= 6) < Y41 (= 8), and Y23 (= –15) < Y43 (= 12). Y21 and Y41 differ only by 2, as is reflected in Figure 3.6a, where E1 is close to the perpendicular line.

3.3.4 Visual Comparison of Two Column Factors for Each of the Rows

Analogous to Figure 3.6a, Figure 3.6b demonstrates graphical comparison of two column factors with regard to each of the rows. To compare column factors E1 and E2, we draw a connector line between them, and draw a perpendicular line that passes through the biplot origin. It becomes obvious that G2 and G3 are on the same side of the perpendicular line as E2, indicating that E2 has greater values than E1 in these two rows. On the contrary, G1 and G4 are on the same side of the perpendicular line as E1, indicating that E1 has greater values than E2 in rows G1 and G4. These statements can be easily verified from Table 3.2.

Theory of Biplot

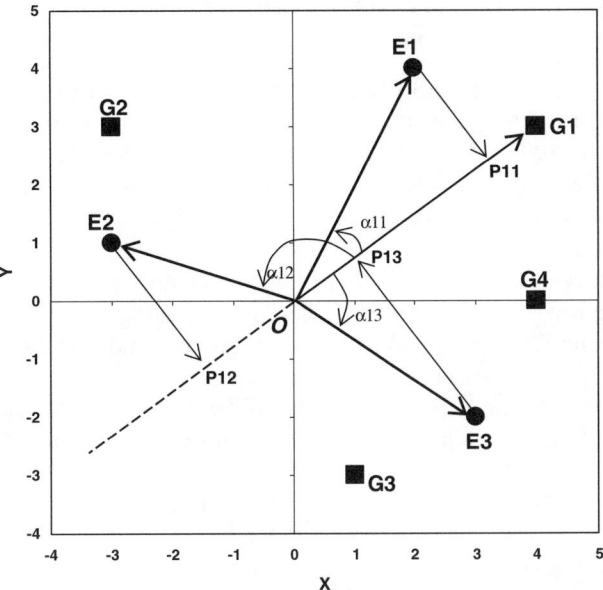

FIGURE 3.5 Comparison among the elements in the first row of the product matrix by visualizing their projections onto the vector of G1. Since P11 > P13 > 0 > P12; Y11 > Y13 > 0 > Y12.

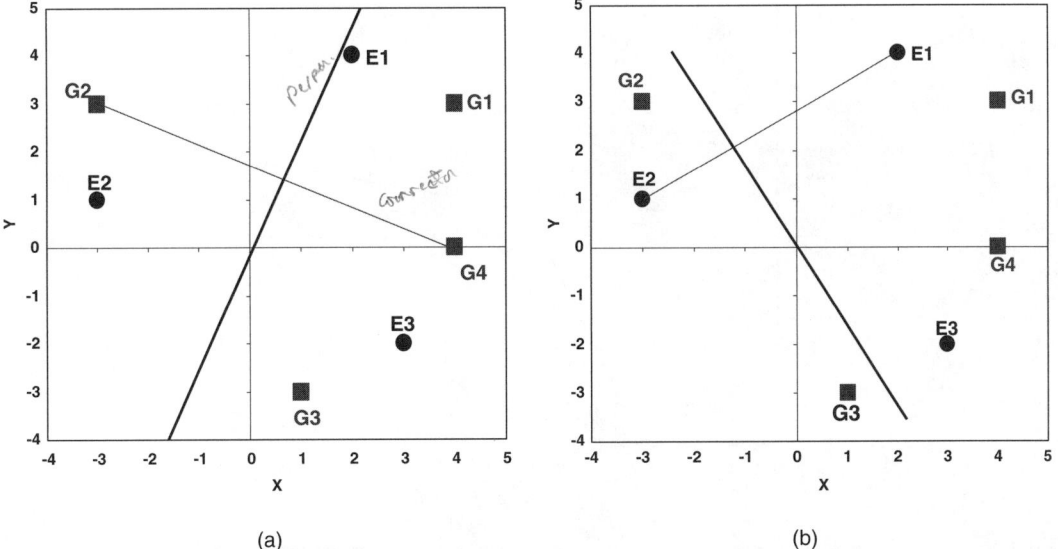

FIGURE 3.6 Visual comparison of two rows or genotypes (a) and of two columns or environments (b) based on the biplot.

3.3.5 Visual Identification of the Largest Column Factor within a Row

As described above, if two column factors are to be compared, they are connected with a connector line, and a perpendicular line is drawn to facilitate the comparison. To identify the column with the largest value within a row, comparisons must be made for all possible pairs of the three column factors. Thus, three connector lines are drawn, which form a triangle, and three respective perpendicular lines are drawn, starting from the biplot origin, which divide the biplot into three sectors (Figure 3.7). The following comparisons can be made from Figure 3.7:

- The line connecting E1 and E2 and its perpendicular line indicate that E1 has greater values than E2 in rows G1 and G4, and E2 has greater values than E1 in rows G2 and G3
- The line connecting E2 and E3 and its perpendicular line indicate that E2 has a greater value in G2, but E3 has greater values in G1, G3, and G4
- The line connecting E1 and E3 and its perpendicular line indicate that E1 has greater values in rows G1 and G2, but E3 has greater values in rows G3 and G4

Therefore, based on the first two comparisons, E2 has the largest value in row G2; in the last two comparisons, E3 has the largest values in rows G3 and G4; and based on the first and last comparisons, E1 has the largest value in G1. These conclusions can be easily verified from Table 3.2.

These statements can be generalized as follows: the perpendicular lines divide the biplot into three sectors, and the four row factors fall into different sectors. The column factor at the top of the triangle in each sector has the largest value in rows falling in that sector.

3.3.6 Visual Identification of the Largest Row Factor within a Column

To find the row factor with the largest value in a column, all possible pairs of rows must be compared. Since there are four rows in this sample, six possible pair-wise comparisons exist. The

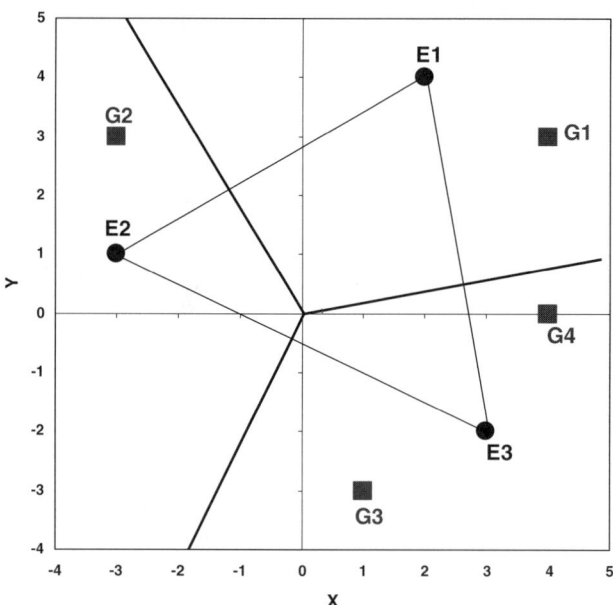

FIGURE 3.7 Visual identification of columns with the largest values for the rows. The column with the largest value for row G2 is E2, the column with the largest value for row G1 is E1, and the column with the largest values for rows G3 and G4 is E3.

Theory of Biplot

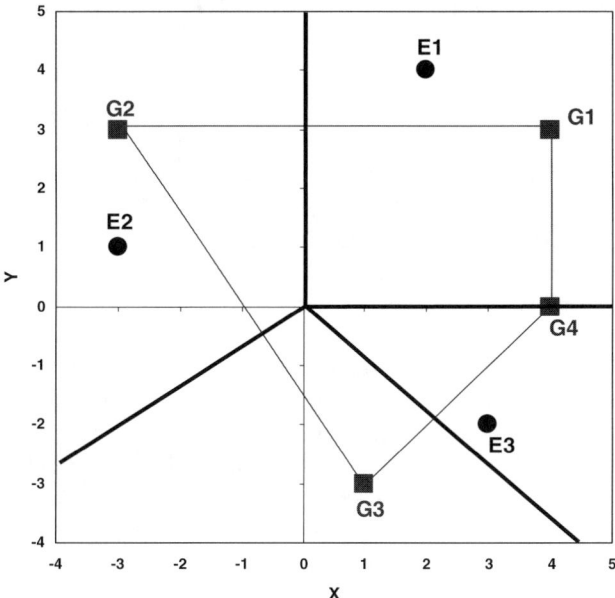

FIGURE 3.8 Visual identification of rows with the largest values for the columns. The row with the largest value for column E2 is G2, the row with the largest value for column E1 is G1, and row with the largest value for column E3 is G4. Row G3 does not have the largest values for any of the columns.

number of pairs quickly becomes unmanageable as the number of rows increases. Using a biplot, however, rows with the largest values for different columns can be easily visualized, no matter how many rows there are in a dataset.

To visualize, markers of the four rows are connected with straight lines to form a polygon, and lines are drawn from the biplot origin, which are perpendicular to each side of the polygon. The perpendicular lines divide the biplot into four sectors (Figure 3.8).

The perpendicular line between G1 and G2 indicates that G2 is greater than G1 in E2. The perpendicular line between G2 and G3 indicates that G2 is greater than G3 in E2. Although there is no direct comparison between G2 and G4, the perpendicular line between G1 and G4 indicates G1 > G4 in E2, and that between G3 and G4 indicates G3 > G4 in E2. Therefore, G2 has the largest value in E2. The largest row factor for other columns can be similarly determined. In general, the row factor at the vertex of the polygon in a sector has the largest values in all column factors that fall in that sector. Thus:

- G1 has the largest value in E1
- G2 has the largest value in E2
- G4 has the largest value in E3
- G3 does not have the largest value in any of the columns

Again, these statements can be easily verified from Table 3.2. The polygon view of a biplot makes identification of the largest row factor for each column, or identification of the largest column factor for each row, extremely convenient. Its advantage becomes more pronounced as the numbers of rows and columns in a matrix increase. A polygon can always be drawn regardless of the number of rows and columns.

3.4 RELATIONSHIPS AMONG COLUMNS AND AMONG ROWS

3.4.1 Relationships among Columns

Relationships among the three columns can be visualized by the angles among their vectors (Figure 3.9). The rule is that the cosine of the angle between the vectors of two columns approximates the correlation between them. The angle between the vectors of E1 and E2 is 98°, and the cosine of this angle is − 0.141. The angle between E1 and E3 is 97°, and cos97° = − 0.124. The angle between E2 and E3 is 165°, and cos165° = − 0.965.

Correlation coefficients and the angles in degrees among the columns calculated based on matrix Y are shown in Table 3.3. The angle between two columns is estimated as the arccosine of the correlation coefficient.

Comparison of the values in Table 3.3 with those presented in Figure 3.9 reveals that the two sets of values, although not exactly identical, correspond quite well. Note that the cosine of the angle between the vectors of two columns is determined solely by values in matrix E and has nothing to do with the values in matrix G (Table 3.1), whereas the correlation coefficient calculated based on matrix Y is dependent on both E and G. Consequently, angles between the columns of E in the biplot should be somewhat related to the correlation coefficients among the columns in Y, but no strict

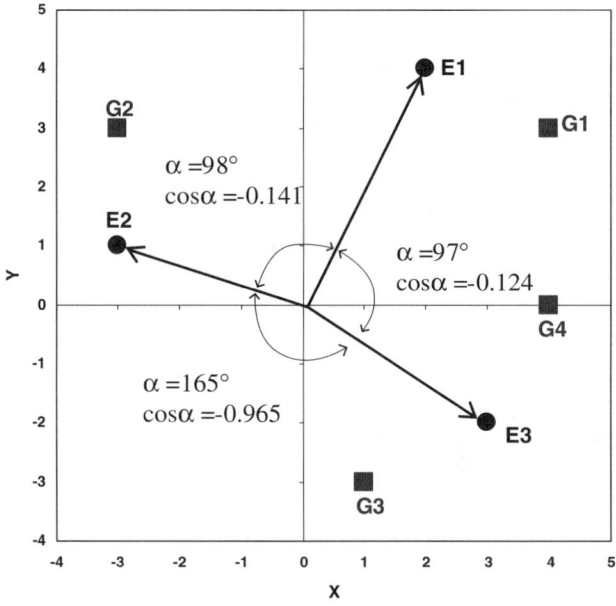

FIGURE 3.9 The cosine of the angle between the vectors of two columns represents an estimate of the correlation coefficient between the two columns.

TABLE 3.3
Correlation Coefficients among Columns and Predicted Angles among Column Vectors (in Parentheses)

	E2	E3
E1	−0.407 (114°)	−0.251 (105°)
E2		−0.977 (168°)

Theory of Biplot

correspondence should be expected. Closer or near-perfect correspondence is expected if matrix G has many rows (a large number of observations) and the rows are randomly scattered in the biplot.

3.4.2 Relationships among Rows

Relationships among rows also can be similarly visualized from the angles among their vectors (Figure 3.10). From the angles (Table 3.4), it can be predicted that rows G1 and G4 in Y are highly positively correlated (38°, acute angle), and rows G2 and G3 in matrix Y are highly negatively correlated (156°, obtuse), etc. These predictions are confirmed by the correlation coefficients among rows calculated based on matrix Y (Table 3.4).

Again, values in Table 3.4 should not be expected to match exactly those in Figure 3.10, since values in Figure 3.10 are based on matrix G alone, and values in Table 3.4 are based on matrix Y, which is dependent on both matrices G and E. Nevertheless, the angles among rows based on matrix G and angles among columns based on matrix E (Figures 3.9 and 3.10) are highly correlated with the corresponding angles based on Y (Figure 3.11). This property is useful for visualizing via a biplot the interrelationships among row factors and those among column factors of a two-way dataset.

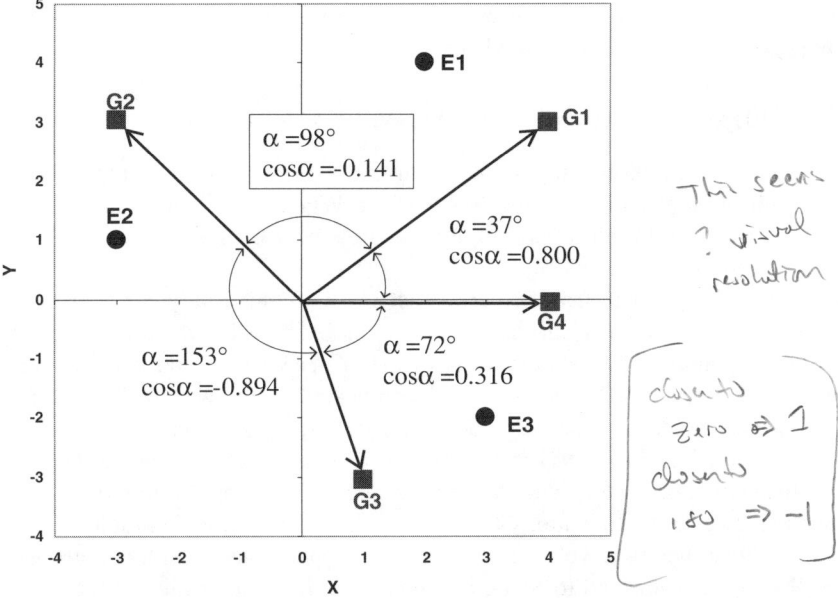

FIGURE 3.10 The cosine of the angle between the vectors of two rows represents an estimate of the correlation coefficient between the two rows.

TABLE 3.4
Correlation Coefficients among Rows and Predicted Angles among Row Vectors (in Parentheses)

	G2	G3	G4
G1	−0.231 (103°)	−0.180 (100°)	0.790 (38°)
G2		−0.915 (156°)	−0.779 (141°)
G3			0.461 (63°)

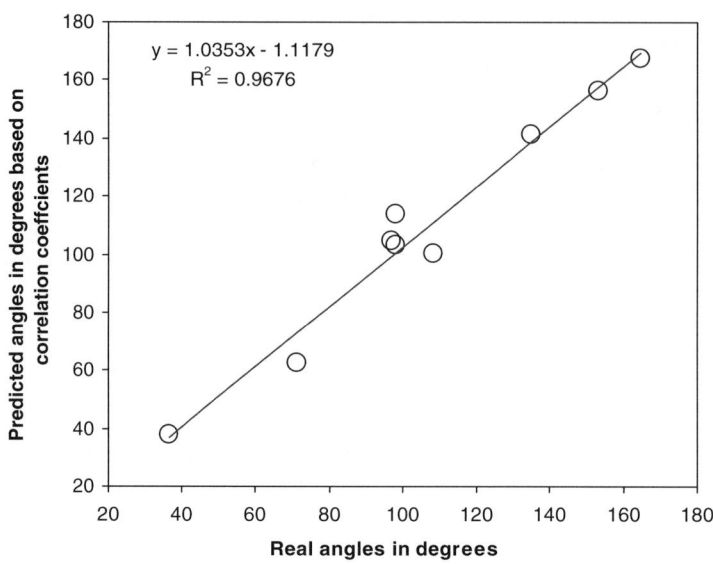

FIGURE 3.11 The relationship between true angles among rows and those among columns and their respective estimates based on correlation coefficients.

3.5 BIPLOT ANALYSIS OF TWO-WAY DATA

We have thus far assumed two known matrices, G, with two columns, and E, with two rows; their multiplication generates a rank-two matrix Y. When both G and E are plotted in a two-dimensional plot, which is referred to as a biplot, each element of matrix Y can be visualized and compared across rows and columns.

All two-way data from experimental research, such as yield data of a number of genotypes tested in a number of locations, can be visualized easily when they are displayed in a biplot. To display a genotype-by-location two-way data Y in a biplot, a prerequisite is to find its two component matrices G and E. The process of decomposing a two-way table (matrix) into two component matrices is called singular value decomposition (SVD), which is essentially the reverse process of matrix multiplication. SVD is a highly specialized mathematical subject and is beyond the scope of this book. Fortunately, due to its importance, SVD is built into most statistical analysis packages such as SAS (SAS Institute, 1996), and its implementation is simple.

Assume that we have yield data of n genotypes tested in m locations, which is an $n \times m$ matrix Y. When Y is subjected to SVD, k principal components result, where k is equal to the smaller of n and m (Equation 3.9). Each principal component is made of three parts: a singular value (λ), a set of genotype eigenvectors (ξ_i), and a set of location eigenvectors (η_j). Therefore, each element of Y is recovered by:

$$Y_{ij} = \sum_{l=1}^{k} \lambda_l \xi_{il} \eta_{lj} \tag{3.9}$$

where:

$i = 1\ldots n, j = 1\ldots m, k = \min(m,n),$
$Y_{ij} =$ yield of genotype i in location j,
$l =$ the rank of a principal component, $l = 1\ldots k$,
$\lambda_l =$ singular value of lth principal component, with $\lambda_1 > \lambda_2 > \ldots > \lambda_k$. The square of λ_l is the sum of squares explained by the lth principal component.

η_{lj} = the eigenvector or singular vector of location j for the lth principal component, and
ξ_{il} = the eigenvector or singular vector of genotype i for the lth principal component.

All k principal components are orthogonal to one another for both the genotypes and the environments:

$$\sum_{i=1}^{n} \xi_{il}\xi_{il'} = 0; \sum_{j=1}^{m} \eta_{lj}\eta_{l'j} = 0 \quad (3.10)$$

where l and l' represent different principal components. For each principal component, the genotype eigenvectors and the environment eigenvectors are both normalized and meet the following restrictions:

$$\sum_{i=1}^{n} \xi_{il} = 1; \sum_{j=1}^{m} \eta_{lj} = 1. \quad (3.11)$$

To present a genotype-by-environment two-way data in a k-dimensional biplot, the singular value for each principal component must be partitioned into the genotype and location eigenvectors, resulting in a row matrix G with n rows and k columns and a column matrix E with k rows and m columns. G and E are not unique, because there are numerous ways to partition the single values. This topic will be discussed in more detail in Chapter 4.

By subjecting a two-way data set to SVD, the resultant G and E matrices can be, theoretically, displayed in a k-dimensional biplot. Only two-dimensional biplots are easy to visualize and interpret, however. A two-dimensional biplot can be constructed using the first two principal components. Such a biplot is meaningful, however, when the first two principal components sufficiently approximate the yield matrix. This is one reason why a GGE biplot is preferred, which will be described in Chapter 4.

4 Introduction to GGE Biplot

SUMMARY

GGE refers to genotype main effect (G) plus genotype-by-environment interaction (GE). It is the part of the variation that is relevant to cultivar evaluation. Included in this chapter are: 1) the general formula for a GGE biplot, 2) two models (regression vs. singular value decomposition) for generating a GGE biplot, 3) three methods of data transformation for generating a GGE biplot, 4) four methods of singular value partitioning between the genotypic and environmental scores, and 5) a step-by-step instruction to construct a GGE biplot using SAS and Microsoft Office software. We trust the reader will realize at the conclusion of this chapter that generating a biplot is just the beginning, rather than the end, of biplot analysis.

4.1 THE CONCEPT OF GGE AND GGE BIPLOT

Everyone who studies biology should be familiar with the fact that the observed phenotypic variation (P) consists of variations of the environment (E), genotype (G), and genotype-by-environment interaction (GEI).

$$P = E + G + GE. \tag{4.1}$$

E may be further partitioned into year (Y), location (L), and year-by-location interaction (LY), and GE can be partitioned into genotype-by-year interaction (GY), genotype-by-location interaction (GL), and genotype-by-location-by-year interaction (GLY) (Comstock and Moll, 1963). For single-year multi-environment trials (MET), no GY, LY, and GLY can be estimated. Therefore E is composed only of L, and GE is composed only of GL.

Regardless of whether the data are from a single year MET or multi-year MET, a universal phenomenon in all regional yield trials is that E is always the predominant source of yield variation, and G and GE are relatively small (e.g., Gauch and Zobel, 1996). This is illustrated by data from winter wheat (Table 4.1) and soybean (Table 4.2) performance trials conducted in Ontario, Canada. The environment main effect E averaged 80% of the total yield variation across 13 years for winter wheat and 59% across 10 years for soybean. The environmental variation E for soybean was relatively small because 60 to 120 cultivars were tested in only 3 or 4 locations, as compared with the winter wheat trials in which 10 to 40 cultivars were tested in 8 to 10 locations.

The large environment main effect, however, is not relevant to cultivar evaluation. Only G and GE are relevant to cultivar evaluation. Therefore, for cultivar evaluation, it is essential to remove E from data and to focus on G and GE. Gauch and Zobel (1996) provided a vivid description of this idea. They compared cultivar evaluation based on MET data with taking pictures of a small cat sitting in a large window. It is logical to zoom in on the cat only rather than to include the frame of the window in the picture, if the purpose is not to show how small the cat is relative to the window.

Another important point in cultivar evaluation is that G and GE must be considered simultaneously to make any meaningful selection decisions. For the past 20 to 30 years, genotype-by-environment interaction (GEI) has been a hot topic among breeders, and particularly among

TABLE 4.1
Relative Magnitude of Genotype, Location, and Genotype-by-Location Interaction in Ontario Winter Wheat Performance Trials

	Percentage of Total Yield Variation		
Year	Genotype	Location	Interaction
1989	4.5	82.7	12.8
1990	4.2	83.6	12.2
1991	2.0	90.7	7.3
1992	9.9	78.2	11.9
1993	11.7	74.2	14.2
1994	4.5	80.4	15.1
1995	2.5	84.7	12.8
1996	1.8	86.5	11.6
1997	6.7	83.5	9.8
1998	28.5	55.4	16.1
1999	5.6	86.3	8.0
2000	14.2	73.2	12.6
2001	4.2	85.2	10.6
Mean	7.7	80.4	11.9

TABLE 4.2
Relative Magnitude of Genotype, Location, and Genotype-by-Location Interaction in Ontario Soybean Performance Trials

	Percentage of Total Yield Variation		
Year	Genotype	Location	Interaction
1991	20.2	56.9	22.9
1992	25.2	61.1	13.7
1993	5.6	86.6	7.8
1994	19.9	49.6	30.6
1995	32.8	26.0	41.2
1996	30.8	33.2	36.0
1997	15.4	64.4	20.1
1998	4.6	86.0	4.8
1999	15.9	63.2	20.9
2000	11.8	66.3	21.9
Mean	18.2	59.3	22.0

quantitative geneticists and biometricians. Total variation of yield (or any other trait) was partitioned into E, G, and GE, and these components were studied separately. Particularly, GE received much attention because G is so much more straightforward to visualize and use; G is the arithmetic mean of genotypes across environments. The emphasis on GE is evidenced by the numerous measures of stability index (Lin and Binns, 1994; Kang, 1998; Chapters 1 and 2 of this book). Some publications dealt with GE in such a manner as if GE were the only purpose of MET data analysis. In reality, however, many decisions are made solely based on G, which is equivalent to the mean yields of cultivars across environments. Indeed, selection based on G alone may be justified if GE is known to be random and cannot be exploited. On the contrary, meaningful decisions will never be achieved based on GE only.

The investigation of GE interaction began to make much sense following the advent of the concept of crossover interaction (Baker, 1990) or rank change (Hühn, 1996). That is, only crossover GEI, which causes cultivar rank change in different environments, affects genotype selection. To determine crossover interaction or rank change of cultivars in different environments, both G and GE must be considered. This implies that investigation into GE is much more meaningful when it is treated in conjunction with G. The points we want to make are: 1) E is large but irrelevant to cultivar evaluation, and therefore it should be removed from the data, and 2) only G and GE are relevant to meaningful cultivar evaluation, and they must be considered simultaneously in making selection decisions. The concept can be represented by the formula:

$$P - E = G + GE. \tag{4.2}$$

The term GGE is the contraction of G + GE. A biplot that displays the GGE of an MET dataset is called a GGE biplot. It allows the researcher to concentrate on the part of the MET data that is most useful for cultivar evaluation (Figure 4.1).

4.2 THE BASIC MODEL FOR A GGE BIPLOT

Equations 4.1 and 4.2 are in terms of variance components. When presented as effects, which have the unit of the originally measured values, they become:

$$\hat{Y}_{ij} = \mu + \alpha_i + \beta_j + \Phi_{ij} \tag{4.3}$$

or

$$\hat{Y}_{ij} - \mu - \beta_j = \alpha_i + \Phi_{ij} \tag{4.4}$$

where

$\hat{Y}_{ij}$ = the expected yield of genotype i in environment j,
μ = the grand mean of all observations,
α_i = the main effect of genotype i,
β_j = the main effect of environment j, and
Φ_{ij} = the interaction between genotype i and environment j.

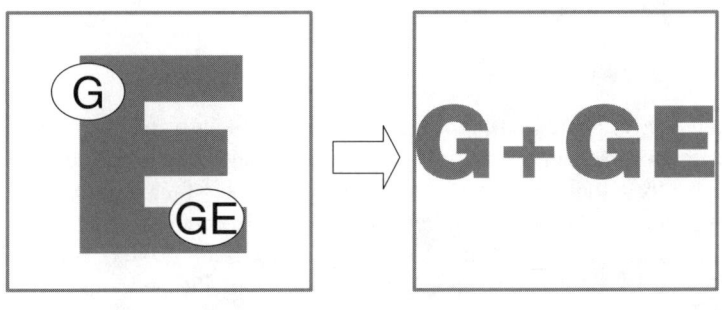

Full model   GGE model

FIGURE 4.1 A diagram comparing the full model and the GGE model. The full model contains E, which is typically many times larger than G and GE but is not relevant to cultivar evaluation. The GGE model allows the analysis to be focused only on the useful part of the data, i.e., G and GE.

Instead of trying to separate G and GE, a GGE biplot model keeps G and GE together and partitions this mixture GGE into two multiplicative terms:

$$\hat{Y}_{ij} - \mu - \beta_j = g_{i1}e_{1j} + g_{i2}e_{2j} + \varepsilon_{ij} \tag{4.5}$$

where g_{i1} and e_{1j} are called the primary scores for genotype i and environment j, respectively; g_{i2} and e_{2j}, the secondary scores for genotype i and environment j, respectively; and ε_{ij} is the residue not explained by the primary and secondary effects. A GGE biplot is constructed by plotting g_{i1} against g_{i2}, and e_{1j} against e_{2j} in a single scatter plot. The primary scores can be obtained through singular value decomposition (SVD) of the GGE (Section 4.3) or through regression of GGE against genotype main effects (Section 4.4).

4.3 METHODS OF SINGULAR VALUE PARTITIONING

The most common way to implement Equation 4.5 is by subjecting the GGE data to SVD, as mentioned in Chapter 3:

$$\hat{Y}_{ij} - \mu - \beta_j = \lambda_1 \xi_{i1} \eta_{1j} + \lambda_2 \xi_{i2} \eta_{2j} + \varepsilon_{ij} \tag{4.6}$$

where λ_1 and λ_2 are the singular values of first and second largest principal components, PC1 and PC2, respectively; the square of the singular value of a PC is the sum of squares explained by the PC; ξ_{i1} and ξ_{i2} are the eigenvectors of genotype i for PC1 and PC2, respectively; and η_{1j} and η_{2j} are the eigenvectors of environment j for PC1 and PC2, respectively.

PC1 and PC2 eigenvectors cannot be plotted directly to construct a meaningful biplot before the singular values are partitioned into the genotype and environment eigenvectors. The partition can be generalized by the following formula:

$$g_{il} = \lambda_l^{f_l} \xi_{il} \text{ and } e_{lj} = \lambda_l^{1-f_l} \eta_{lj} \tag{4.7}$$

where l can be 1 or 2, referring to the principal component number, and f is the partition factor. Theoretically, the partition factor (f_l) can be any value between 0 and 1. Within this range, the choice of f_l does not alter the relative relationships or interactions between the genotypes and the environments, although the appearance of the biplot will be different. Different f_l values, however, do affect visualization of the interrelationships among genotypes and those among environments. This area has rarely been discussed in the literature.

4.3.1 Genotype-Focused Scaling

When $f_{sl} = 1$, i.e., $g_{il} = \lambda_l \xi_{il}$ and $e_{\ell j} = \eta_{\ell j}$, the singular values are partitioned entirely into the genotype eigenvectors. The genotype scores are in principal coordinates. They have unrestricted means and their vector lengths are equal to the associated singular values that represent the original unit of the data (DeLacy et al., 1996a). On the other hand, the environment scores are in standard coordinates, i.e., they have zero means and unit lengths (DeLacy et al., 1996a). Since all of the singular values are partitioned into the genotype scores, the range of the genotype scores is likely to be much greater than that of the environment scores. Therefore, when a biplot is generated, environments are likely to be crowded relative to genotypes. To generate a biplot in which the ranges of genotypes and environments are comparable, the environment scores of both axes can be divided by a factor d ($d < 1$). Or equivalently, the genotype scores can be multiplied by this factor. This factor can be defined as:

Introduction to GGE Biplot

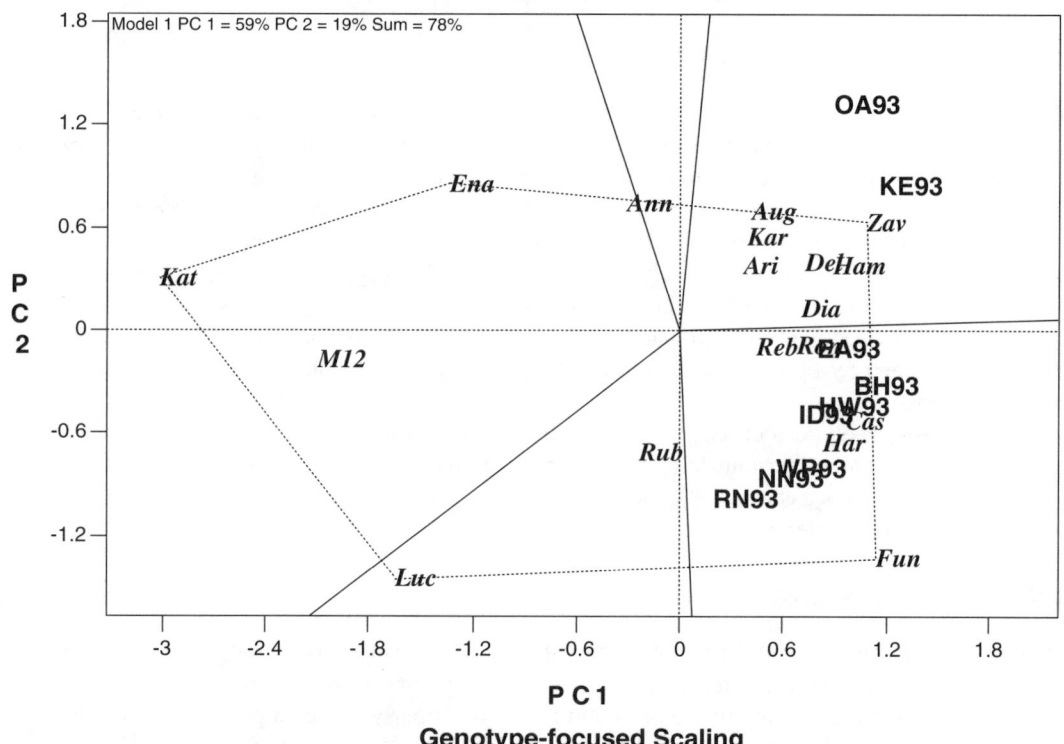

FIGURE 4.2 The GGE biplot based on genotype- or entry-focused singular value scaling. It is based on yield data of 1993 Ontario winter wheat performance trials. The 18 cultivars are displayed in italics and the 9 locations in regular uppercase.

$$d = \max(d_1, d_2)$$

where

$$d_l = \frac{\max(e_{il}) - \min(e_{il})}{\max(g_{lj}) - \min(g_{lj})} \tag{4.8}$$

l denotes the lth principal component.

The GGE biplot based on genotype-focused scaling is presented in Figure 4.2. Multiplying both axes of the environment scores with a positive number is equivalent to multiplying such a number to every element of the environment-centered data matrix, and it will not alter the GE pattern of the data, or the interrelationships among environments or those among genotypes.

4.3.2 Environment-Focused Scaling

When $f_l = 0$, i.e., $g_{il} = \xi_{il}$ and $e_{lj} = \lambda_l \eta_{lj}$, the singular values are partitioned entirely into environment scores. As a result, environment scores are in principal coordinates. Their means are not restricted and their vector lengths are equal to the associated singular values that assume the original unit of the data. On the contrary, the genotype scores are in standard coordinates, and they have zero means and unit lengths. Since the singular value is partitioned into environment scores, the range of environment scores is likely to be much greater than that of genotypes, and the genotypes are likely to be crowded in the biplot when directly plotted. To generate a biplot in which the ranges of genotypes and environments are comparable, the environment scores for both axes can be multiplied by a factor defined by Equation 4.8. Figure 4.3 represents the GGE biplot based on environment-focused scaling.

In both the genotype-focused scaling and the environment-focused scaling, genotype scores assume the same unit, either unit-less or the unit of original data, for both PC1 and PC2. Likewise, the environment scores assume the same units for both axes. The units for genotypes and environments are, however, different.

4.3.3 Symmetric Scaling

Symmetrical scaling occurs when f_l takes the value of 0.5. This type of scaling has a property that both PC1 and PC2 have the same unit, and both genotype and environment scores also have the same unit. It is the square root of the original unit. This property makes it possible to visualize the relative magnitude of PC1 vs. PC2, and the relative magnitude of genotype vs. environment variation for a given PC. To exemplify, the biplot for the 1993 Ontario winter wheat performance trial data using symmetrical scaling (Figure 4.4) seems to indicate that environments had greater variation than genotypes for PC1, but the opposite is true for PC2. This functionality is not available in all other scaling methods. Symmetric scaling is intermediate to entry-focused (entry = genotype) and tester-focused (tester = environment) scaling. Although it is not ideal for visualizing the interrelationships among either the entries or the testers, it is an approximation of both the entry-focused and the tester-focused scaling methods.

4.3.4 Equal-Space Scaling

The equal-space scaling method was devised such that the biplot space used by genotypes and that used by environments is the same (Yan et al., 2001). This is achieved by assigning the singular value partition factor to:

Introduction to GGE Biplot

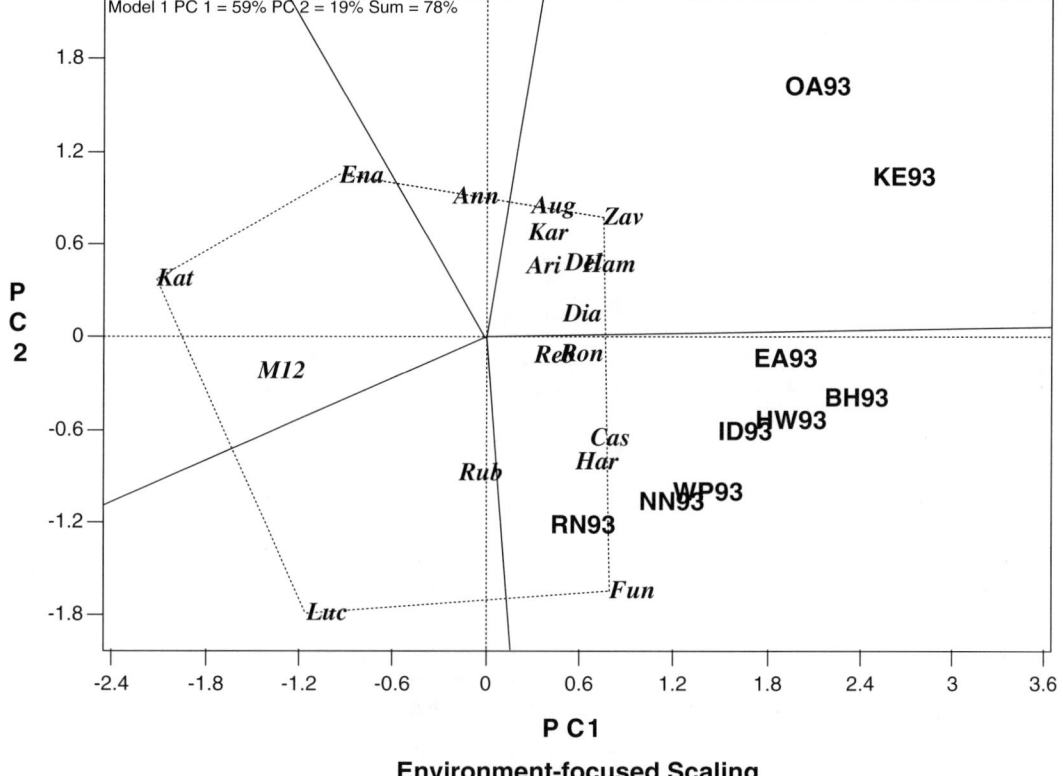

FIGURE 4.3 The GGE biplot based on environment- or tester-focused singular value scaling. Cultivars are in italics and environments are in regular uppercase. Based on yield data of 1993 Ontario winter wheat performance trials.

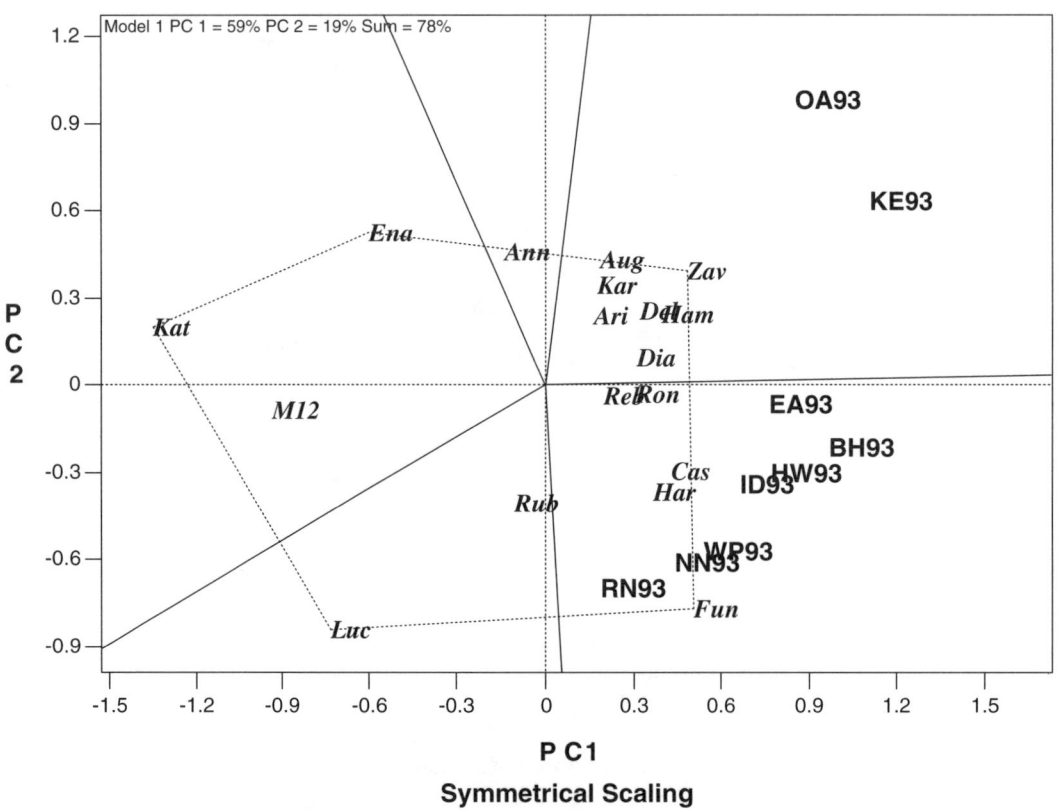

FIGURE 4.4 The GGE biplot based on symmetrical singular value scaling. Cultivars are in italics and environments are in regular uppercase. Based on yield data of 1993 Ontario winter wheat performance trials.

$$f_l = 0.5\left\{1 + \frac{\ln\left(\dfrac{\max(\eta_{lj}) - \min(\eta_{lj})}{\max(\xi_{il}) - \min(\xi_{il})}\right)}{\ln\lambda_l}\right\}. \quad (4.9)$$

In this scaling, the units of the genotype scores and of the environments scores are usually different; the units of PC1 and PC2 are also different. The equal-space scaling is equivalent to the symmetric scaling when $\max(\eta_{lj}) - \min(\eta_{lj}) = \max(\xi_{il}) - \min(\xi_{il})$ for both PC1 and PC2. Other than aesthetic aspects, a biplot based on this scaling (Figure 4.5) does not allow visual comparison of genotype variation vs. environment variation or comparison of PC1 variation vs. PC2 variation. Thus, the equal-space scaling method may be the least useful. It is necessary, however, to construct biplots based on models containing results from methods other than SVD (Section 4.4).

4.3.5 Merits of Different Scaling Methods

All four scaling methods, and any other possible methods, of singular value partitioning do not alter the GEI pattern, which is best represented by the polygon view of a biplot. The polygon view is presented for the example cited (Figures 4.2 to 4.5) because it shows the "which-won-where" pattern of MET data. Interpretation of such a biplot was alluded to in Chapter 3 and will be described in more detail in Chapter 5. Briefly, a polygon is drawn on genotypes located away from the biplot origin so that all other genotypes are contained within the polygon. Lines starting from the biplot origin and perpendicular to each side of the polygon are drawn, which divide the biplot into sectors. Within a sector, the genotype at the vertex of the polygon is the winner in all environments falling in the sector. Thus, regardless of the shape of the biplots, all biplots (Figures 4.2 to 4.5) indicate the same "which-won-where" pattern: genotype *zav* was the highest-yielding cultivar in environments OA93 and KE93, and genotype *fun* was the highest-yielding cultivar in the other seven environments. See Chapter 5 for a more detailed description of this dataset.

Different scaling methods do differ in several aspects, however. The first aspect is their accuracy in displaying the interrelationship among environments and among genotypes. The genotype-focused scaling (Figure 4.2) partitions the entire singular value into genotype scores. It thereby displays the interrelationship among genotypes more accurately than any other method does. It is obvious from Figure 4.2 that PC1 is much more important than PC2 for cultivar evaluation, as implied by the 59 vs.19% GGE variation explained by the two axes, respectively. Figure 4.2 does not correctly display the interrelationship among environments, however, since it displays only the standard coordinates of environments and does not reflect the relative importance or the weights of the two axes. This genotype-focused scaling has rarely been discussed in the literature.

The environment-focused scaling (Figure 4.3) is the most frequently used method and is referred to as the principal component analysis scaling. Since the singular values are fully incorporated into environment scores, this type of scaling is most informative of interrelationships among environments. Citing Kroonenberg (1995), DeLacy et al. (1996a) pointed out the following properties with regard to this type of scaling: 1) the cosine of the angle between any two environments approximates their correlation, with equality if the fit is perfect; 2) the lengths of the environment vectors are approximately proportional to their standard deviations, with exact proportionality if the fit is perfect; and 3) the inner product between two environments approximates their covariance, with equality if the fit is perfect. This is summarized by:

$$Cov(E_j, E_{j'}) = OE_j \cos\alpha_{jj'} OE_{j'} \quad (4.10)$$

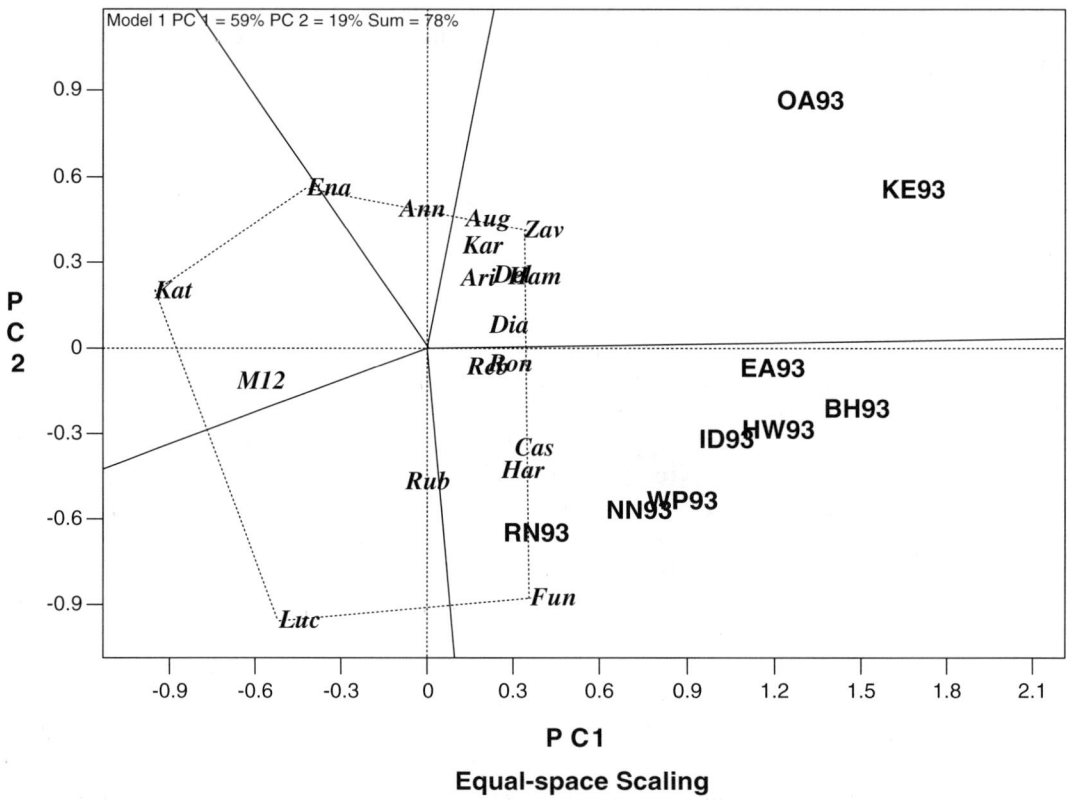

FIGURE 4.5 The GGE biplot based on equal-space singular value scaling. Cultivars are in italics and environments are in regular uppercase. Based on yield data of 1993 Ontario winter wheat performance trials.

where j and j' represent different environments, and OE_j and $OE_{j'}$ are the vector lengths of the two environments. The symmetric scaling method (Figure 4.4) is intermediate in all aspects between the genotype-focused scaling and the environment-focused scaling methods. Therefore, it does not have exact display of either the genotype relations or the environment relations. The equal-space scaling (Figure 4.5) is the least informative method in this respect.

The second differential aspect among the scaling methods is their allowance of visual comparison between PC1 and PC2 for genotypes and for environments. The genotype-focused scaling allows comparison of PC1 and PC2 for genotypes, whereas the environment-focused scaling allows comparison for environments. The symmetric scaling allows comparison for both genotypes and environments. The equal-space scaling method allows comparison neither for genotypes nor for environments.

The third differential aspect is their allowance of comparison of the relative variance between genotypes and environments. Only the symmetric scaling method allows this.

Thus, it seems that the equal-space scaling method is least desired for displaying interrelationships among genotypes and among environments, although it may produce a more presentable, less-crowded biplot. Each of the other three scaling methods has merits, but none of these is perfect in all aspects. Therefore, they should be used complementarily to achieve different purposes. All four singular value partition methods were built into the GGEbiplot software (Chapter 6), and switching between them for a given dataset is just a mouse-click away.

4.4 AN ALTERNATIVE MODEL FOR GGE BIPLOT

One valuable property of a GGE biplot is that it displays the "which-won-where" pattern of an MET dataset (Figures 4.2 to 4.4). It would be ideal if a biplot also could display mean performance and stability of cultivars. A GGE biplot based on Equation 4.6 does have this property as demonstrated via analyses of the yearly data of Ontario winter wheat performance trials (Yan et al., 2000), because PC1 scores of genotypes are highly correlated with their mean yields for most years, ranging from 1.00 to 0.82 for 1992 to 1998 (Yan and Hunt, 2001). This correlation becomes much smaller, however, if multi-year data are combined. For example, the correlation coefficient was only 0.58 when the 1996 to 1999 data for 11 commonly tested cultivars were combined to form an 11-genotype by 34-environment two-way table (Yan et al., 2001). Furthermore, the correlation between genotype PC1 scores and G is dependent on the magnitude of G relative to the total GGE (Figure 4.6). The correlation is greater than 0.90 when G is more than 20% of the total GGE. It drops considerably as the proportion of G is reduced (Figure 4.6). In such cases, the PC1 of the biplot cannot be interpreted as representing G — genotype mean performance.

To account for these exceptions, Yan et al. (2001) proposed an alternative model for generating a GGE biplot, which is written as:

$$\hat{Y}_{ij} - \beta_j = b_j \alpha_i + \lambda_1 \xi_{i1} \eta_{1j} + \varepsilon_{ij} \qquad (4.11)$$

where all symbols are similarly defined as for previous equations, except that b_j is the regression coefficient against G for environment j. In this model, the regression term, $b_j \alpha_i$, corresponds to the primary effects of Equations 4.5 and 4.6, and the PC1 corresponds to the secondary effect.

The singular value partition for the secondary effect in Equation 4.10 can follow Equation 4.7 so that:

$$g_{i2} = \lambda_1^{f_1} \xi_{i1} \text{ and } e_{2j} = \lambda_1^{1-f_1} \eta_{1j}$$

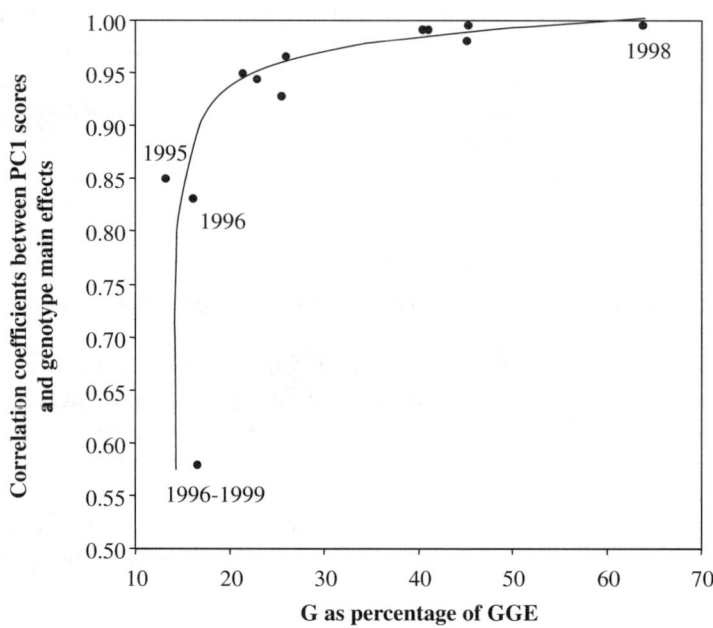

FIGURE 4.6 The correlation between G and genotype PC1 score in a GGE biplot is a function of the relative magnitude of G in total GGE. The correlation is high if G is more than 25% of the total GGE. Adapted from Yan, W. et al., *Crop Sci.*, 41:656–663, 2001.

with f_1 being determined by any of the four methods discussed in Section 4.3 but preferentially using the equal-space scaling (Equation 4.9), to be consistent with the scaling method for the primary effects, which is scaled by:

$$g_{i1} = B^{-1}\alpha_i \text{ and } e_{1j} = Bb_j \tag{4.12}$$

where

$$B = \sqrt{\frac{\max(\alpha_i) - \min(\alpha_i)}{\max(b_j) - \min(b_j)}}. \tag{4.13}$$

In general, GGE biplots, based on the principal component analysis (PCA) model, tend to explain more variation (Table 4.3). The PC1 of the PCA model always explains more variation than the regression term of the regression model. This is expected, since PC1 is obtained using the least squares solution, whereas the regression is restrained on G. This disadvantage of the regression model in accounting for less GGE by the primary effects was somewhat compensated for by its secondary effects. This is again expected, since more variation is left by the regression term in the regression model as compared with that by PC1 in the PCA model (Equation 4.6).

Nevertheless, the two models reveal similar "which-won-where" patterns (Yan et al., 2001), as demonstrated in Figures 4.7 and 4.8, which are GGE biplots based on Equation 4.10 — the regression model and Equation 4.6 — the PCA model, respectively, using the yield data of the 1996 Ontario winter wheat performance trials. This dataset was chosen because it represents datasets for which PC1 scores of the genotype have relatively poor correlations with the genotype main effects ($r = 0.82$, Yan and Hunt, 2001). Both biplots indicate that cultivar *2533* won or was close to winning in environments EA96, WK96, CA96, HW96, and OA96, and that cultivar *2510* won or was close to winning in environments RN96, LN96, WE96, and ID96.

Introduction to GGE Biplot

TABLE 4.3
Proportions of GGE Variation Explained by the PCA Model and the Regression Model for 12 Datasets from 1989 to 1999 Ontario Winter Wheat Performance Trials

| | | | | % of Total GGE Explained By | | | | | |
| | Number of Cultivars | Number of Locations | D.F. | The PCA Model (Equation. 4.6) | | | The Regression Model (Equation 2.10) | | |
Year				PC1	PC2	Total	Regression	PC1	Total
1989	10	9	32	42.5	21.3	63.8	40.7	21.9	62.6
1990	10	7	28	59.7	21.2	80.9	53.5	25.1	78.6
1991	10	9	32	53.3	20.7	74.0	49.1	22.1	71.2
1992	10	10	34	57.0	19.9	76.9	56.4	20.1	76.5
1993	18	9	48	56.8	20.0	76.8	55.4	21.2	76.6
1994	14	11	44	45.6	16.2	61.8	41.6	16.8	58.4
1995	14	14	50	54.2	13.4	67.6	40.8	25.2	66.0
1996	23	9	56	29.6	24.5	54.1	26.7	25.3	52.0
1997	28	8	66	55.0	15.9	70.9	54.0	15.9	69.9
1998	33	8	76	71.5	14.7	86.2	71.0	15.2	86.2
1999	31	9	74	51.5	17.4	68.9	50.7	17.7	68.4
1996–99	11	34	84	24.5	22.7	47.2	23.0	23.9	46.9
Mean				50.1	19.0	69.1	46.9	20.9	67.8

The regression model is designed to indicate explicitly the genotype main effects (Figure 4.9). We defer to Chapter 5 the detailed discussion of interpreting the biplot, but briefly, the abscissa of Figure 4.9 indicates the genotype main effects. Thus, based on mean yield, the cultivars are ranked in the order of *2533* > *2510* > *rub* > *fre* > ... > *dex* > *har*. This rank cannot be directly visualized from the PCA-based biplot (Figure 4.8).

Since the abscissa of the biplot based on the regression model (Figure 4.9) represents G, the ordinate of the biplot should represent stability of the genotypes. Thus, cultivars close to the abscissa, i.e., with small projections onto the *y*-axis, such as *har, reb, ari, car, mar*, and *rub* are relatively stable, whereas cultivars away from the abscissa, i.e., with large projections onto the *y*-axis, such as *dex, fun, dyn*, and *2737* below the biplot origin, and *f93, del*, and *2533* above the biplot origin, are less stable.

Because the two axes in the regression model have different units or scales, it is not possible, however, to visualize the relative magnitude of mean performance vs. stability of cultivars via Figure 4.9. This problem had not surfaced while the work reported in Yan et al. (2001) was in progress.

The PCA model explains more GGE than the regression model, and it also has the advantage of having both axes assume the same scale and units. The only disadvantage is that its abscissa (i.e., PC1) does not necessarily indicate the G, and as a result, its ordinate (PC2) does not necessarily indicate genotype stability.

A recent understanding is that through axis rotation, the GGE biplot based on the PCA model can sufficiently approximate the G and their stabilities (Yan, 2001). This is done by defining an average environment coordinate (AEC) (Figure 4.10). The various steps follow. First, define an average environment as average PC1 and PC2 scores of all environments. Second, draw a line that passes through the biplot origin and the average environment. This line is the abscissa of the AEC and indicates the mean performance of cultivars. The ordinate of the AEC is the line that passes through the biplot origin and is perpendicular to the AEC abscissa. It indicates stability of the cultivars. To verify this, we look at the ranking of the cultivars along the AEC abscissa: *2533* > *2510* = *fre* > *rub* = *mor* > ... > *har* > *dex*. This ranking is very close to the ranking shown in

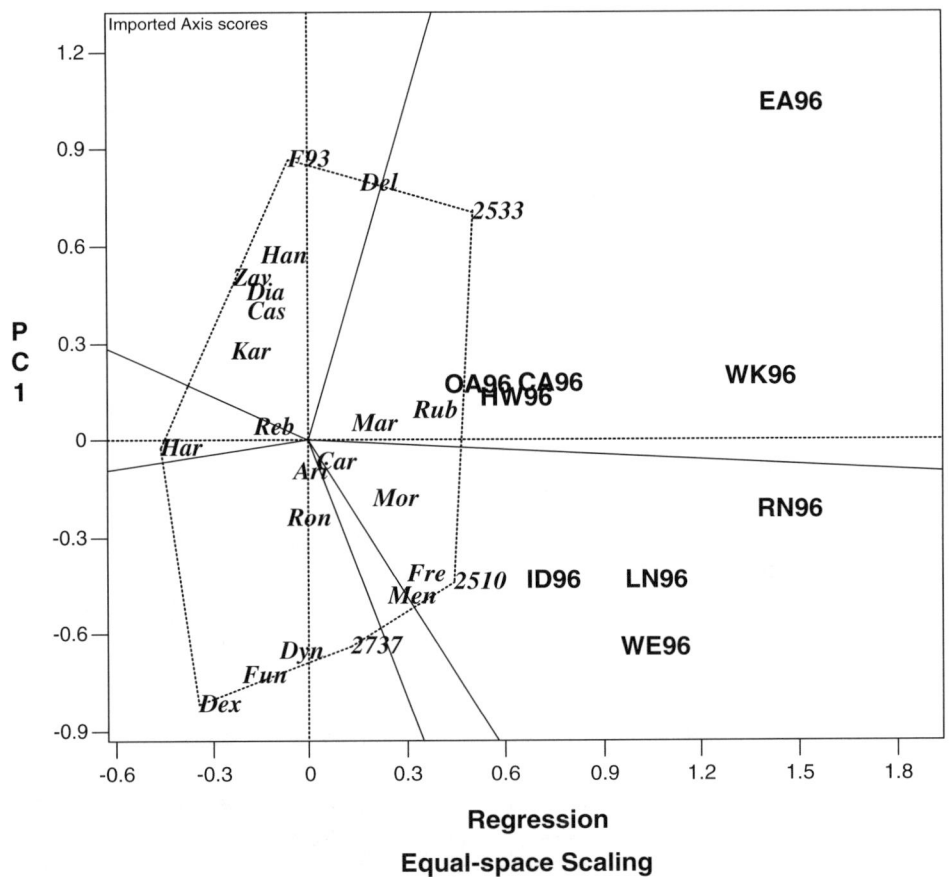

FIGURE 4.7 Polygon view of the GGE biplot based on the regression model (regression of environment-centered data on G, see text). Yield data of 1996 Ontario winter wheat performance trials are used here. Cultivars are in italics and environments are in regular uppercase. Equal-space scaling is used for constructing the biplot.

Introduction to GGE Biplot

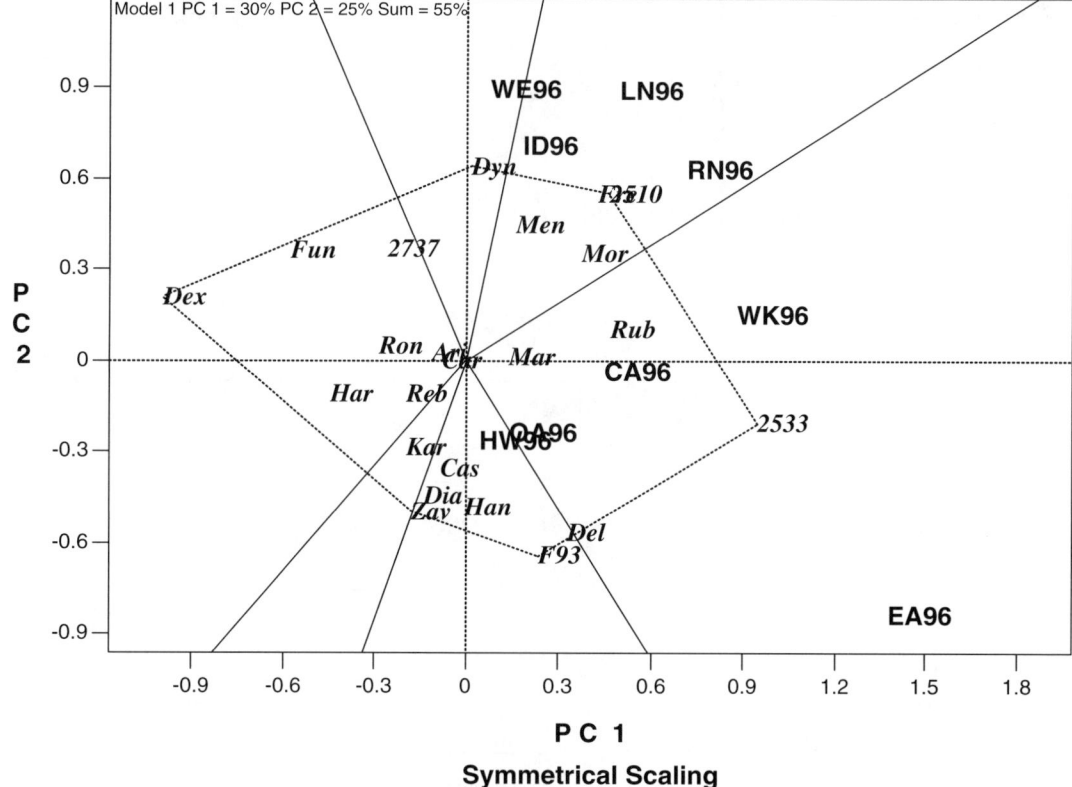

FIGURE 4.8 Polygon view of the GGE biplot based on the PCA model for yield data of 1996 Ontario winter wheat performance trials. Cultivars are in italics and environments are in regular uppercase. Symmetrical scaling is used for constructing the biplot.

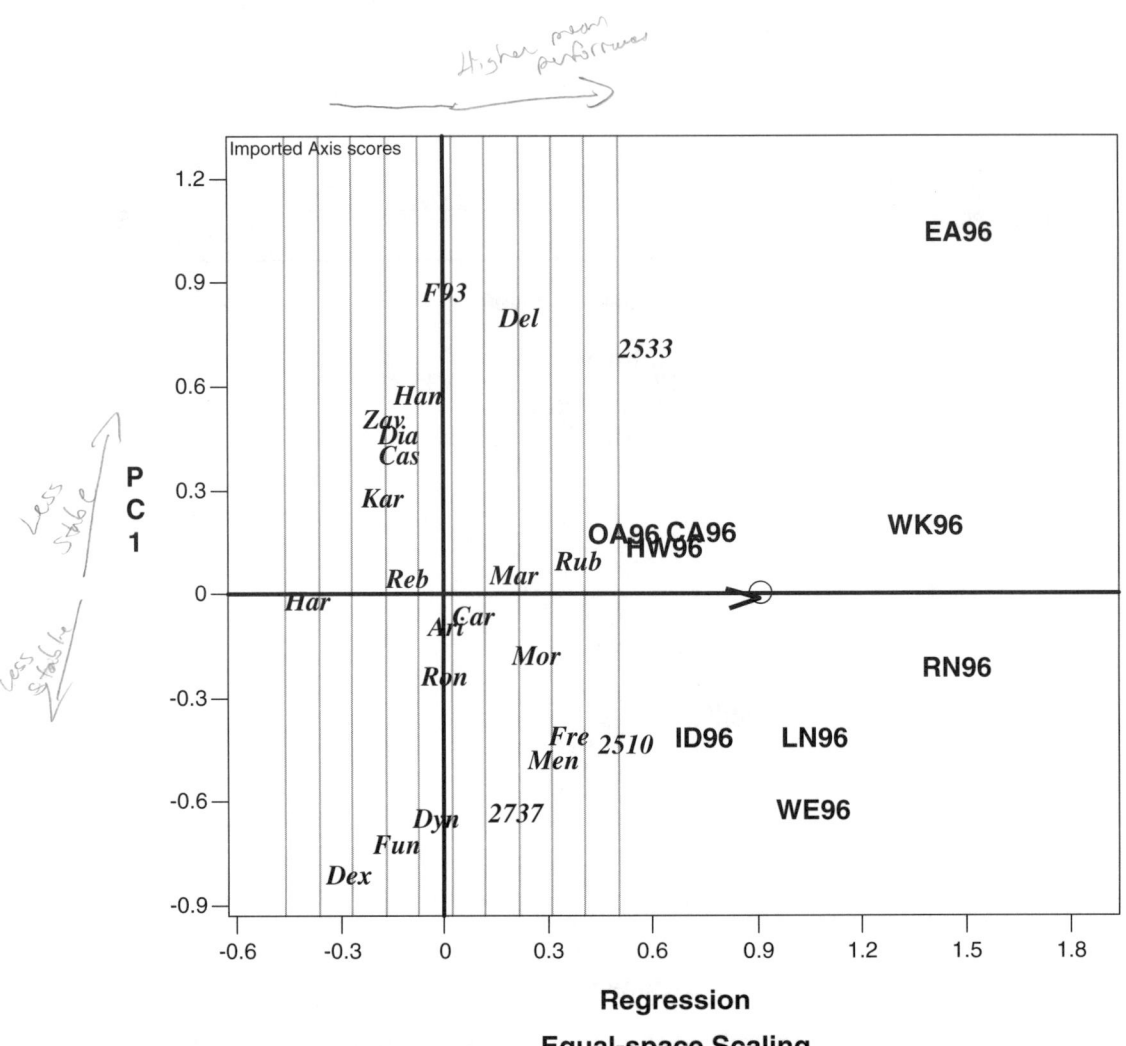

FIGURE 4.9 Average-environment coordinate (AEC) view of the GGE biplot based on the regression model (regression of environment-centered data on G, see text). Yield data of 1996 Ontario winter wheat performance trials are used here. As expected, the AEC abscissa is the abscissa of the biplot, indicating that the genotype scores are the genotype main effects per se.

Introduction to GGE Biplot

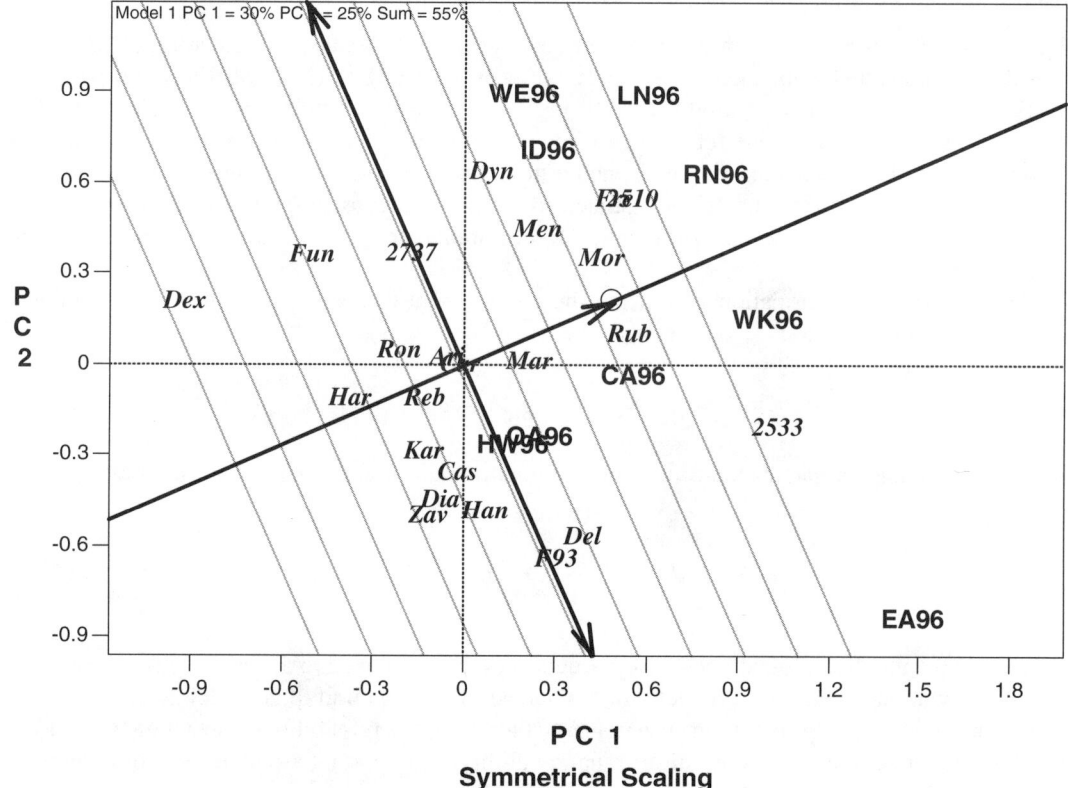

FIGURE 4.10 Average-environment coordinate (AEC) view of the GGE biplot based on the PCA model for yield data of 1996 Ontario winter wheat performance trials. The AEC abscissa, which assumes an angle with the abscissa of the biplot, approximates the G.

Figure 4.9. The projection of the cultivars onto the AEC abscissa is highly correlated with the genotype main effect ($r = 0.964$), as compared with the correlation of 0.82 between PC1 scores and the genotype main effect (Figure 4.6). That the AEC abscissa accurately indicates the genotype main effect is a general property of the GGE biplot.

The seemingly superior GGE biplot based on the regression model is actually not much superior for two reasons. First, its two axes assume different units, which makes it difficult to visualize the relative magnitude of mean performance and stability. Second, indication of the G also can be achieved by the PCA model through axis rotation. Therefore, the alternative GGE biplot based on the regression model will not be discussed hereafter in this book.

4.5 THREE TYPES OF DATA TRANSFORMATION

Equation 4.5 is referred to as an environment-centered model, because the environmental means ($\mu + \beta_j$) are subtracted from each of the observed mean values ($\hat{Y}_{ij}$). This model has two potential problems. First, it can only be used for analyzing MET data for a single trait, e.g., yield, and the trait must be in the same unit for all environments. It will not be appropriate for analyzing a genotype-by-trait two-way data, in which each trait assumes a different unit. Second, for MET data on a single trait, the environments are assumed to be homogeneous. It is, therefore, inappropriate for datasets with heterogeneous environments. To accommodate these scenarios, two data transformations are devised.

The first data transformation is to scale the environment-centered data with the within-environment standard deviation:

$$(\hat{Y}_{ij} - \mu - \beta_j) / d_j = g_{i1}e_{1j} + g_{i2}e_{2j} + \varepsilon_{ij} \qquad (4.14)$$

where d_j is the phenotypic standard deviation in environment j, and is calculated by:

$$d_j = \sqrt{\frac{1}{m-1} \sum_{i=1}^{m} (Y_{ij} - \bar{Y}_j)^2} \qquad (4.15)$$

where m is the number of genotypes tested. Equation 4.14 is referred to as the within-environment standard deviation-standardized model. For MET data of a single trait, this model assumes that all environments are equally important. It precludes, therefore, the possibility of detecting any differences among environments in their discriminating ability, an important aspect for test-environment evaluation. This topic is discussed in more detail in Chapter 5.

The second data transformation is to scale the environment-centered data with the within-environment standard error (s_j):

$$(\hat{Y}_{ij} - \mu - \beta_j) / s_j = g_{i1}e_{1j} + g_{i2}e_{2j} + \varepsilon_{ij} \qquad (4.16)$$

where s_j is the standard error in environment j, and is calculated by:

$$s_j = \sqrt{\frac{1}{m(r-1)} \sum_{i=1}^{m} \sum_{k=1}^{r} (Y_{ijk} - \bar{Y}_{ij.})^2} \qquad (4.17)$$

where r is the number of replicates in environment j. Equation 4.16 is referred to as the within-environment standard error-standardized model. Obviously, this model can only be used for replicated data, since s_j cannot be estimated from nonreplicated data where $r = 1$.

Introduction to GGE Biplot

GGE biplots based on the 1993 Ontario winter wheat performance trials using Equations 4.14 and 4.16 are presented in Figures 4.11 and 4.12, respectively. They can be compared with Figure 4.4, which used Equation 4.5. All these biplots used the symmetric scaling method. Note that the shapes of the biplots are somewhat different, although the "which-won-where" patterns are similar. In Figure 4.11, which is based on Equation 4.14, all environments have similar vector lengths. This is expected if the biplot sufficiently approximates the standard deviation-transformed data, because this model assumes all environments to be equally discriminating. Figures 4.4 and 4.12 do reveal differences among environments in discriminating among genotypes. Figure 4.4 reveals that OA93 and KE93 were more discriminating, whereas Figure 4.12 reveals that RN93, ID93, HW93, and EA93 were more discriminating. The latter is probably more valid since it takes into account the heterogeneity among environments.

4.6 GENERATING A GGE BIPLOT USING CONVENTIONAL METHODS

This section will provide step-by-step guidance for generating a GGE biplot based on actual data. The sample data represent yield data from the 1993 Ontario winter wheat performance trials, in which 18 genotypes were tested in 9 locations in 1993. Table 4.4 lists the mean yield of each genotype at each location.

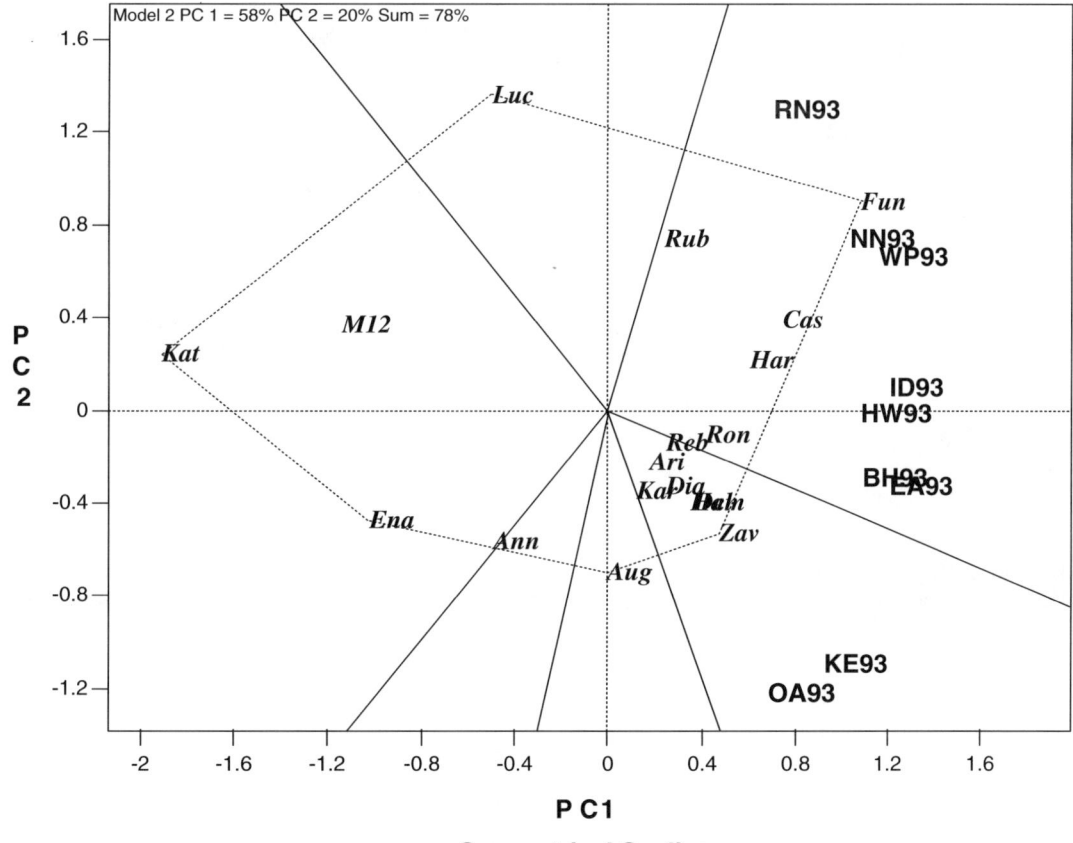

FIGURE 4.11 GGE biplot based on within-environment standard deviation-scaled data. Cultivars are in italics and environments are in regular uppercase. Based on yield data of 1993 Ontario winter wheat performance trials.

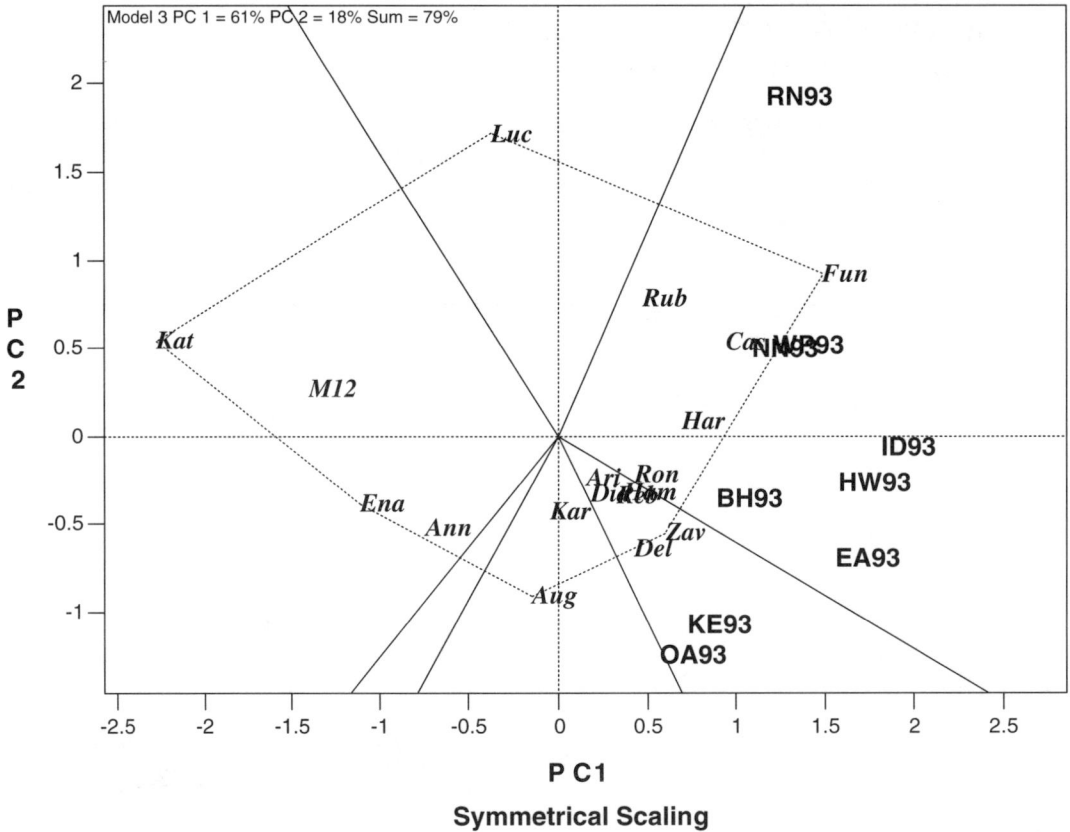

FIGURE 4.12 GGE biplot based on within-environment standard error-scaled data. Cultivars are in italics and environments are in regular uppercase. Based on yield data of 1993 Ontario winter wheat performance trials.

Introduction to GGE Biplot

TABLE 4.4
Yield Data of 18 Cultivars Tested in Nine Environments (from the 1993 Ontario Winter Wheat Performance Trials)

Names	BH93	EA93	HW93	ID93	KE93	NN93	OA93	RN93	WP93
Ann	4.460	4.150	2.849	3.084	5.940	4.450	4.351	4.039	2.672
Ari	4.417	4.771	2.912	3.506	5.699	5.152	4.956	4.386	2.938
Aug	4.669	4.578	3.098	3.460	6.070	5.025	4.730	3.900	2.621
Cas	4.732	4.745	3.375	3.904	6.224	5.340	4.226	4.893	3.451
Del	4.390	4.603	3.511	3.848	5.773	5.421	5.147	4.098	2.832
Dia	5.178	4.475	2.990	3.774	6.583	5.045	3.985	4.271	2.776
Ena	3.375	4.175	2.741	3.157	5.342	4.267	4.162	4.063	2.032
Fun	4.852	4.664	4.425	3.952	5.536	5.832	4.168	5.060	3.574
Ham	5.038	4.741	3.508	3.437	5.960	4.859	4.977	4.514	2.859
Har	5.195	4.662	3.596	3.759	5.937	5.345	3.895	4.450	3.300
Kar	4.293	4.530	2.760	3.422	6.142	5.250	4.856	4.137	3.149
Kat	3.151	3.040	2.388	2.350	4.229	4.257	3.384	4.071	2.103
Luc	4.104	3.878	2.302	3.718	4.555	5.149	2.596	4.956	2.886
m12	3.340	3.854	2.419	2.783	4.629	5.090	3.281	3.918	2.561
Reb	4.375	4.701	3.655	3.592	6.189	5.141	3.933	4.208	2.925
Ron	4.940	4.698	2.950	3.898	6.063	5.326	4.302	4.299	3.031
Rub	3.786	4.969	3.379	3.353	4.774	5.304	4.322	4.858	3.382
Zav	4.238	4.654	3.607	3.914	6.641	4.830	5.014	4.363	3.111

When the data are prepared in a text format, e.g., Table 4.4, and named w93.prn, running the SAS program given in Table 4.5 will lead to the output shown in Table 4.6. PRIN1 and PRIN2 in the listing denote PC1 and PC2 scores for the 9 environments and 18 genotypes. In the SAS procedure PRINCOMP, the singular value for each principal component is incorporated into the genotype (row) scores. Thus, a GGE biplot based on the genotype-focused scaling can be easily constructed.

The maximum and minimum PC1 scores for the environments were found to be 0.494 and 0.082, respectively, and they were 1.134 and –3.012 for the genotypes (Table 4.6). Thus, based on Equation 4.8, we have:

$$d_1 = \frac{0.494 - 0.082}{1.134 - (-3.012)} = \frac{0.41}{4.15} = 0.10$$

TABLE 4.5
SAS Program for PCA Analysis

```
data ds1;
    infile 'w93.prn' firstobs = 2;
    input geno $ BH93 EA93 HW93 ID93 KE93 NN93 OA93 RN93 WP93;
proc princomp n = 2 out = pcaout;
proc print data = pcaout;
    var prin1 prin2;
    ID GENO;
RUN;
```

TABLE 4.6
Part of the Output from Running the SAS Program in Table 4.5 Using Data in Table 4.4

```
The SAS System                     16:41 Thursday, August 30, 2001  2
                        Principal Component Analysis
                        Total Variance = 2.5076058497
                     Eigenvalues of the Covariance Matrix
                 Eigenvalue      Difference      Proportion     Cumulative
PRIN1            1.47693         0.996739        0.588979       0.588979
PRIN2            0.48019           .             0.191492       0.780471
                                Eigenvectors
                                  PRIN1             PRIN2
                    BH93         0.431198          0.138700
                    EA93         0.341176          0.048408
                    HW93         0.343467          0.189017
                    ID93         0.296789          0.211609
                    KE93         0.494251         -.364467
                    NN93         0.195934          0.373329
                    OA93         0.380041         -.568809
                    RN93         0.083052          0.424326
                    WP93         0.240023          0.349437
          The SAS System            16:41 Thursday, August 30, 2001  3
                    GENO          PRIN1             PRIN2
                    ann          -0.30821          -0.74380
                    ari           0.37302          -0.37252
                    aug           0.42221          -0.69256
                    cas           0.96125           0.53011
                    del           0.72381          -0.39324
                    dia           0.70500          -0.12507
                    ena          -1.33783          -0.85436
                    fun           1.13795           1.33273
                    ham           0.88603          -0.37953
                    har           0.82575           0.65541
                    kar           0.39031          -0.54781
                    kat          -3.01243          -0.30521
                    luc          -1.64126           1.45231
                    m12          -2.09507           0.17680
                    reb           0.44509           0.09603
                    ron           0.68358           0.08650
                    rub          -0.23910           0.71537
                    zav           1.07987          -0.63115
```

Similarly, the minimum and maximum PC2 scores for both the genotype and the environment can be found from Table 4.6 and d_2 calculated as:

$$d_2 = \frac{0.426 - (-0.567)}{1.453 - (-0.855)} = \frac{1.00}{2.31} = 0.43.$$

We therefore use $d = \max(d_1, d_2) = 0.43$ to divide the environment scores or to multiply the genotype scores. In this example, we divided the environment scores of both axes by 0.43 (see resultant environment and genotype scores in Table 4.7). Plotting PC1 against PC2 for both environments and genotypes leads to Figure 4.13. Note that both PC1 and PC2 are made to assume the same physical scale, which is crucial for valid visualization.

TABLE 4.7
Environment and Genotype Scores Used for Generating a GGE Biplot

Environments	PC1	PC2
BH93	0.960	0.308
EA93	0.757	0.108
HW93	0.763	0.418
ID93	0.658	0.473
KE93	1.099	−0.808
NN93	0.433	0.830
OA93	0.847	−1.260
RN93	0.183	0.947
WP93	0.533	0.779

Genotypes	PC1	PC2
Ann	−0.307	−0.745
Ari	0.378	−0.371
Aug	0.425	−0.694
Cas	0.958	0.530
Del	0.723	−0.393
Dia	0.704	−0.122
Ena	−1.338	−0.855
Fun	1.134	1.328
Ham	0.891	−0.379
Har	0.827	0.650
Kar	0.391	−0.547
Kat	−3.012	−0.308
Luc	−1.643	1.453
M12	−2.097	0.173
Reb	0.446	0.097
Ron	0.684	0.092
Rub	−0.242	0.718
Zav	1.078	−0.628

It is obvious that Figure 4.13 is identical to Figure 4.2 except for two things: 1) the PC2 axis in Figure 4.13 is reversed compared with that in Figure 4.2, and 2) the genotypes and the environments in Figure 4.13 are not labeled with names. These differences are created on purpose to make two points. First, the biplot pattern will not be altered when both the genotype and the environment scores of a principal component are multiplied by −1, which reverses the direction of the axis. Second, it is a tedious job to manually label the genotype and environment names on a biplot, particularly when there are many of them.

Figure 4.13 is a genotype-focused GGE biplot since the singular values in the PRINCOMP procedure of SAS are partitioned into the genotype singular vectors by default. Table 4.6 contains enough information to generate an environment-focused GGE biplot, however. To do this, we first calculate the sum of squares (SS) explained by each PC, which is the product of the eigenvalue for each PC multiplied by the number of genotypes (referred to as number of observations in SAS). The SS for PC1 is thus $1.477 \times 18 = 26.58$. The singular value for PC1 is the square root of 26.58, which is 5.156. Multiplying the environment scores and simultaneously dividing the genotype scores of PC1 using this value will re-partition the whole PC1 singular value to the environment scores. Likewise, for PC2, the environment scores should be multiplied, and genotype scores

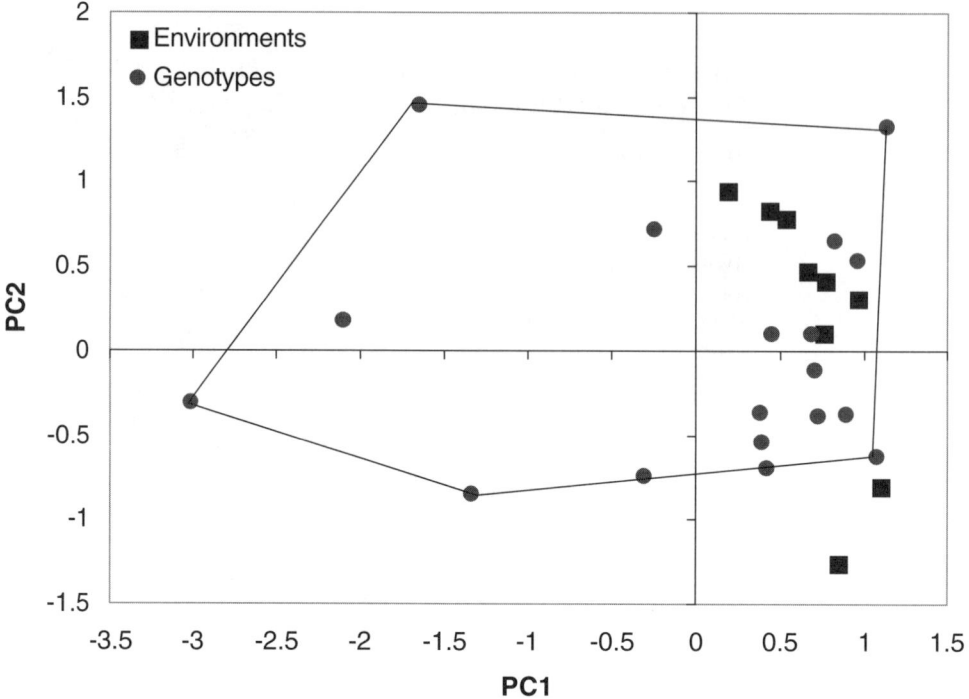

FIGURE 4.13 Hand-made GGE biplot using Microsoft Excel from scores generated by an SAS program. Based on yield data of 1993 Ontario winter wheat performance trials. Cultivars or genotypes are represented by circles and environments by squares.

divided, by the singular value, which is 2.94. Once the singular values are re-partitioned into the environment scores, an environment-focused GGE biplot can be easily constructed following the steps described for the genotype-focused GGE biplot. Once we know how to determine the singular values from the SAS output, a symmetrically scaled GGE biplot also can be easily constructed.

A point we now want to make is that although GGE biplot analysis is elegant and effective, generating a GGE biplot using conventional tools is quite tedious. You must run SAS and get SAS output; you must open the SAS output file with Microsoft Excel and convert the PC1 and PC2 scores into values required for generating a biplot; you must draw the plot carefully and make sure the two axes are in the same physical scales; and you must label the genotype and environment names on the biplot so that the biplot can be properly visualized. Furthermore, you may need to generate different types of biplots since different biplots have different utilities. Those sufficiently trained in SAS can have SAS generate reasonably good biplots, with biplot scales exactly adjusted and all the names precisely labeled. Nevertheless, biplot analysis is still a tedious job. As will be seen in Chapter 5, having a biplot generated is just the beginning, rather than the end, of biplot analysis. Therefore, there is a real need for highly specialized software to assist researchers interested in biplot analysis. Such software is now available. A preliminary version is described in Yan (2001), and a more advanced version is described in Chapter 6. The GGEbiplot software is continually evolving.

5 Biplot Analysis of Multi-Environment Trial Data

SUMMARY

This chapter deals with analysis of multi-environment trial (MET) data and identifies its four major tasks: 1) mega-environment investigation for understanding the target environment, 2) cultivar evaluation for each mega-environment, 3) test-location evaluation for each mega-environment, and 4) understanding causes of genotype-by-environment interaction (GEI). This chapter details how the genotype and genotype-by-environment (GGE) biplot technique can be used to achieve each of these goals. You will be amazed at how powerful the GGE biplot analysis is and how it can help you understand your data more fully.

5.1 OBJECTIVES OF MULTI-ENVIRONMENT TRIAL DATA ANALYSIS

One may think it a kindergarten-level question to ask what is the purpose of a MET data analysis. Of course, it is to identify the best cultivars. There are other issues, however, that must be addressed before one can effectively address the problem of cultivar evaluation. The good news is that information exists in, and can be extracted from, MET data, which may be beneficial for cultivar evaluation in the short as well as long run.

GEI is an issue that cannot be avoided in MET data analysis. Actually, GEI is the primary reason why MET rather than a single-environment trial must be conducted for cultivar evaluation. We wish to emphasize that GEI is not the only factor that must be considered in cultivar evaluation and selection: the other factor, which may be more important than GEI, is G. Analysis of MET data may be equated to GGE analysis, which is the gist of this book, particularly this chapter. The flow chart (Figure 5.1), modified from Yan and Hunt (1998), summarizes the tasks of MET data analysis. For a breeding program, the first question to ask is, "Is there any GEI?" If no interaction exists, any single environment should suffice for selecting best cultivars; if interaction does exist, the second question to ask is, "Is there any crossover GEI?" The concept of crossover interaction (Baker, 1990) is important because it implies that both G and GEI must be considered simultaneously in cultivar evaluation. This concept has led to studies to identify homogeneous groups of environments with negligible crossovers (Crossa and Cornelius, 1997). A further development of this concept is the emphasis on the "which-won-where" pattern (Gauch and Zobel, 1997), which implies that only the crossover interactions that involve the best cultivars are meaningful in cultivar evaluation. The GGE biplot methodology (Yan et al., 2000) is the most effective, useful, and elegant way to reveal the "which-won-where" pattern of an MET dataset. If there is no important crossover interaction, i.e., if a single genotype won in all environments, any environment should suffice to select the best cultivar, but a best test environment exists, which is most discriminating (Yan et al., 2000, 2001). If there are important crossovers, a third question to ask would be, "Is the 'which-won-where' pattern repeatable across years?" This is a crucial question to determine if the target environment should be divided into different mega-environments (Cooper et al., 1993, Yan and Hunt, 1998). If the answer is no, we are facing a complex mega-environment that has nonrepeatable crossover GEI (Yan and Rajcan, 2002). When this is the case, cultivar evaluation must be based

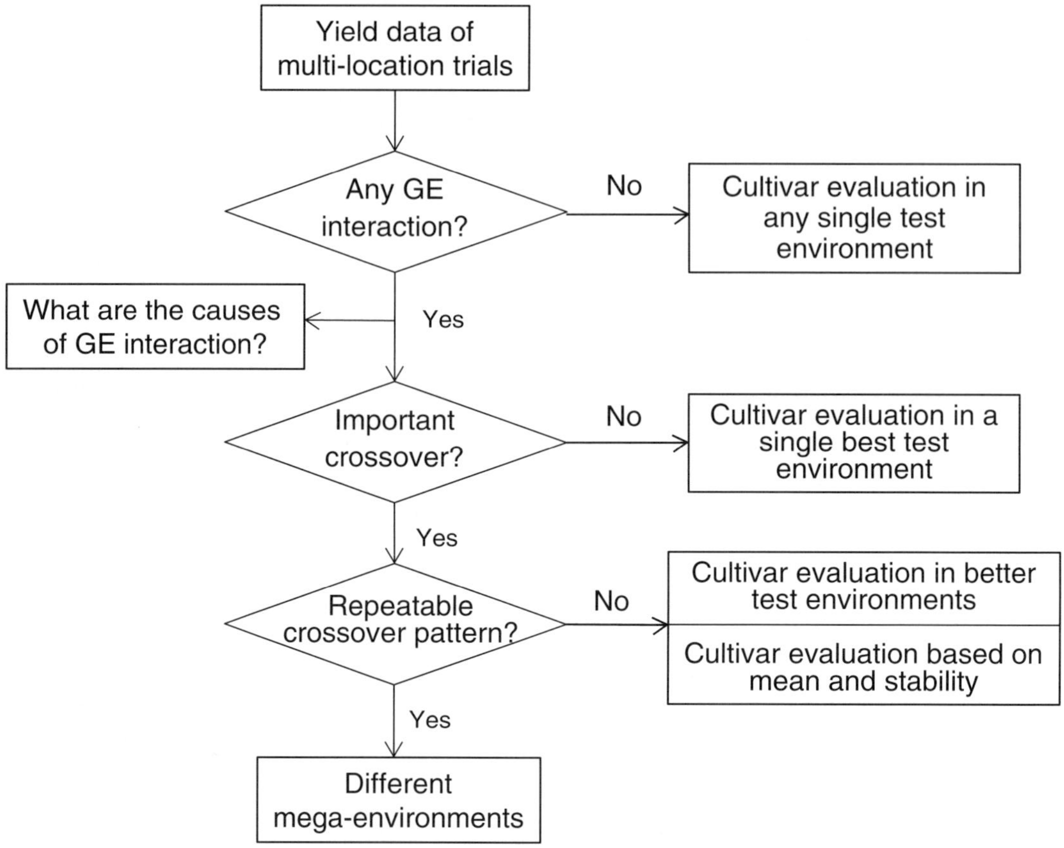

FIGURE 5.1 A schematic chart showing tasks of an MET data analysis.

on both mean performance and stability, which is a perennial topic for almost all breeding programs (Kang, 1993, 1998; Lin and Binns, 1994). If the "which-won-where" pattern is repeatable across years, different mega-environments are suggested, and cultivar evaluation must be specific to each of them. In addition, whenever GEI is detected, one must try to ascertain its environmental and/or gentic cause. Understanding of GEI can have long-term benefits in cultivar evaluation as well as in evaluation of test environments (Yan and Hunt, 1998, 2001). Thus, a MET data analysis should address four major issues: 1) the presence of different mega-environments; 2) cultivar evaluation within each mega-environment; 3) test-environment evaluation within each mega-environment; and 4) investigation of the causes of GEI. This chapter demonstrates how these four objectives can be achieved using GGE biplot analyses.

We use the 1993 Ontario winter wheat performance trial data as an example to demonstrate biplot analysis of MET data. The dataset consists of 18 cultivars grown in 9 locations, with 4 blocks or replicates at each location. Table 4.4 lists only the mean values of each cultivar–location combination. Biplots based on different models and scaling methods for this particular dataset were presented in Chapter 4. The biplots were presented in a polygon view, which is best in revealing the general GEI patterns of a dataset. We show in this chapter that there are numerous ways to visualize a GGE biplot, each revealing a different piece of information from the data. Unless explicitly stated, the biplot is based on the environment-centered model (Equation 4.5) and the tester-focused scaling method (Section 4.3).

A simple GGE biplot is presented in Figure 5.2, where cultivars are shown in italics, to differentiate them from environments, which are in regular uppercase. The first principal component (PC1) scores are used as the abscissa or *x*-axis and the second principal component (PC2) scores

Biplot Analysis of Multi-Environment Trial Data

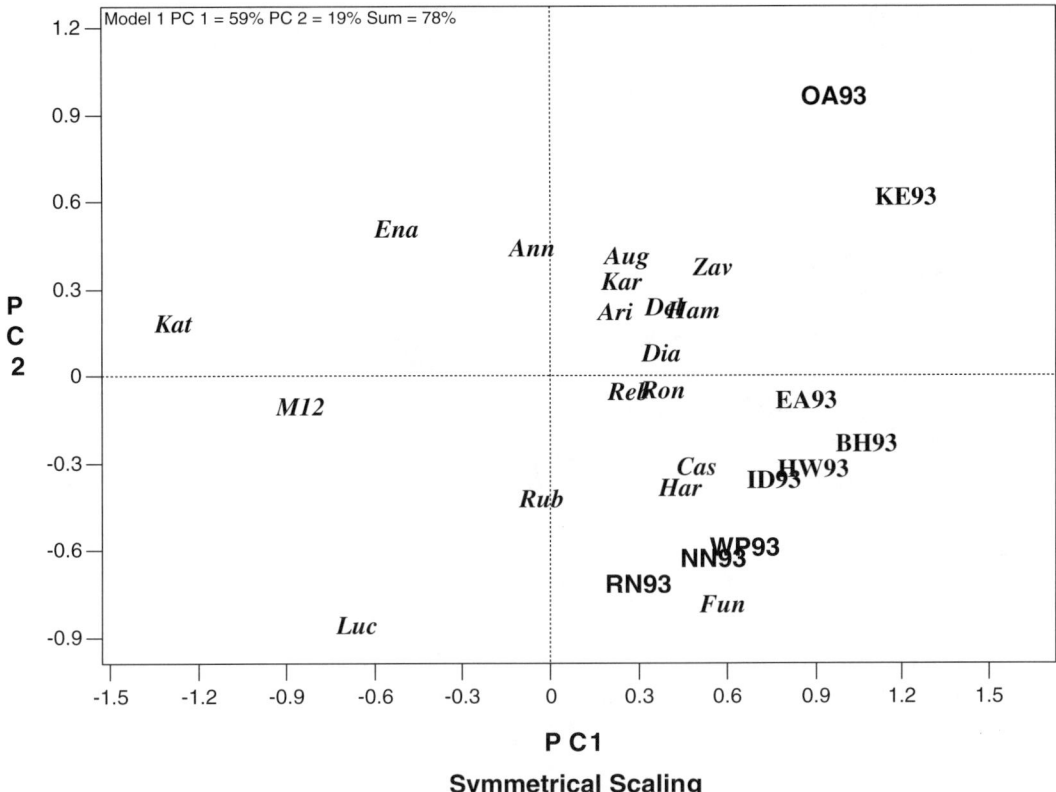

FIGURE 5.2 The GGE biplot based on yield data of 1993 Ontario winter wheat performance trials. It is based on model 1, i.e., singular value decomposition of environment-centered yield. The scaling method used is symmetrical singular value partitioning. The biplot consists of PC1 scores plotted against PC2 scores for both genotypes and environments. The two PCs explained 59 and 19% of the total GGE variation, respectively. The genotypes are in italics and environments are in regular uppercase. The guidelines indicate zero values for the two axes, respectively. The accurate position of each environment or genotype is at the beginning of its label.

as the ordinate or y-axis. The model used (model 1, the environment-centered model) and the percentage of GGE explained by each axis (PC1 = 59% and PC2 = 19%) are presented in the upper-left corner of the biplot. The biplot thus explained 78% of the total variation relative to G plus GEI. The two guidelines indicate the zero point for PC1 and PC2. Scales are indicated for the axes, but it is the relative relationships among the genotypes, among the environments, and between the genotypes and the environments, that are important. The two axes use the same physical scale, which is an important requirement for correct visualization of a biplot.

5.2 SIMPLE COMPARISONS USING GGE BIPLOT

5.2.1 Performance of Different Cultivars in a Given Environment

If you are a crop grower or an extension specialist, you are most likely interested in identifying cultivars most adapted to your area. This, of course, can be achieved by examining the data per se, but it can be more easily achieved via a GGE biplot. For example, to visualize the performance of different genotypes in a given environment, i.e., BH93, simply draw a line that passes through the biplot origin and the marker of BH93, which may be referred to as the BH93 axis. The genotypes can be ranked according to their projections onto the BH93 axis based on their performance in

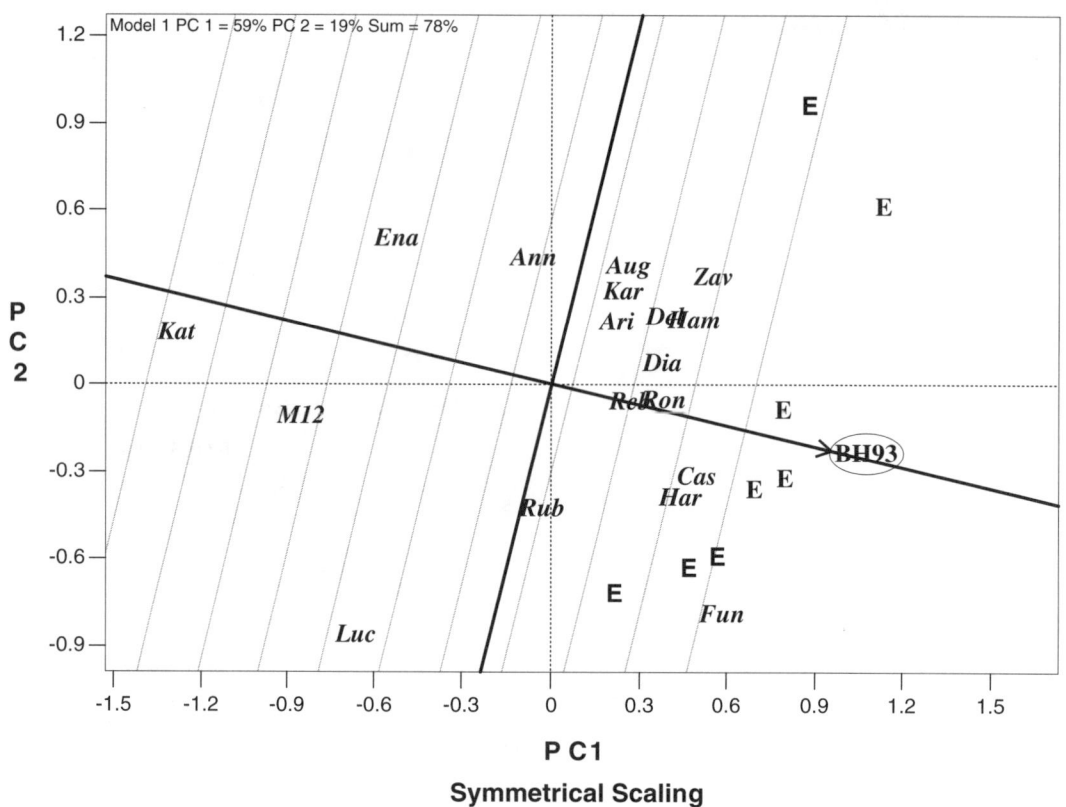

FIGURE 5.3 Comparison of cultivar performance in a selected environment (BH93). The line passing through the biplot origin and the marker of BH93 is called the BH93 axis. The yields of the cultivars in BH93 are approximated by their projections onto the BH93 axis. The thick line passing through the biplot origin and perpendicular to the BH93 axis separates cultivars with below-average yield from those with above-average yield in BH93. The parallel lines help rank the cultivars. Environments other than BH93 are presented by the letter "E" since they are not of interest here.

BH93, in the direction pointed by the arrow (Figure 5.3). Thus, at BH93, the highest-yielding cultivar was *fun*, and the lowest-yielding cultivar was *kat*, and the order of the cultivars is: *fun* > *cas* ≈ *har* > *zav* ≈ *ham* ≈ *ron* > *dia* ≈ *del* > *reb* > *aug* ≈ *kar* ≈ *ari* > *rub* > *ann* > *luc* > *ena* > *m12* > *kat*. The line that passes through the biplot origin and is perpendicular to the BH93 axis separates genotypes that yielded above the mean from *fun* to *rub* and those that yielded below the mean from *ann* to *kat* in BH93. The lighter lines that are parallel to the perpendicular line divide the whole range of the genotypes into ten equal segments and help in ranking the cultivars.

To verify, the determination coefficient between these ranks and the actual yield of the cultivars at BH93 was 75.5% (Figure 5.4). This value is about the order of the total GGE explained by the biplot (78%). Although the determination is not perfect, it is good enough to separate good cultivars from poor cultivars in the selected environment. Further, it is arguable whether the rank based on the biplot, which contains information from all test environments, or the observed rank, which is based on BH93 alone, is more reliable.

The distance from the biplot origin to the marker of an environment is called that environment's vector. The length of the vector is a measure of the environment's ability to discriminate among cultivars. A short vector, relative to the biplot size, implies that all cultivars tend to have similar yield in the associated environment. Thus, cultivar differences based on projections onto vectors of such an environment may not be reliable; they may only reflect noise.

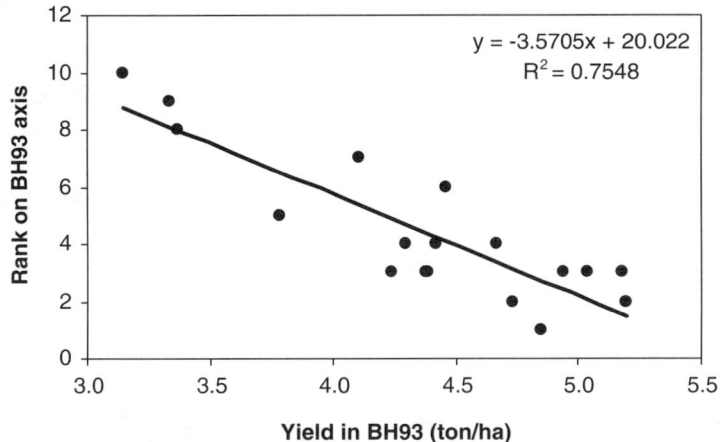

FIGURE 5.4 Association between yield in environment BH93 and cultivar rank on the BH93 axis on the biplot.

5.2.2 Relative Adaptation of a Given Cultivar in Different Environments

Breeders would like to know which environment is most suited for a cultivar. Figure 5.5 illustrates how to visualize the relative adaptation of cultivar *fun* in different environments. A line is drawn that passes through the biplot origin and the marker of *fun*, which may be called the *fun* axis. The environments are ranked along the *fun* axis in the direction indicated by the arrow. Thus, the relative performance of *fun* in different environments was WP93 ≈ NN93 ≈ BH93 ≈ RN93 ≈ ID93 ≈ HW93 > EA93 > KE93 > OA93. The line passing through the biplot origin and perpendicular to the *fun* axis separates environments in which *fun* yielded below the mean, e.g., OA93 from those in which *fun* yielded above the mean i.e., all the other environments. Cultivar *fun* was almost equally well adapted to RN93, NN93, WP93, ID93, HW93, BH93, and EA93, but not to OA93 and KE93. The lighter parallel lines, which divide the whole range of the environments into ten equal segments, help in ranking the environments.

The determination coefficient between the rank of the environments on the biplot and their environment-centered yield of *fun* in different environments is 65% (Figure 5.6). This determination is not perfect but is sufficient to differentiate between the well-suited environments and the not well-suited environments for *fun*. Again, the ranking based on the biplot has taken into account information from other cultivars, whereas the observed values were for cultivar *fun* only. Hence, which of the two rankings is more reliable remains a question.

The distance from the biplot origin to the marker of a cultivar is called the cultivar's vector. The length of the vector is a measure of the cultivar's responsiveness to environment. A short vector, relative to the biplot size, implies that the associated cultivar tends to rank the same in all environments. Thus, the environment ranking based on their projections onto the vector of such a cultivar may not be reliable; it may reflect just the noise.

5.2.3 Comparison of Two Cultivars

To compare two cultivars, for example, *zav* and *fun*, draw a connector line to connect them and draw a perpendicular line that passes through the biplot origin and is perpendicular to the connector line (Figure 5.7). We see two environments, OA93 and KE93, are on the same side of the perpendicular line as *zav*, and the other seven environments are on the other side of the perpendicular line, together with cultivar *fun*. This indicates that *zav* yielded more than *fun* in OA93 and KE93, but *fun* yielded more than *zav* in the other seven environments.

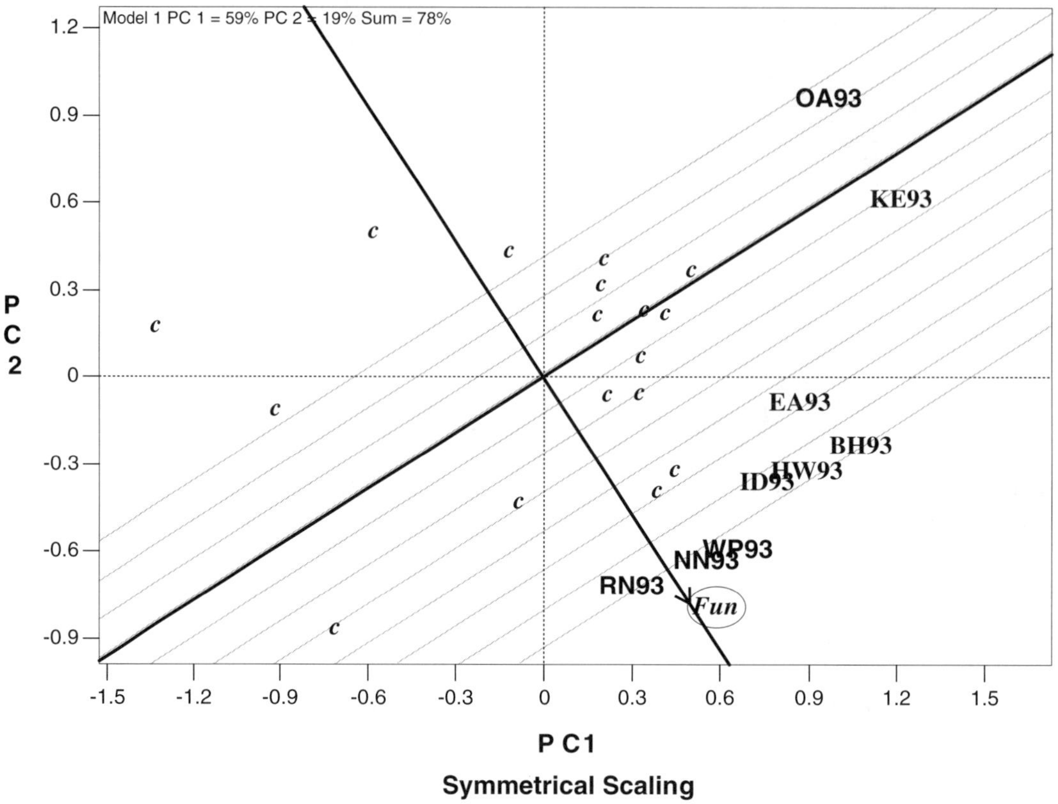

FIGURE 5.5 Comparison of the performance of cultivar *fun* in different environments. The line passing through the biplot origin and the marker of *fun* is called the *fun* axis. The relative yield of *fun* in different environments is approximated by the projections of the environments onto the *fun* axis. The thick line passing through the biplot origin and perpendicular to the *fun* axis, separates environments where *fun* yielded below the mean from those where *fun* yielded above the mean. The parallel lines help rank the environments based on the adaptation of *fun*. Cultivars other than *fun* are presented by the letter "c" since they are not of interest here.

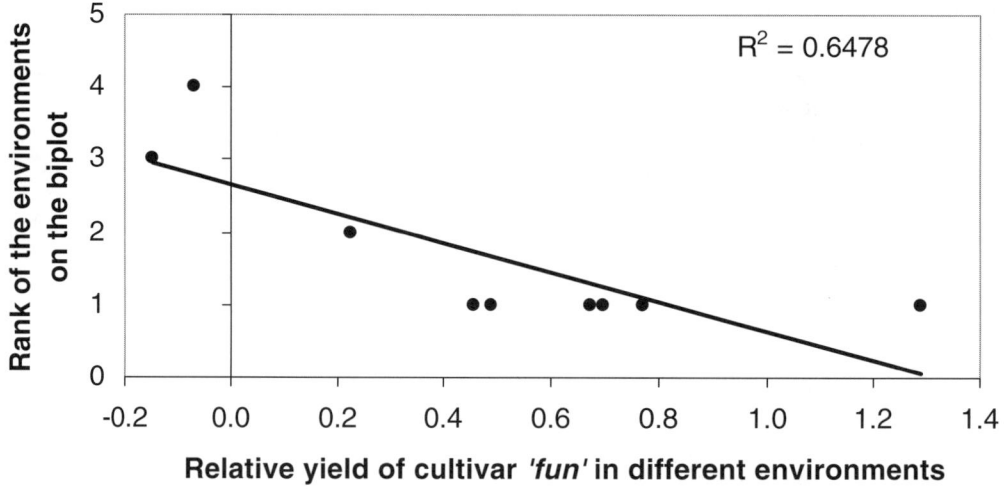

FIGURE 5.6 Association between the environment-centered yield of cultivar *fun* in different environments and the rank of the environments on the *fun* axis in the biplot.

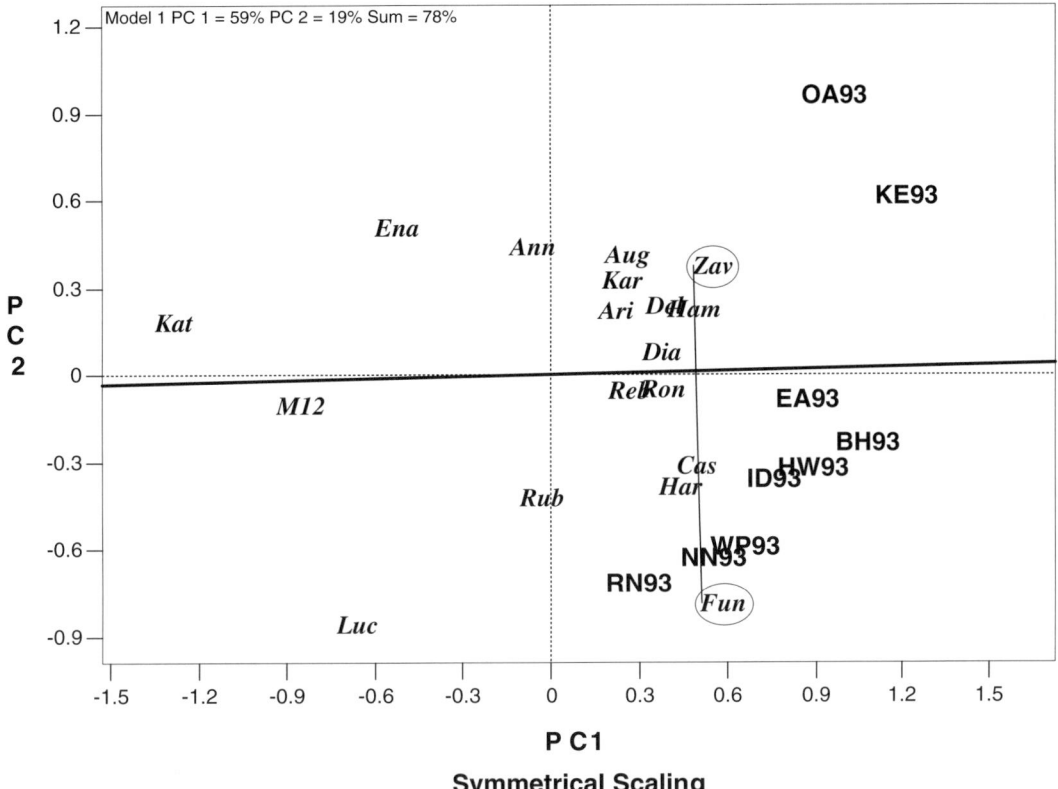

FIGURE 5.7 Comparison of two cultivars, *zav* and *fun*, based on the GGE biplot. The two cultivars to be compared are connected with a straight line, and a perpendicular line passing through the biplot origin separates environments in which *zav* performed better from those in which *fun* performed better.

This type of interpretation, which was first documented in Yan (1999), is an extended use of the "inner-product" property. Based on the inner-product property, cultivars *fun* and *zav* should have yielded the same in any virtual environments that are located on the perpendicular line, because their projections onto the perpendicular line are equal. Furthermore, it is not possible for these two cultivars to have equal projection length or yield in any other environments. Cultivar *zav* had a longer projection onto the vector of any environment that is on its side of the perpendicular line but a shorter projection onto the vector of any environment on the other side of the perpendicular line. A longer projection means higher yield, and vice versa.

To verify, environment-centered yields of the two cultivars in each of the environments are presented in Figure 5.8. Cultivar *fun* (the *x*-axis) had above-average yield in all environments except OA93 and KE93, and *zav* (the *y*-axis) had above-average yield in all environments except NN93, BH93, and RN93. Two environments, OA93 and KE93, are above the equality line, and the other seven environments are below the equality line, indicating that *zav* had higher yield in OA93 and KE93 whereas *fun* had higher yield in the other seven environments. The above interpretation based on the biplot is, thus, validated.

Cultivar *cas* is on the connector line between *zav* and *fun* (Figure 5.7), indicating that it had yield intermediate between the yields of *zav* and *fun* in all environments. To validate this statement, we first compare *cas* and *zav*, and then compare *cas* and *fun*. The comparison of *cas* and *zav* (Figure 5.9) indicates that *zav* had higher yield in OA93 and KE93, whereas *cas* had higher yield in the other seven environments. The comparison of *cas* and *fun* (Figure 5.10) indicates that *cas* had higher yield in OA93 and KE93, whereas *fun* had higher yield in the other seven environments. Therefore, *cas* was neither the highest-yielding nor the lowest-yielding cultivar in any of the nine environments.

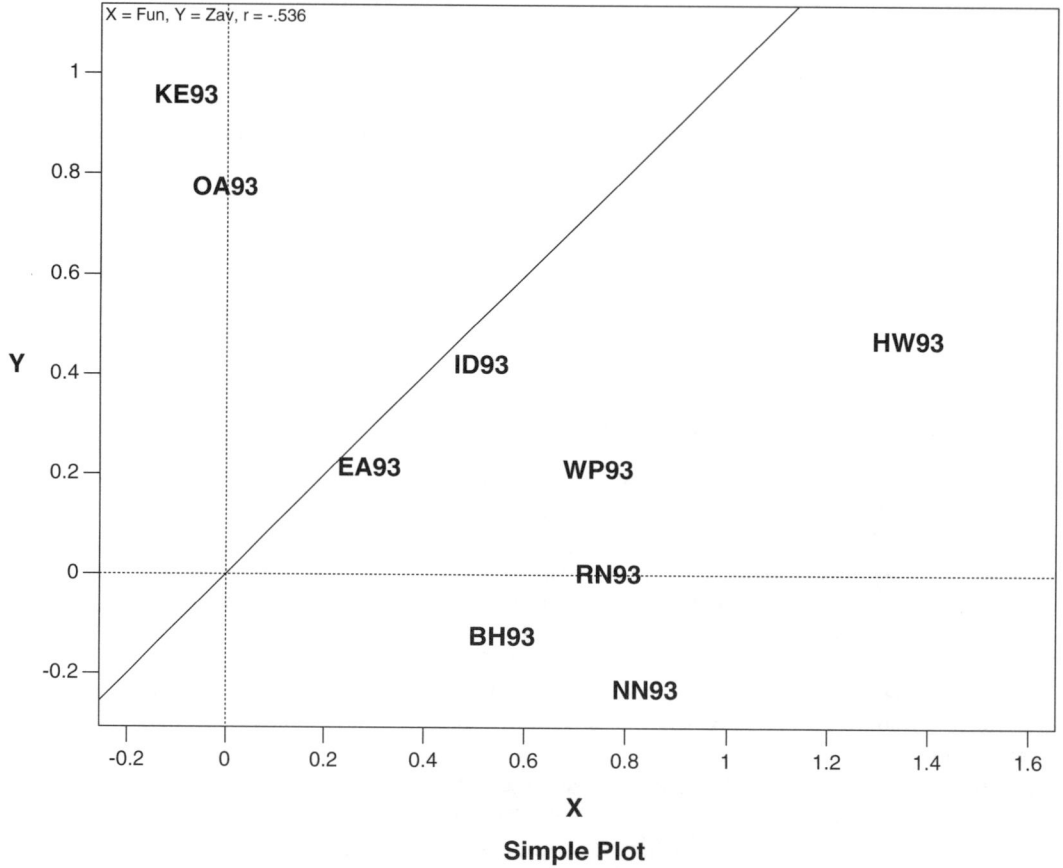

FIGURE 5.8 Environment-centered yield of cultivars *fun* (*x*-axis) and *zav* (*y*-axis) in various environments. The guidelines indicate zero or average yield in each environment. The equality line represents environments where the two cultivars yielded the same.

Biplot Analysis of Multi-Environment Trial Data

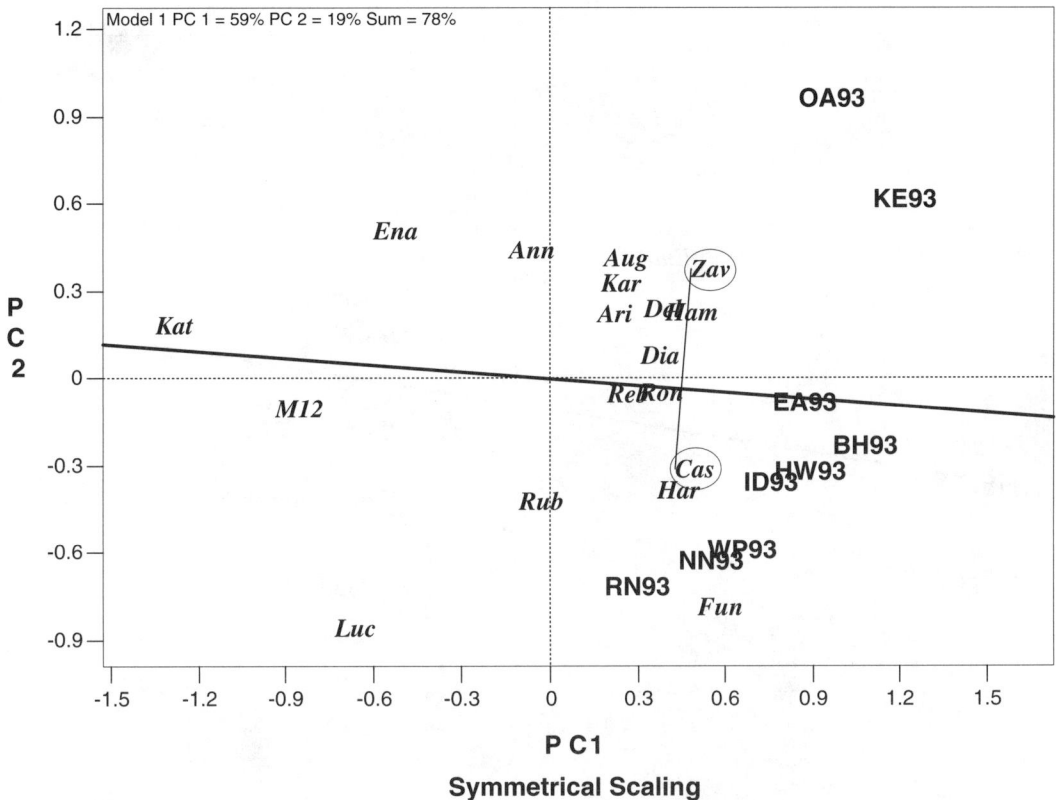

FIGURE 5.9 Comparison of two cultivars, *zav* and *cas*, based on the GGE biplot. The two cultivars to be compared are connected with a straight line, and a perpendicular line passing through the biplot origin separates environments in which *zav* performed better from those in which *cas* performed better.

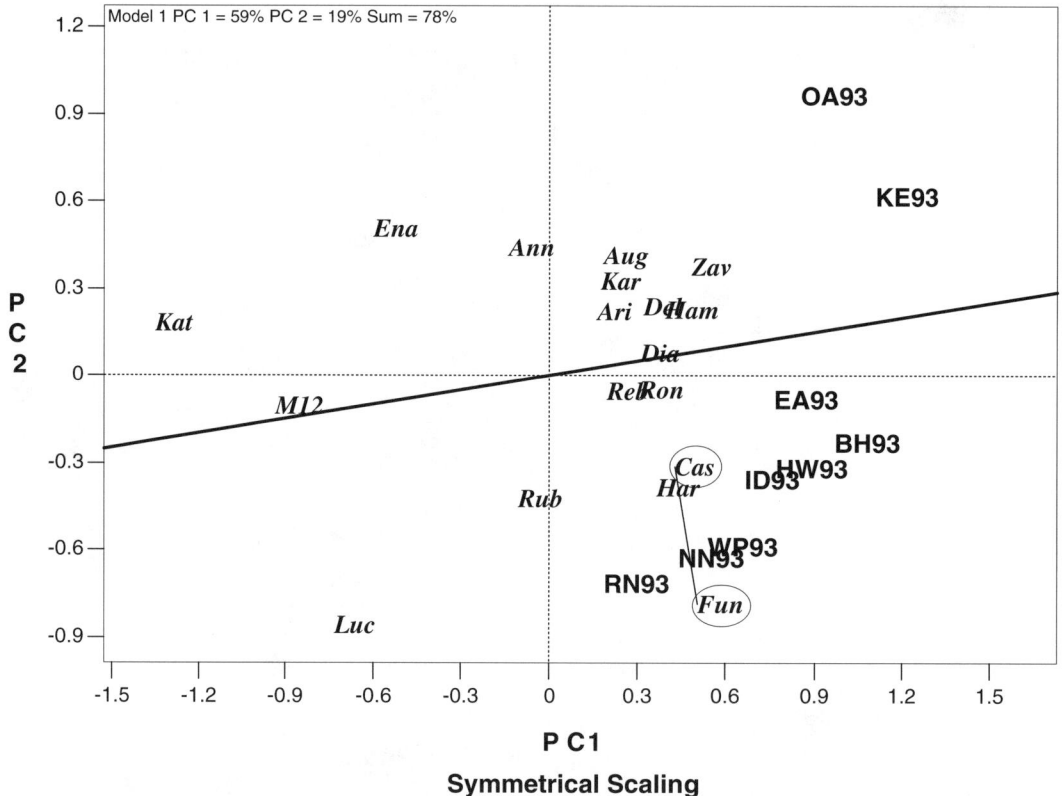

FIGURE 5.10 Comparison of two cultivars, *cas* and *fun*, based on the GGE biplot. The two cultivars to be compared are connected with a straight line, and a perpendicular line passing through the biplot origin separates environments in which *cas* performed better from those in which *fun* performed better.

Compared with Figures 5.7 and 5.9, Figure 5.10 is different in that the two cultivars to be compared, *cas* and *fun,* are on the same side of the perpendicular line. In such cases, the cultivar that is closer to the perpendicular line (i.e., *cas*) is better in environments on the other side of the perpendicular line (i.e., OA93 and KE93), and the cultivar that is farther away from the perpendicular line (i.e., *fun*) is better in environments on the same side of the perpendicular line (i.e., the environments other than OA93 and KE93).

Figure 5.11, which compares cultivars *ena* and *zav,* represents another type of scenario, in which all environments are on the same side of the perpendicular line. This indicates that the cultivar that is on the same side of, or closer to, the perpendicular line (i.e., *zav*), is better than the other cultivar (i.e., *ena*) in all environments. This statement can be validated from the real data (Figure 5.12). While *zav* was above average in yield in all environments, except NN93, BH93, and RN93, *ena* had below-average yield in all environments. Moreover, all environments are above the equality line, indicating that the *y*-axis (*zav*) had higher values than the *x*-axis (*ena*) for all environments.

When making pairwise comparisons, the length of the connector line is important for making inferences. The longer it is, the more reliable the comparison. A short connector line implies that the two cultivars to be compared are similar in all environments, and therefore, the comparison may be meaningless.

5.3 MEGA-ENVIRONMENT INVESTIGATION

5.3.1 "Which-Won-Where" Pattern of an MET Dataset

The polygon-view of a GGE biplot, first documented in Yan (1999), provides an effective and elegant means of visualizing the "which-won-where" pattern of an MET dataset (Figure 5.13). Notice that Figure 5.13 is exactly a duplication of Figure 4.3, where little interpretation was provided. The polygon view was used extensively, although not fully explained, in Chapter 4 because it is the most succinct way of summarizing the GGE pattern of a dataset.

The polygon is drawn joining the cultivars (*fun, zav, ena, kat* and *luc*) that are located farthest from the biplot origin, so that all other cultivars are contained in the polygon. The polygon can be called a convex hull, and the cultivars at the corner of the polygon can be called the vertex cultivars. Thus, the vertex cultivars are those located farthest from the origin. They have the longest vectors, in their respective directions, which is a measure of responsiveness to environments. The vertex cultivars are, therefore, among the most responsive cultivars; all other cultivars are less responsive in their respective directions. A cultivar located at the origin would rank the same in all environments and is not at all responsive to the environments.

In Figure 5.13, a line perpendicular to each side of the polygon is drawn. The polygon is used as an instrument to compare adjacent vertex cultivars. Comparison of *zav* with *ena,* and also with *ann* and *aug,* both located on the polygon side that connects *zav* and *ena,* indicates that *zav* had higher yields in environments OA93 and BH93 (see also Figure 5.11 and associated discussion in the previous section). Comparison of *zav* with *fun,* and also with *cas* and *ham,* indicates that *zav* had higher yields in OA93 and KE93 (see also Figure 5.7 and associated discussion in the previous section). Although no direct comparison between *zav* and *kat* is made, *kat* is obviously inferior to *ena* in OA93 and KE93. Similarly, although no direct comparison between *zav* and *luc* is made, *luc* is obviously inferior to *fun* in OA93 and KE93. Consequently, *zav* is the highest-yielding cultivar in OA93 and KE93.

Interestingly, OA93 and KE93 are in the same sector, separated from the rest of the biplot by two perpendicular lines, and *zav* is the vertex cultivar in this sector. The perpendicular lines to the sides of the polygon divide the biplot into sectors. Each sector has a vertex cultivar; a sector can, therefore, be named after the vertex cultivar. For instance, the sector with the vertex cultivar *zav* may be referred to as the *zav* sector; and two environments, OA93 and KE93, fell in this sector. As a rule, the vertex cultivar is the highest-yielding cultivar in all environments that share the sector with it. As another example, *fun* is the highest-yielding cultivar in its sector, which contains seven environments, namely, EA93, BH93, HW93, ID93, WP93, NN93, and RN93.

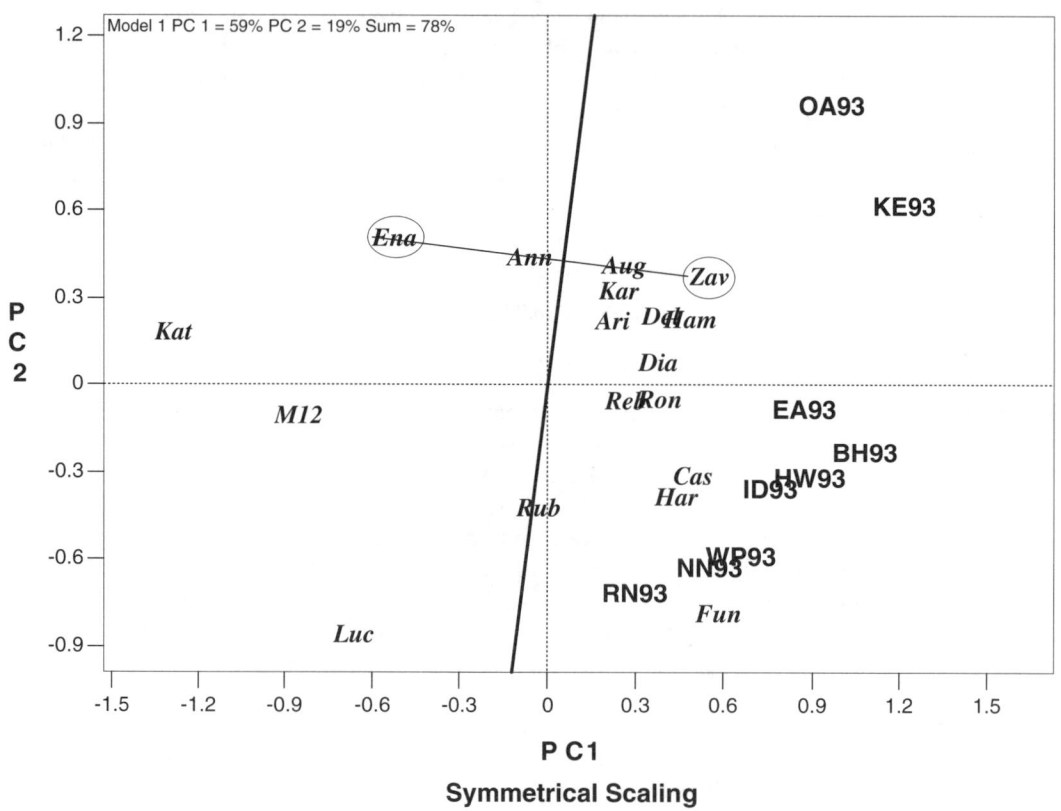

FIGURE 5.11 Comparison of two cultivars, *zav* and *ena*, based on the GGE biplot. The two cultivars to be compared are connected with a straight line, and a perpendicular line passing through the biplot origin separates environments in which *zav* performed better from those in which *ena* performed better. *Zav* was better than *ena* in all environments.

Biplot Analysis of Multi-Environment Trial Data

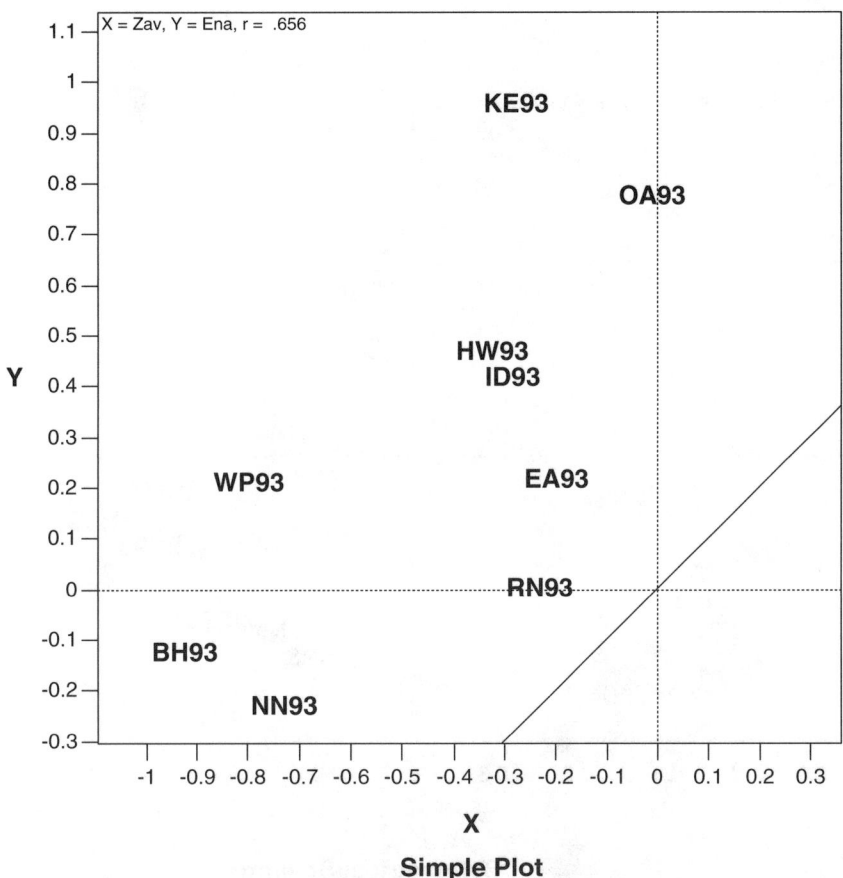

FIGURE 5.12 Environment-centered yield of cultivars *zav* (*x*-axis) and *ena* (*y*-axis) in all environments. The guidelines indicate zero or average yield in each environment. The equality line represents environments where the two cultivars yielded the same. This figure indicates that *zav* was better than *ena* in all environments, even though it was below average in environments NN93 and BH93.

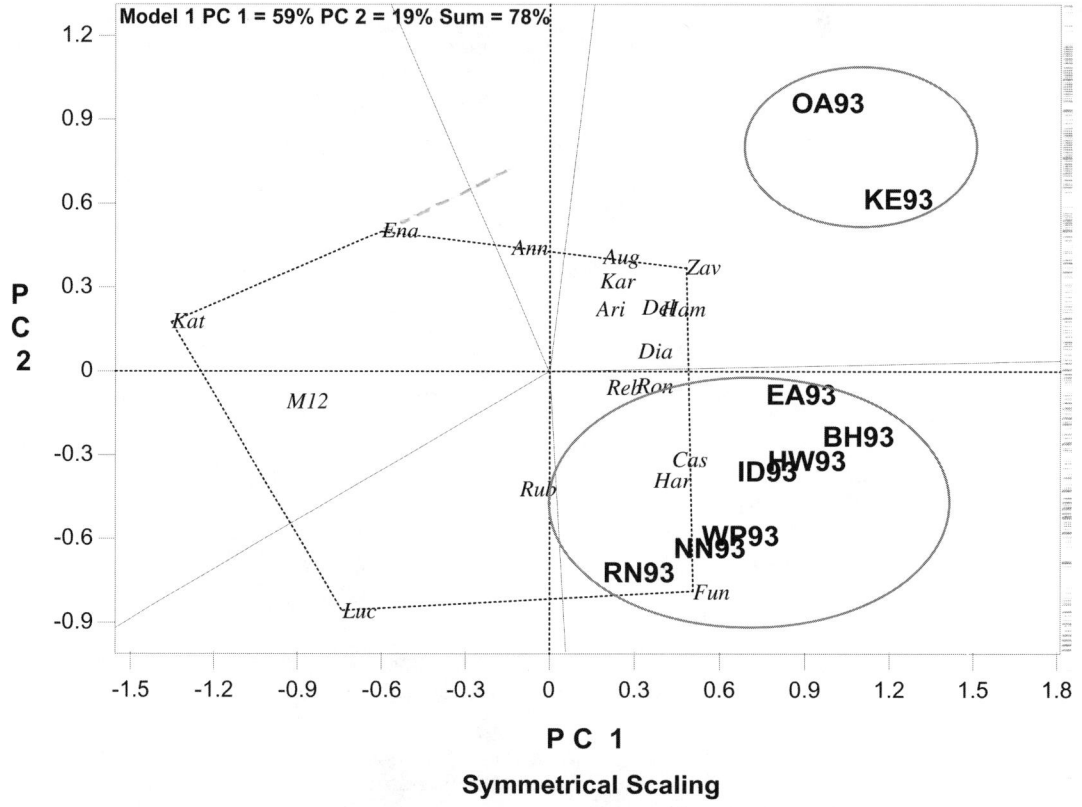

FIGURE 5.13 Polygon view of the GGE biplot, showing which cultivar yielded best in which environments. The polygon was drawn on cultivars located away from the biplot origin so that all other cultivars are contained in the polygon. Perpendicular lines are drawn to each side of the polygon, which divide the biplot into sectors. The vertex cultivar in each sector is the highest-yielding cultivar in environments that fell in the particular sector.

No environments fell in the sectors with *ena*, *kat*, and *luc* as vertex cultivars. This indicates that these vertex cultivars were not the best in any of the test environments. Moreover, this indicates that these cultivars were the poorest in some or all of the environments. To generalize, vertex cultivars are the most responsive cultivars; they are the best or else the poorest cultivars in some or all of the test environments. By reversing the sign of all test environments and redrawing the polygon, the poorest cultivars for each of the test environments will be indicated (Figure 5.14). As can be expected from Figure 5.13, Figure 5.14 shows that *kat* was the poorest in all environments except OA93 and RN93. These two environments are outside the *kat* sector. The poorest cultivars for these two environments were *luc* and *ena*, respectively. *Ena* was the vertex cultivar for the sector that contained environment RN93.

5.3.2 MEGA-ENVIRONMENT INVESTIGATION

The polygon view of a GGE biplot not only shows the best cultivar for each test environment but also divides the test environments into groups. In Figure 5.13, two groups of environments are obvious: OA93 and KE93 in the *zav* sector, and the other seven environments in the *fun* sector. Figure 5.13, thus, suggests that there are two winter wheat mega-environments in Ontario.

Two criteria are required to suggest existence of different mega-environments. First, there are different winning cultivars in different test environments (Gauch and Zobel, 1997). Second, the among-group variation should be significantly greater than the within-group variation, a common criterion for clustering. Graphically, different mega-environments should consist of groups of test environments that are apparently separated in a biplot. Both criteria are met in the present case (Figure 5.13). Interestingly, the two-mega-environment suggestion in Figure 5.13 coincides well with the geographical distribution of the locations (Figure 5.15). Location OA (= Ottawa) and KE (= Kemptville) belong to Eastern Ontario; BH (= Bath) also belongs to Eastern Ontario but is much warmer than OA and KE. The other six locations belong to Western or Southern Ontario.

Any suggestion of the existence of different mega-environments must be validated by multiple-year data. Specifically, the "which-won-where" pattern observed in one year should be largely repeatable in other years to conclude existence of different mega-environments. The pattern we saw in Figure 5.13 was often seen in other years, though not exactly. Therefore, it is justifiable to say that there exist two winter-wheat mega-environments in Ontario, as depicted in Figure 5.15. The Ontario winter wheat-growing regions had been thought to consist of four sub-regions based on heat resources.

Note that similar location grouping in different years does not necessarily indicate existence of different mega-environments. For example, 28 soybean cultivars were tested from 1997 to 1999 at 3 to 4 locations in the 2800 Crop Heat Unit area of Ontario. When analyzed separately, location WIN (= Winchester) was always different, whereas the other locations, namely, STP (= St Paul), EXE (= Exeter), and WOO (= Woodstock), were always similar (Figures 5.16 to 5.18). When three-year data are analyzed jointly, however, no clear differentiation of mega-environments is suggested (Figure 5.19): Cultivar *Cm401028* was the best in almost all year–location combinations except WIN99 and STP98. Thus, the observed yearly genotype-by-location interactions are essentially random genotype × location × year interactions. This example indicates that a joint multi-year data analysis is mandatory to declare existence of different mega-environments.

Mega-environment differentiation is an important concept (Gauch and Zobel, 1997). Dividing the target environment into different mega-environments and deploying different cultivars in different mega-environments is the best way to utilize GEI. However, unjustified mega-environment division can be counterproductive because limited resources are stretched across sub-regions (Atlin et al., 2000a,b). Here we emphasize that mega-environment division must meet two requirements. First, there must be clear crossover genotype-by-location interactions that suggest different groups of locations, not environments. Second, any suggestions of mega-environments must be critically validated by joint analysis of multiple-year data.

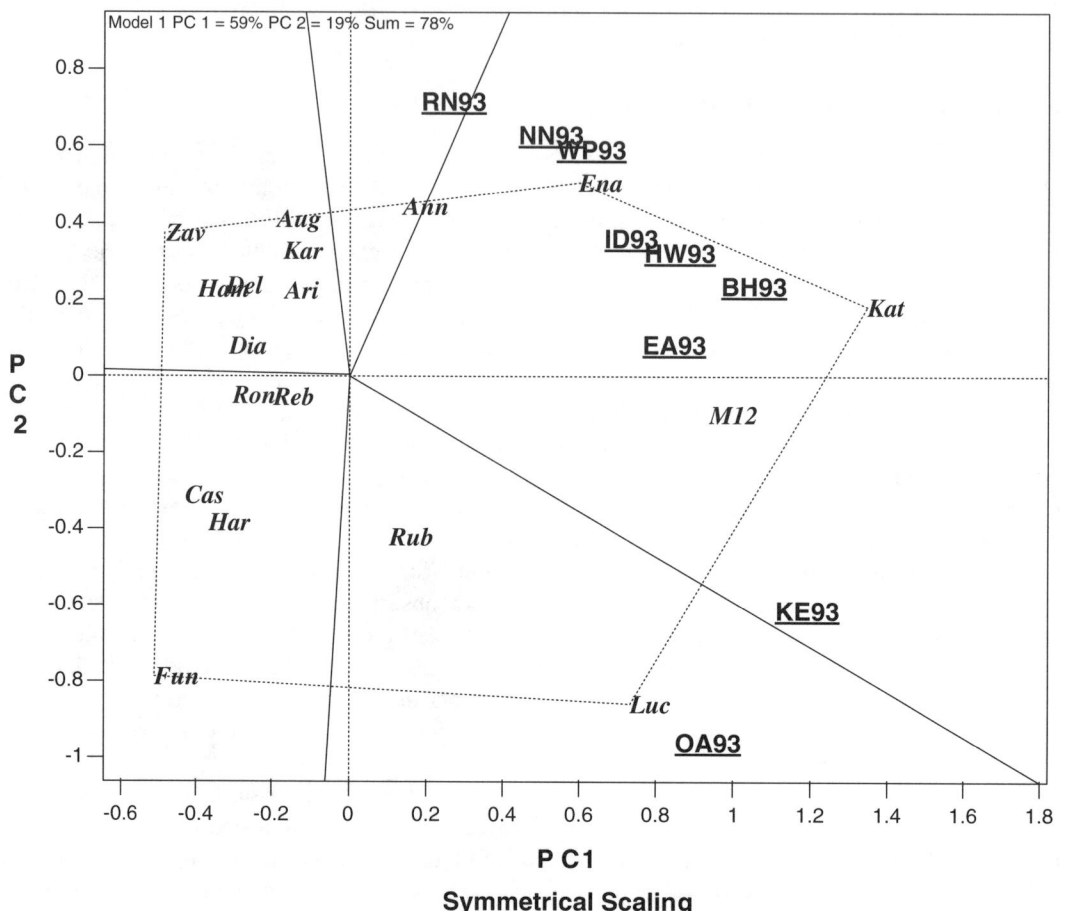

FIGURE 5.14 Polygon view of the GGE biplot after the signs of the environments are reversed, showing which cultivar yielded lowest in which environments. The polygon was drawn on cultivars located away from the biplot origin so that all other cultivars are contained in the polygon. Perpendicular lines are drawn to each side of the polygon, which divide the biplot into sectors. The vertex cultivar in each sector was the lowest-yielding cultivar in environments falling in the particular sector.

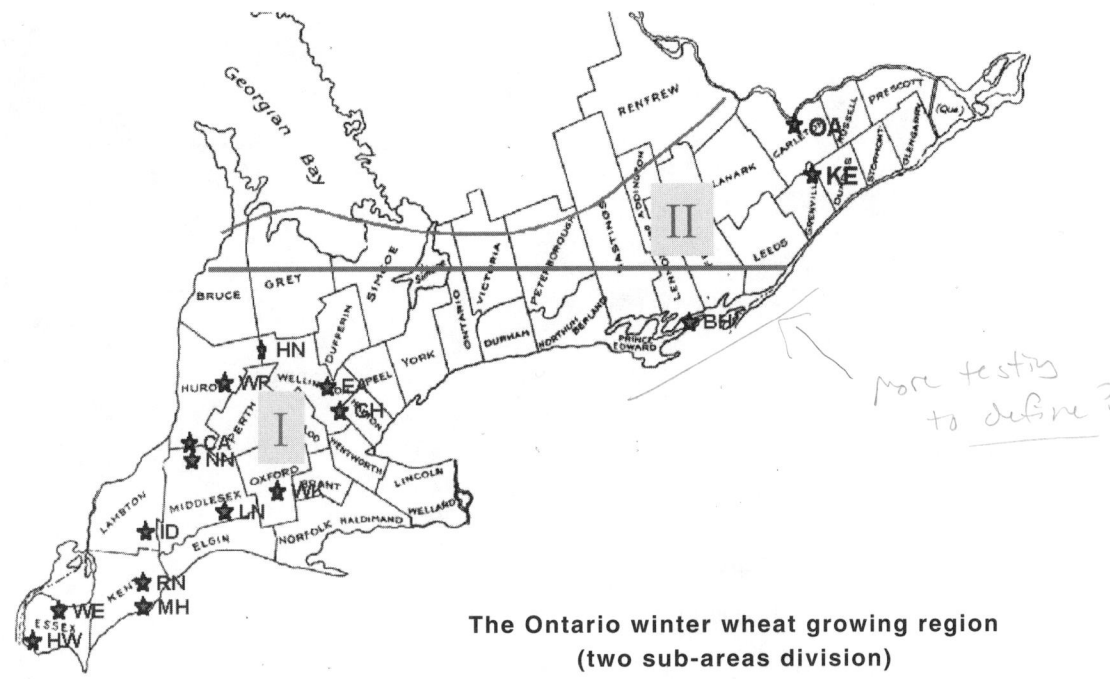

FIGURE 5.15 Geographical distribution of the test locations used in Ontario winter wheat performance trials.

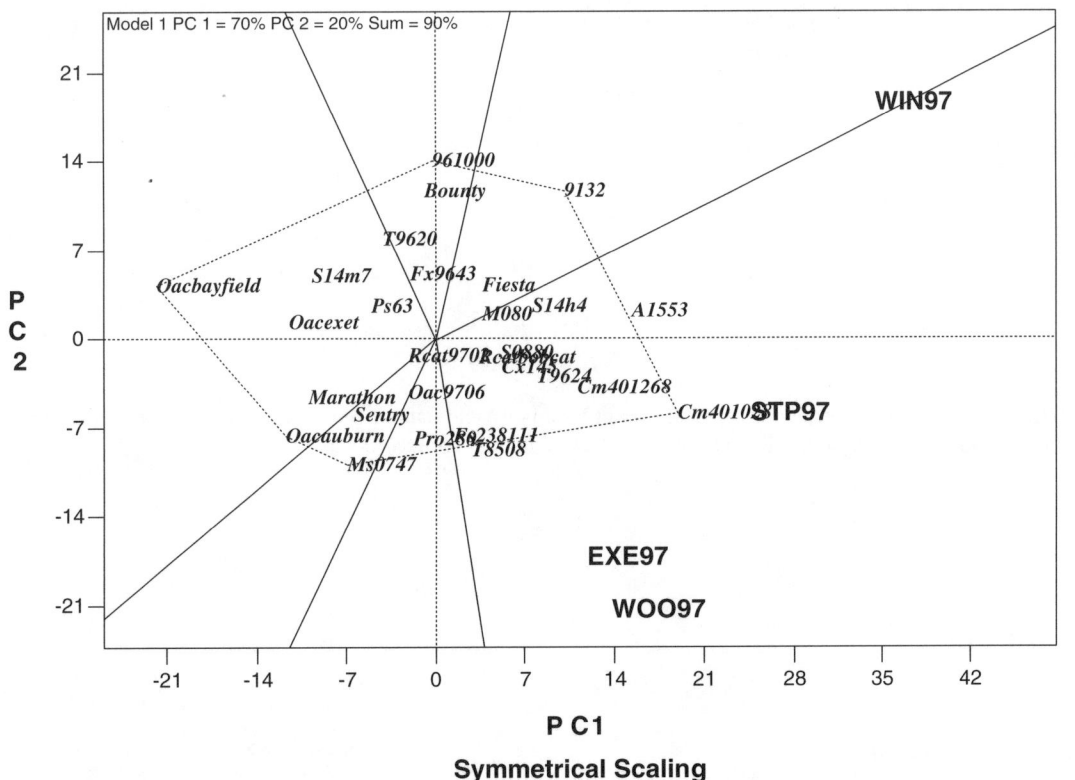

FIGURE 5.16 Polygon view of the GGE biplot based on yield data of selected soybean genotypes tested at four locations in 1997 in the 2800 Crop Heat Unit area in Ontario, Canada.

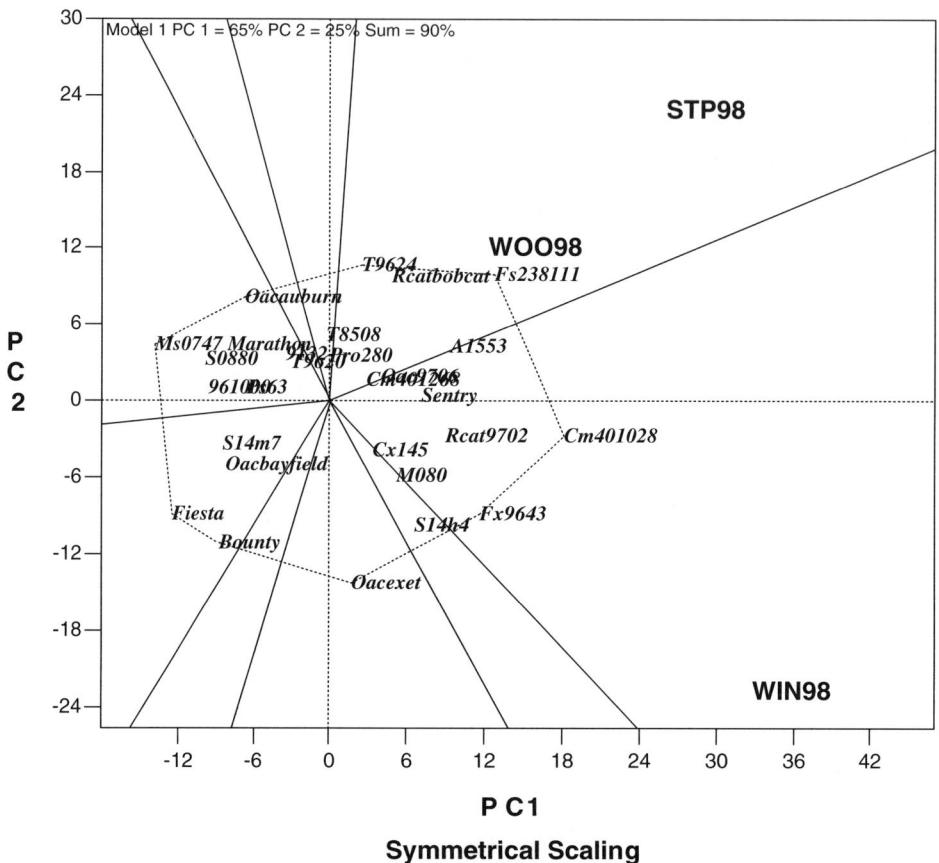

FIGURE 5.17 Polygon view of the GGE biplot based on yield data of selected soybean genotypes tested at three locations in 1998 in the 2800 Crop Heat Unit area in Ontario, Canada.

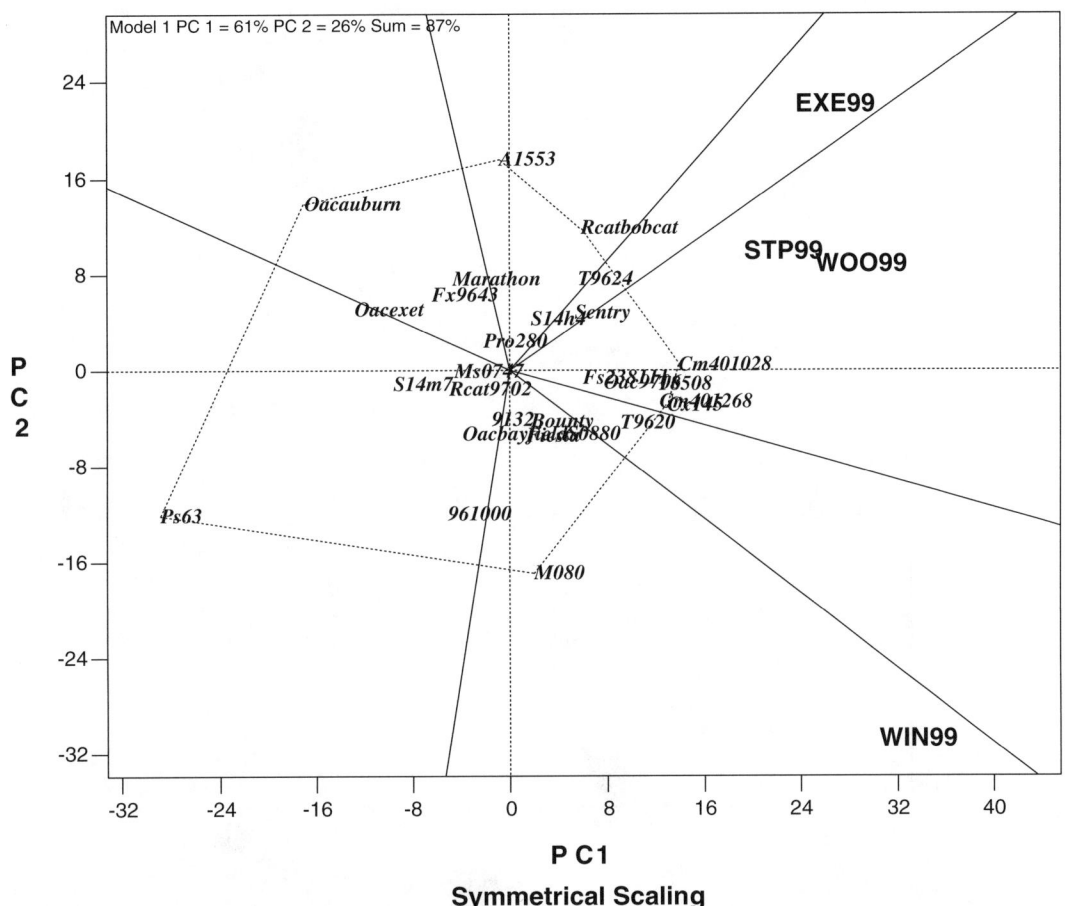

FIGURE 5.18 Polygon view of the GGE biplot based on yield data of selected soybean genotypes tested at four locations in 1999 in the 2800 Crop Heat Unit area in Ontario, Canada.

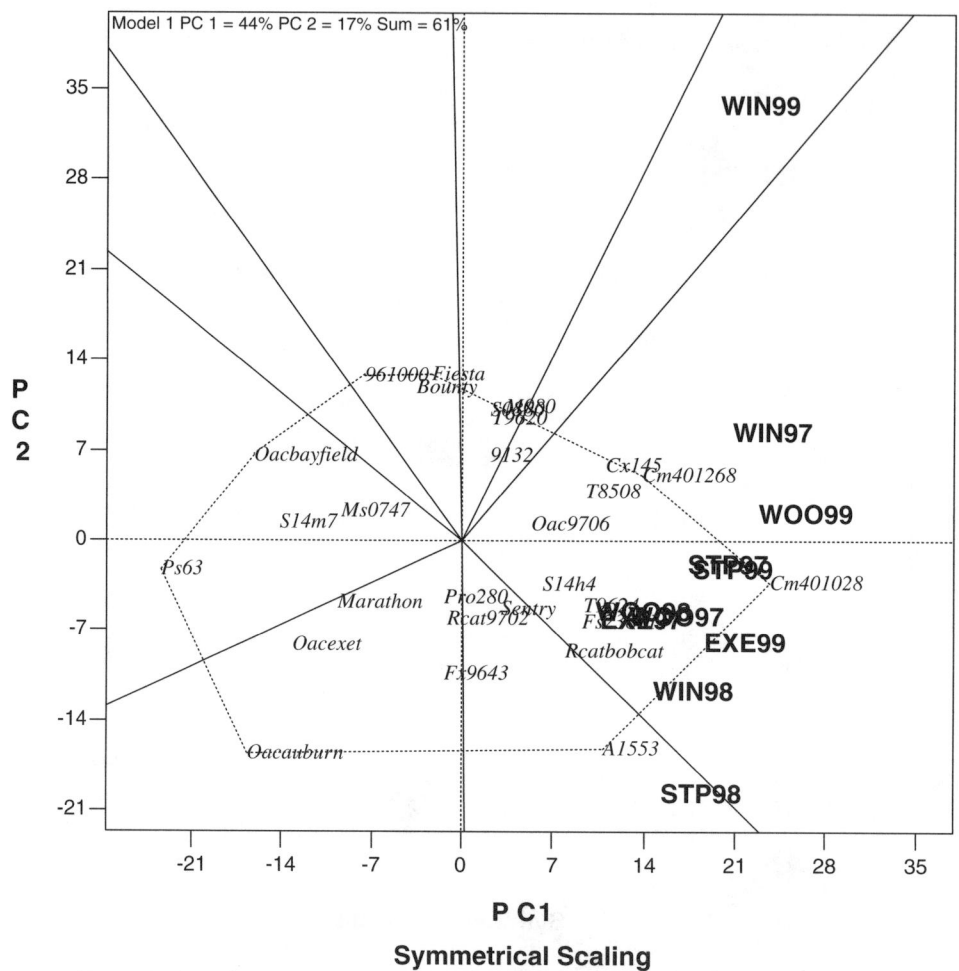

FIGURE 5.19 Polygon view of the GGE biplot based on yield data of selected soybean genotypes tested in 11 year–location combinations during 1997 to 1999 in the 2800 Crop Heat Unit area in Ontario, Canada.

As pointed out in the beginning of this chapter, a mega-environment can be simple or complex (Yan and Rajcan, 2002). A simple mega-environment is one within which there are no major crossover GEI. For such a mega-environment, one or a few test locations are sufficient to identify the very best cultivars that can be recommended for everywhere within the mega-environment. A complex mega-environment has large but unpredictable crossover genotype-by-location interactions, and when analyzed across years, the yearly genotype-by-location interactions turn into genotype × location × year three-way interactions, and the across-year genotype-by-location interaction disappears or becomes insignificant (Yan, unpublished research). For such mega-environments, METs are essential, and cultivar evaluation must be based on both mean performance and stability. In reality, most breeding programs face complex mega-environments with unpredictable GEI and cultivar evaluation based on mean performance and stability has been a perennial problem and challenge.

5.4 CULTIVAR EVALUATION FOR A GIVEN MEGA-ENVIRONMENT

5.4.1 Cultivar Evaluation Based on Mean Performance and Stability

Referring to the winter wheat trials in Ontario, the "which-won-where" pattern in Figure 5.13 suggests that there exist two mega-environments: Eastern Ontario represented by OA and KE, and Western and Southern Ontario represented by the other seven test locations. An examination of MET data from many other years supported this mega-environment differentiation (Yan, 1999; Yan et al., 2000). Our next task is to select cultivars for each of the mega-environments.

This task appears to be straightforward, since Figure 5.13 clearly suggests that *zav* was the best cultivar for the two locations in Eastern Ontario, and *fun* was the best cultivar for the other locations. But experienced breeders will never try to select a single cultivar. There is a need to have a closer look at the individual mega-environments and to evaluate all tested cultivars for mean performance and stability.

Removal of environments OA93 and KE93 from Figure 5.13 results in Figure 5.20. It shows that BH93 had different winning cultivars, *har* and *dia*, than the other six locations. This relatively small crossover was suppressed in Figure 5.13 by the larger difference between the two mega-environments. Location BH is geographically closer to OA and KE than to the southern locations (Figure 5.15), but its interaction with the cultivars is frequently more similar to the southern locations than to the eastern locations. Therefore, the crossover pattern shown in Figure 5.20 is regarded as unrepeatable (i.e., unpredictable). Accepting that the seven locations in Figure 5.20 belong to the same mega-environment implies that all locations will be viewed as random effects of the mega-environment, and no adaptation to specific locations will be attempted. In other words, cultivar evaluation within a mega-environment should be based on both mean performance and stability to avoid the random GEI rather than trying to exploit it.

Since mean performance is simply the arithmetic mean, most research in the past has focused on stability, hence the numerous measures of stability (Lin and Binns, 1994; Kang, 1998; Yan, 1999; and Chapters 1 and 2 of this book). The genotypic interaction principal component (IPCA1) scores of the AMMI model (Gauch and Zobel, 1988), and the genotypic PC1 scores of the regression-based GGE biplot model (Yan et al., 2001; Section 4.4, Equation 4.11) are additional measures of stability. Nevertheless, stability has rarely been used by plant breeders for various reasons. One reason is that it is difficult to weigh between mean performance and stability, although some attempts were made (reviewed by Kang, 1998). Here we show that GGE biplot methodology, facilitated by the GGEbiplot software (Yan, 2001, and Chapter 6 of this book) may be the solution to help breeders make selection decisions based on both mean performance and stability.

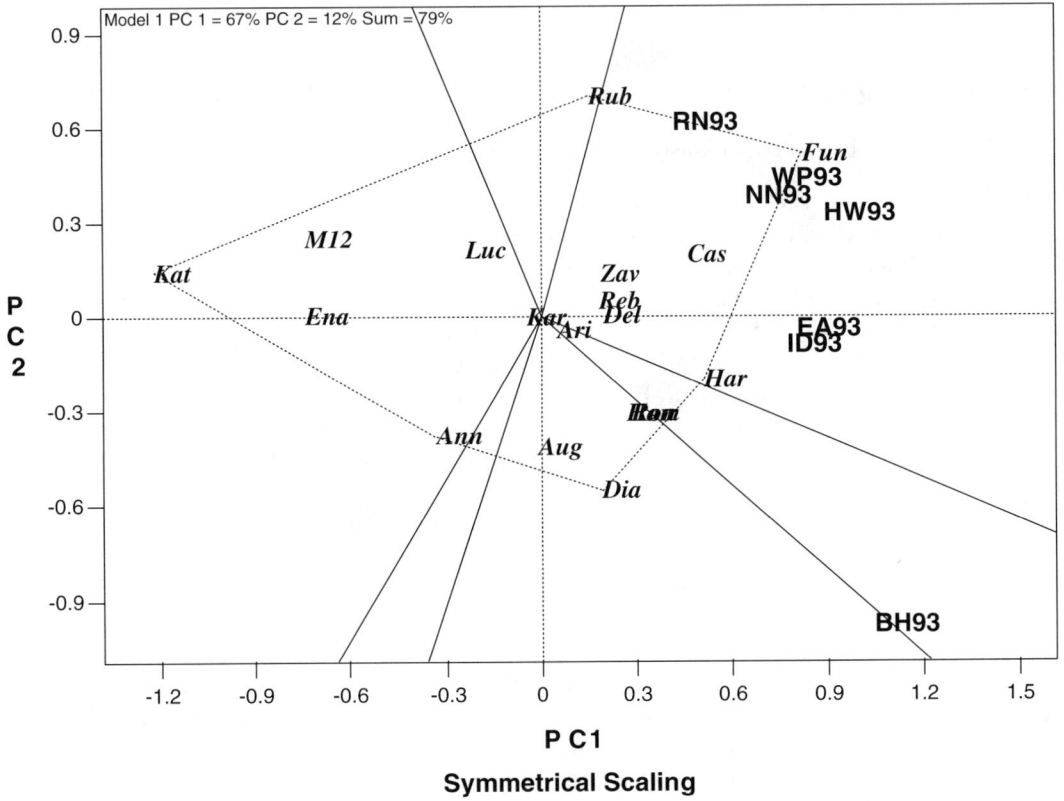

FIGURE 5.20 Polygon view of the GGE biplot based on yield data of 1993 Ontario winter wheat performance trials after two locations (OA93 and KE93) were removed.

5.4.2 THE AVERAGE ENVIRONMENT COORDINATE

Here we want to reiterate that the GGE biplot based on the principal component analysis (PCA) model (Equation 4.6), particularly its genotype-focused scaling form, provides a superior means for visualizing both mean performance and stability of the tested genotypes. This is because, in a GGE biplot based on genotype-focused scaling method, the unit of both axes for the genotypes is the original unit of the data. This property is desirable for visualizing the relative importance of PC1 and PC2 scores, hence the mean performance and stability of genotypes. Figures 5.21 and 5.20 are exactly the same except that the former uses the genotype-focused singular value scaling method (Chapter 4).

Visualization of the mean and stability of genotypes is achieved by drawing an average environment coordinate (AEC) on the genotype-focused biplot (Figure 5.21). First, an average environment, represented by the small circle, is defined by the mean PC1 and PC2 scores of the environments. Analogous to visualizing cultivar performances in a given environment (Figure 5.3), the line that passes through the biplot origin and the average environment may be called the average environment axis — it is the abscissa of the AEC. Projections of genotype markers onto this axis should, therefore, approximate the mean yield of the genotypes. Indeed, the r-squared value between these projections and the G was 0.998 (Figure 5.22). Thus, the cultivars are ranked along the AEC abscissa, with the arrow pointing to higher mean performance. Cultivar *fun* was clearly the highest-yielding cultivar, on average, in this mega-environment, followed by *cas* and *har*, followed by *rub*, *zav*, *reb*, *del*, *ron*, *ham*, and *dia*, followed by *ari*, *kar*, and *aug*, etc.

The AEC ordinate is the double-arrowed line that passes through the biplot origin and is perpendicular to the AEC abscissa (Figure 5.21). Perpendicular means orthogonal. Therefore, if the AEC abscissa represents the G, the AEC ordinate must approximate the GEI associated with each genotype, which is a measure of variability or instability of the genotypes. The double arrow indicates that a greater projection onto the AEC ordinate, regardless of the direction, means greater instability. Therefore, *rub* near the top and *dia* near the bottom of the biplot are more variable and less stable than other cultivars. The environment near the bottom is BH93, a test site located in Eastern Ontario characterized by colder winters. Similarly, the environment near the top is RN93, a test site in Southern Ontario characterized by warm and wet climate in most years. This information helps understand the greater instabilities of *rub* and *dia*. Cultivar *rub* was unstable because it is early and has relatively poor winter hardiness. It therefore performed well in RN93 but poor in BH93. Cultivar *dia* is late and tall; it performed relatively well in BH93 but was more prone to lodging in RN93 (more detailed discussion on how to understand GEI is available in Section 5.7 of this chapter). Cultivars placed close to the AEC abscissa, namely, *cas*, *zav*, *reb*, *del*, *ari*, and *kar*, were more stable than others.

Thus, the GGE biplot in the form of Figure 5.21 allows visualization of both mean performance and stability of the genotypes in the unit of yield per se. The common unit allows the two measures to be combined into, and visualized by, a single measure (Figure 5.23). The small circle in Figure 5.23, which is located on the AEC abscissa and with an arrow pointing to it, represents the ideal cultivar. It is defined by two criteria: 1) it has the highest yield of the entire dataset; and 2) it is absolutely stable, as indicated by being located on the AEC abscissa. Such an ideal genotype rarely exists in reality. Nevertheless, it can be used as a reference for cultivar evaluation. The plot distance between any cultivar and this ideal cultivar can be used as a measure of its desirability. The concentric circles, taking the ideal cultivar as the center, help in visualizing the distance between all cultivars and the ideal cultivar (Figure 5.23). Hence, *fun* is closet to the ideal cultivar, and therefore, most desirable of all the tested cultivars. It is followed by *cas* and *har*, which are in turn followed by *ron*, *ham*, *rub*, *zav*, *del*, and *reb*, etc.

It is interesting to note that the cultivar rankings in Figure 5.21, based on mean performance alone, and the cultivar rankings in Figure 5.23, based on both mean performance and stability, are almost identical. This is because the G is much greater than GEI in this dataset. The large G relative to GE is indicated by the shape of the biplot per se, as the range of PC1 is much greater than that of PC2, even though the magnitude of G and GE are not explicitly indicated.

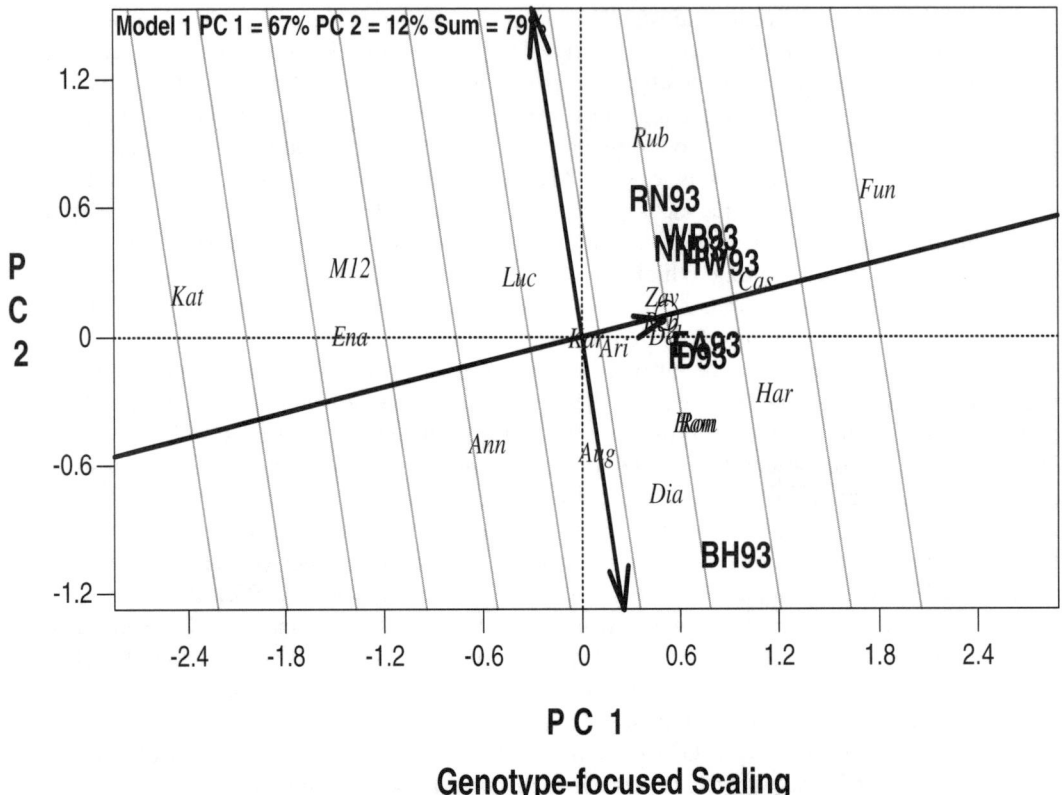

FIGURE 5.21 The average environment or tester AEC or ATC view of the GGE biplot based on the genotype-focused scaling, showing the mean yield and stability of genotypes. The AEC-abscissa, which is the single-arrowed line passing through the biplot origin and the average environment represented by the small circle, approximates the mean yield of the genotypes, and the AEC-ordinate, which is the double-arrowed line passing through the biplot origin and is perpendicular to the AEC-abscissa, approximates the GE interaction or (in)stability of the genotypes. The parallel lines help rank the genotypes based on mean yield. The arrow of the AEC-abscissa points toward the direction of increasing mean yield, and the arrow of the AEC-ordinate point to greater GE interaction or lower stability.

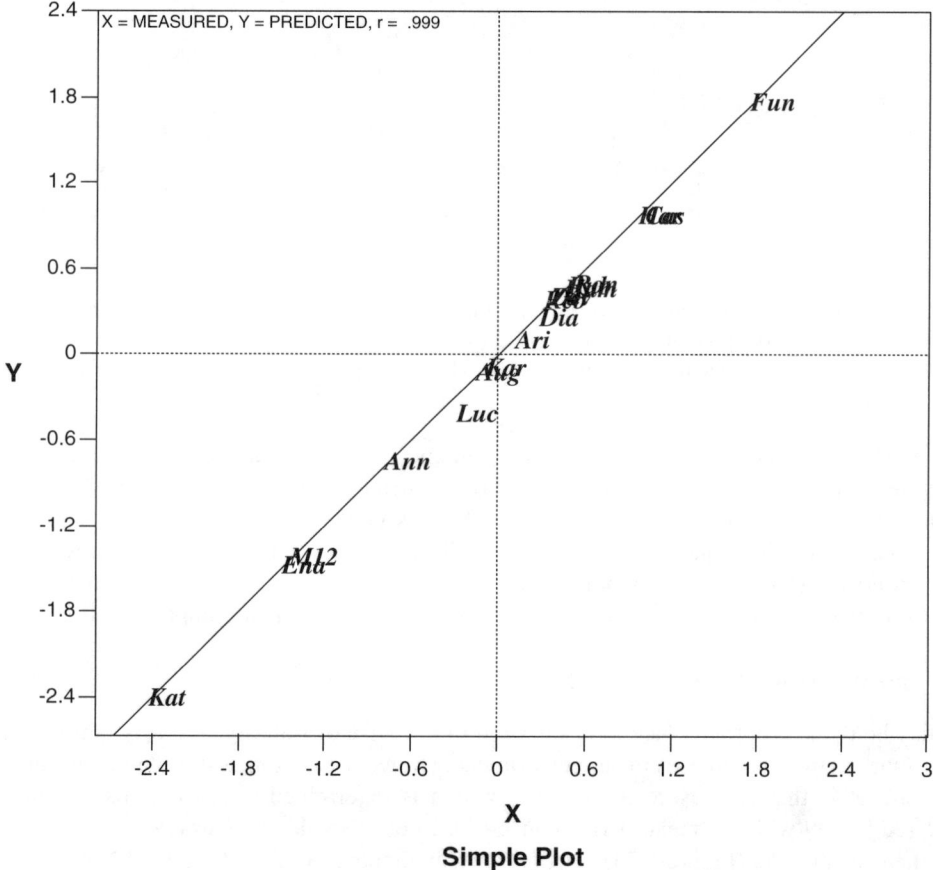

FIGURE 5.22 Relationship between the measured mean yield of the genotypes and their projections onto the abscissa of the average environment coordinate (AEC). It is to show that the latter provides a good approximation of the genotype main effects.

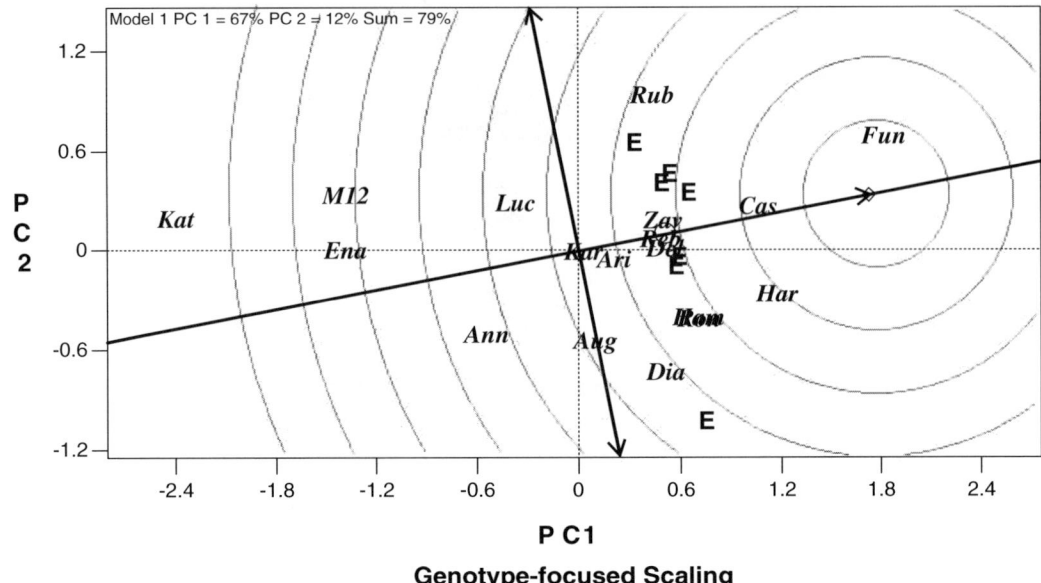

FIGURE 5.23 Comparison of all genotypes with the ideal cultivar. The ideal cultivar, represented by the small circle with an arrow pointing to it, is defined as having the highest yield in all environments. That is, it has the highest mean yield and is absolutely stable. The genotypes are ranked based on their distance from the ideal cultivar.

The ranking of genotypes based on both mean and stability, as measured by the distance from the markers of the genotypes to the ideal cultivar on the GGE biplot, was found to be highly correlated ($r = 0.97$) with the ranking based on Kang's yield-stability (YS_i) statistic (Kang, 1993) for the dataset used in Kang and Magari (1995). This suggests that GGE biplot provides similar information as that provided by popular stability analysis methods such as YS_i statistic. The GGE biplot, however, is much more comprehensive and its visual effects are unparalleled.

5.4.3 How Important is Stability?

This may be the appropriate place to discuss the importance of stability. Stability is important, but it is no more than a modifier of mean performance (Kang, 1993; Yan, 1999). One corollary of the GGE concept is that a measure of stability, which is determined by GEI, is useful only when considered jointly with the mean performance of G. For example, cultivar *ena* is apparently more stable than *rub* and *dia* (Figure 5.23). Does this mean that *ena* is more desirable? Not at all. It only indicates that *ena* yielded consistently but poorly at most locations. Stability is a dimensionless multiplier. It makes a high mean performance cultivar better but makes a low mean performance cultivar worse. Stability alone is only an academic exercise.

Another argument against the over-emphasis on stability follows. Stability cannot be reliably estimated from a few environments; but when it can be estimated reliably, i.e., when results from a rather large number of environments become available, it becomes redundant since, by that time, the mean performance alone should contain all the needed information (Yan, 1999). Stability was reported to have lower heritability than mean performance (Eskridge, 1996). In conclusion, stability is useful only when considered jointly with mean performance. The GGE biplot methodology provides a superior way to integrate both mean performance and stability into a single measure, which can be assessed visually.

Eisemann et al. (1990) listed three possible researcher attitudes or strategies toward GEI: exploiting it, ignoring it, or avoiding it. Exploitation of GEI means that the target environment is meaningfully divided into mega-environments and cultivar recommendation is made for each mega-environment (Figure 5.13). Ignorance of GEI means that one tests only in a few environments and selects on the basis of mean yield only (Figure 5.21). Avoidance of GEI means that one tests widely and selects on the basis of both mean yield and stability (Figure 5.23). Each of these strategies can be materialized via the GGE biplot methodology.

5.5 EVALUATION OF TEST ENVIRONMENTS

METs are conducted to evaluate cultivars. They are also equally useful for evaluating test environments. Evaluation of test environments includes four aspects. The first aspect is to see if the target region belongs to a single mega-environment or consists of different mega-environments, as was discussed in Section 5.3. The second aspect is to identify better test locations or environments. The third is to identify redundant test locations or environments that give no additional information about the cultivars. It is possible that fewer but better test locations can provide equally or more informative data for cultivar evaluation. The fourth aspect is to identify environments that can be used for indirect selection. This section shows how the GGE biplot is used to visualize the interrelationships among test environments and to identify better test locations or environments.

5.5.1 INTERRELATIONSHIPS AMONG ENVIRONMENTS

Figure 5.24 is referred to as the vector view of the GGE biplot, in which the environments are connected with the biplot origin via lines called *vectors*. Genotypes are represented by a single letter "c" since they are not the focus of discussion here. This view of the biplot helps understand the interrelationships among the environments. One interesting interpretation is that the cosine of the angle between the vectors of two environments approximates the correlation coefficient between them. For example, NN93 and WP93 have an angle of about 7° between their vectors; therefore, they should be closely correlated. Indeed, the correlation coefficient between NN93 and WP93 is 0.834 (Table 5.1). The angle between RN93 and OA93 is about 111°, slightly larger than 90°; therefore, they should be slightly negatively correlated: The correlation between them is –0.184 (Table 5.1). Comparison of the angles in Figure 5.24 and the correlation coefficients in Table 5.1 reveals a high correspondence between them. The cosine of the angles does not precisely translate into correlation coefficients, since the biplot does not explain all of the variation in a dataset. Nevertheless, the angles are informative enough to allow a whole picture about the interrelationships among the test environments. Such an understanding is not possible from Table 5.1. In Section 5.3, we indicated that biplot could be used to identify different mega-environments, although we did not explicitly point out that test environments from different mega-environments should have large angles; hence weak or negative correlations. Such relationships become obvious only when the environment vectors are drawn (Figure 5.24).

The vector view of a biplot (e.g., Figure 5.24) helps identify redundant test environments. If some environments have small angles and are, therefore, highly positively correlated, information on genotypes obtained from these environments must be similar. If this similarity is repeatable across years, these environments are redundant; a single environment should suffice. Obtaining the same or better information by using fewer test environments will reduce the cost and increase breeding efficiency. The soybean data provide an example for this. In all three years, locations Woodstock, St. Paul, and Exeter were closely correlated (Figures 5.16 to 5.18), suggesting that these three locations provided redundant information about genotypes. Therefore, it was suggested that one of the three locations be dropped to reduce the cost of testing (Yan and Rajcan, 2002).

The vector view of a biplot also has been used to identify environments that can be used in indirect selection (Cooper et al., 1997). One obvious example is disease screening in greenhouse conditions for selecting disease-resistant cultivars. Other examples include selecting for drought adaptation, acid-soil adaptation, etc., in artificial conditions. The vector view of the biplot helps identify correspondence between natural conditions and artificial conditions for indirect selection.

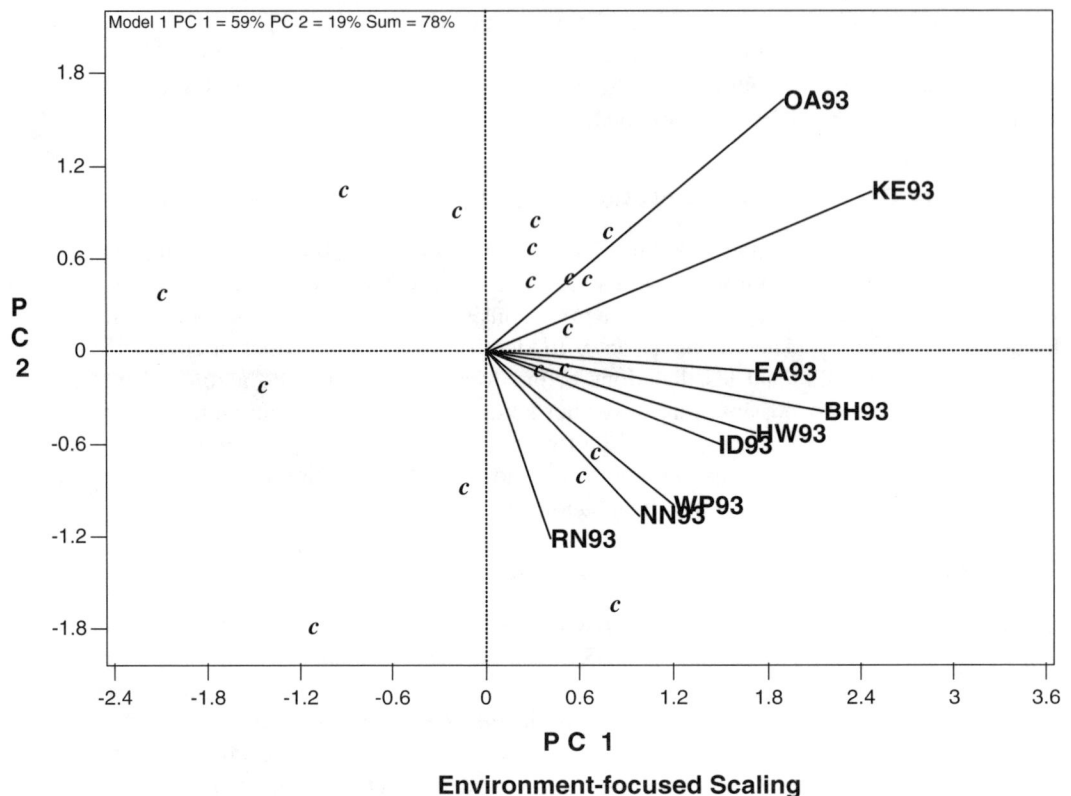

FIGURE 5.24 The vector view of the GGE biplot. It shows the interrelationships among the test environments. The cosine of the angle between the vectors of two environments approximates the correlation coefficient between them.

TABLE 5.1
Correlation Coefficients Among Environments

Names	BH93	EA93	HW93	ID93	KE93	NN93	OA93	RN93	WP93
BH93	1	0.651	0.552	0.746	0.736	0.523	0.358	0.287	0.584
EA93	0.651	1	0.689	0.753	0.659	0.613	0.644	0.339	0.711
HW93	0.552	0.689	1	0.620	0.482	0.559	0.464	0.429	0.661
ID93	0.746	0.753	0.620	1	0.646	0.697	0.324	0.513	0.717
KE93	0.736	0.659	0.482	0.646	1	0.204	0.636	−0.103	0.357
NN93	0.523	0.613	0.559	0.697	0.204	1	0.113	0.542	0.834
OA93	0.358	0.644	0.464	0.324	0.636	0.113	1	−0.184	0.223
RN93	0.287	0.339	0.429	0.513	−0.103	0.542	−0.184	1	0.713
WP93	0.584	0.711	0.661	0.717	0.357	0.834	0.223	0.713	1

5.5.2 Discriminating Ability of the Test Environments

Another interesting observation from the vector view of the biplot is that the length of the environment vectors approximates the standard deviation within each environment (Table 5.2), which is a measure of their discriminating ability. Thus, the three Eastern Ontario locations, OA, KE, and BH, are most discriminating; the southern location RN is the least discriminating (Figure 5.24; Table 5.2). The greater discriminating ability of BH93 is more prominent when OA93 and KE93 are removed (Figure 5.25). Other environments have similar vector lengths, i.e., discriminating ability. All environments within this mega-environment are positively correlated (Figure 5.25 and Table 5.1).

TABLE 5.2
Mean Yield and Standard Deviation Within Each Environment

Environments	Mean (Mg ha^{-1})	Standard Deviation
BH93	4.36	.62
EA93	4.44	.47
HW93	3.14	.54
ID93	3.49	.43
KE93	5.68	.71
NN93	5.06	.41
OA93	4.24	.68
RN93	4.36	.36
WP93	2.90	.42

5.5.3 Environment Ranking Based on Both Discriminating Ability and Representativeness

Discriminating ability is an important measure of a test environment. A test environment's lack of discriminating ability provides no information about the cultivars and, therefore, the test environment is useless. Another equally important measure of a test environment is its representativeness of the target environment. If a test environment is not representative of the target environment, it is not only useless but also misleading since it may provide biased information about the tested cultivars.

The representativeness of an environment is difficult to measure, since it is not possible to sample all possible environments within a mega-environment and subsequently determine the representativeness of each individual environment. The biplot way of measuring representativeness is to define an average environment and use it as a reference or benchmark. The average environment is indicated by the small circle in Figure 5.26. The line that passes through the biplot origin and the average environment is AEC, as discussed earlier. The angle between the vector of an environment (not drawn in Figure 5.26) and the AEC axis is a measure of the representativeness of the environment. Hence, EA93 and ID93 are most representative, whereas RN93 and BH93 are least representative of the average environment.

An ideal test environment should be both discriminating and representative. The small circle on the AEC axis, with an arrow pointing to it, represents the ideal environment. It is defined to be the most discriminating and absolutely representative (on the AEC axis) of all test environments. Anything that is preceded by the adjective "ideal" probably does not exist in reality. Nonetheless, it can be used as a reference point. In Figure 5.26, this ideal environment is used as the center of

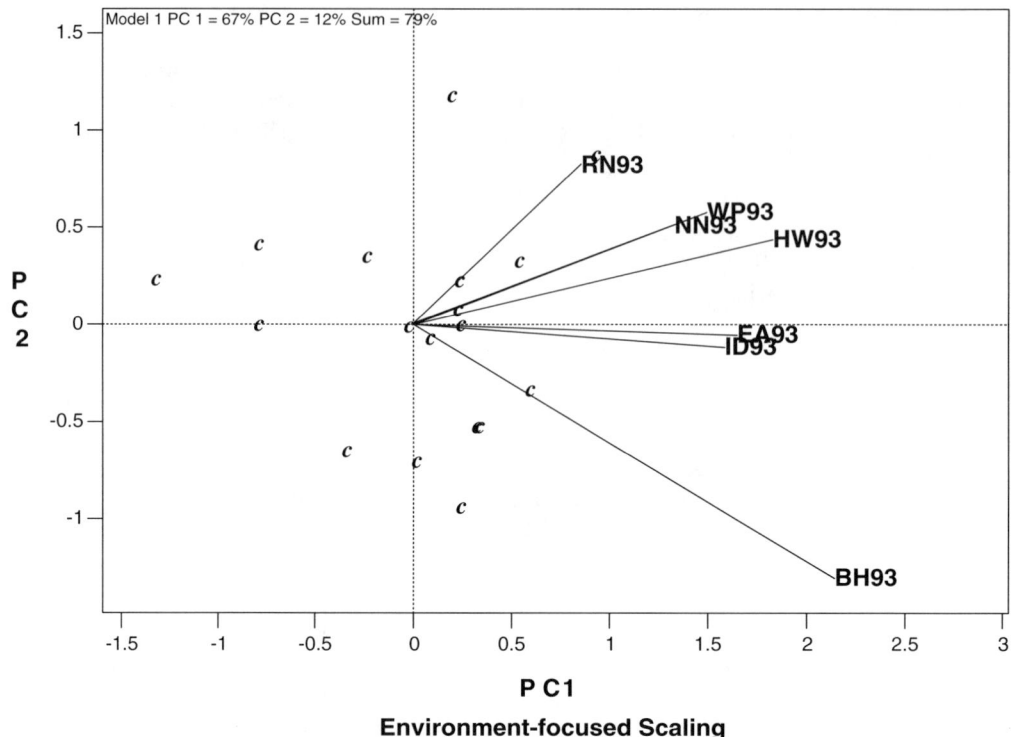

FIGURE 5.25 The vector view of the GGE biplot based on data after two environments were removed. It shows the interrelationships among the remaining test environments. The cosine of the angle between the vectors of two environments approximates the correlation coefficient between them.

a set of concentric lines that serve as a ruler to measure the distance between an environment and the ideal environment. Note that Figures 5.26 and 5.23 are very similar. The difference is that the main focus in Figure 5.23 was genotypes and the genotype-focused scaling was used, whereas the main focus in Figure 5.26 is environments and the environment-focused scaling is used. From Figure 5.26, it can be seen that HW93 is the closest to the ideal environment, and, therefore, is most desirable of all seven environments. HW93 is followed by EA93 and ID93, which are followed in turn by WP93 and NN93. RN93 and BH93 were the least desirable test environments. Also, EA93 and ID93, and NN93 and WP93 are pairs of highly similar environments, which can be better visualized from Figure 5.25.

5.5.4 ENVIRONMENTS FOR POSITIVE AND NEGATIVE SELECTION

When we talk about desirable environments, we mean environments that are most effective in identifying the best cultivars. Environment HW93, e.g., may be most effective for selecting superior cultivars. In essence, all environments are useful, provided that they are discriminative. Environments RN93 and BH93 are not representative; they are therefore not good environments for selecting superior cultivars. They can be used for culling inferior genotypes, however. For example, the unstable cultivar *rub* can easily be discarded based on performance in BH93, and another unstable cultivar *dia* can be easily discarded based on performance in RN93. Discriminating but unrepresentative environments are sometimes called killer or culling environments, which are useful for selecting stable cultivars.

To summarize, a GGE biplot can help identify 1) the more desirable environments for positive selection, 2) the killer environments for negative culling, 3) the redundant environments that can

Biplot Analysis of Multi-Environment Trial Data

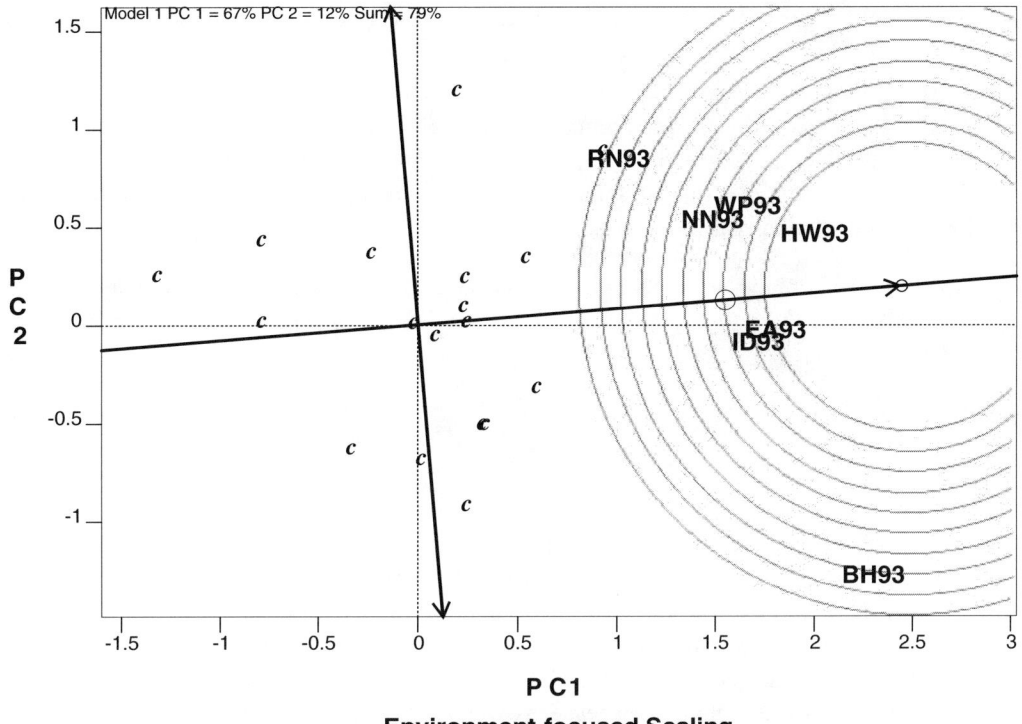

FIGURE 5.26 Comparison of all environments with the ideal environment. The ideal environment, represented by the small circle with an arrow pointing to it, is the most discriminating of genotypes and yet representative of the other test environments. The environments are ranked based on their distance from the ideal environment.

be removed without adversely influencing cultivar evaluation, and 4) artificial environments that can be effectively used in indirect selection.

5.6 COMPARISON WITH THE AMMI BIPLOT

While presenting the methodology of GGE biplot to various audiences, Yan was frequently asked to compare the GGE biplot to the AMMI biplot. Therefore, it seems appropriate to briefly discuss the AMMI biplot. AMMI has become one of the most popular methods for analyzing MET data. Numerous papers on the application of AMMI in MET data analysis have been published following the pattern of Zobel et al. (1988), which compared AMMI with the complete multiplicative model, the variance analysis, and the regression on environmental means model (Finlay and Wilkinson, 1963). The most attractive thing about AMMI, in our opinion, is that it generates a biplot consisting of means vs. interaction PC1 (IPC1), which displays both the mean yield and interaction scores of the genotypes and of the environments in a single plot.

The AMMI model is written as:

$$\hat{Y}_{ij} = \mu + \alpha_i + \beta_j + \sum_{n=1}^{k} \lambda_n \xi_{in} \eta_{nj} \qquad (5.1)$$

where

$\hat{Y}_{ij}$ = the expected yield, or any other trait, of genotype i in environment j,
μ = the grand mean of all observations,
α_i = the main effect of genotype i,
β_j = the main effect of environment j, and
λ_n, ξ_{in}, and η_{nj} are, respectively, singular value, gentotype eigenvectors, and environmental eigenvectors for the nth interaction principal component derived from subjecting the interaction matrix to SVD.

An AMMI1 model is made of the main effects and one interaction principal component, i.e.:

$$\hat{Y}_{ij} = \mu + \alpha_i + \beta_j + \lambda_1 \xi_{i1} \eta_{1j} + \varepsilon_{ij}$$

or

$$\hat{Y}_{ij} = \mu + \alpha_i + \beta_j + g_{i1} e_{1j} + \varepsilon_{ij} \qquad (5.2)$$

with $g_{i1} = \lambda_1^{0.5} \xi_{i1}, e_{1j} = \lambda_1^{0.5} \eta_{1j}$, i.e., the singular value is symmetrically partitioned into the genotype and environment scores (Chapter 4). The AMMI biplot results from plotting α_i against g_{i1} for the genotypes and plotting β_j against e_{1j} for the environments. The unit of the main effects is the original unit of the trait, whereas the unit of the IPC1 scores is square root of the original unit. Consequently, the shape of such a biplot is highly subjective. To amend this problem, square root of the main effects can be taken while keeping their respective signs unchanged (Richard Zobel, 1999, personal communication with W. Yan). Figure 5.27 is an AMMI biplot based on the yield data of the 1993 Ontario winter wheat performance trials we have used extensively in illustrating the GGE biplot methodology. Here we only want to make two points. First, Figure 5.27 does not have the inner-product property of a normal biplot and, therefore, most functions we discussed for the GGE biplot are not applicable. Second, Figure 5.27 can be misleading regarding the "which-won-where" issue. For example, Figure 5.27 may suggest that cultivars *kat*, *m12*, and *luc* won in locations WP93, HW93, and ID93. Actually this is untrue (Figure 5.13). One great thing about the AMMI biplot generated by the GGE biplot software (Chapter 6) is that it displays the relative magnitude of the G, environment main effect, and IPC1 (upper-left corner of Figure 5.27).

The take-home message is this: although a GGE biplot and an AMMI biplot may display similar information, most properties or functions of a GGE biplot described in this chapter do not apply to an AMMI biplot. Moreover, if your purpose is to find which cultivar won in which environments, you should be cautious that the AMMI biplot could be misleading. We have little doubt that most, if not all, researchers will immediately agree that the GGE biplot is the better choice.

5.7 INTERPRETING GENOTYPE-BY-ENVIRONMENT INTERACTION

5.7.1 THE GENERAL IDEA

The mathematical model for GGE biplots (Equation 4.5), which is given below:

$$\hat{Y}_{ij} - \mu - \beta_j = g_{i1} e_{1j} + g_{i2} e_{2j} + \varepsilon_{ij}$$

is ideal for studying causes of the GEI. In this model, GGE is decomposed into two or more principal components, each of which consists of a set of genotype scores multiplied by a set of environment scores. That is, the GGE of an MET data is presented in the form of GEI. If the genotype scores can be related to genes or traits, and if the environment scores can be related to physical factors, the observed GEI can be interpreted in terms of gene or trait-by-environmental factor interaction (Figure 5.28).

Biplot Analysis of Multi-Environment Trial Data

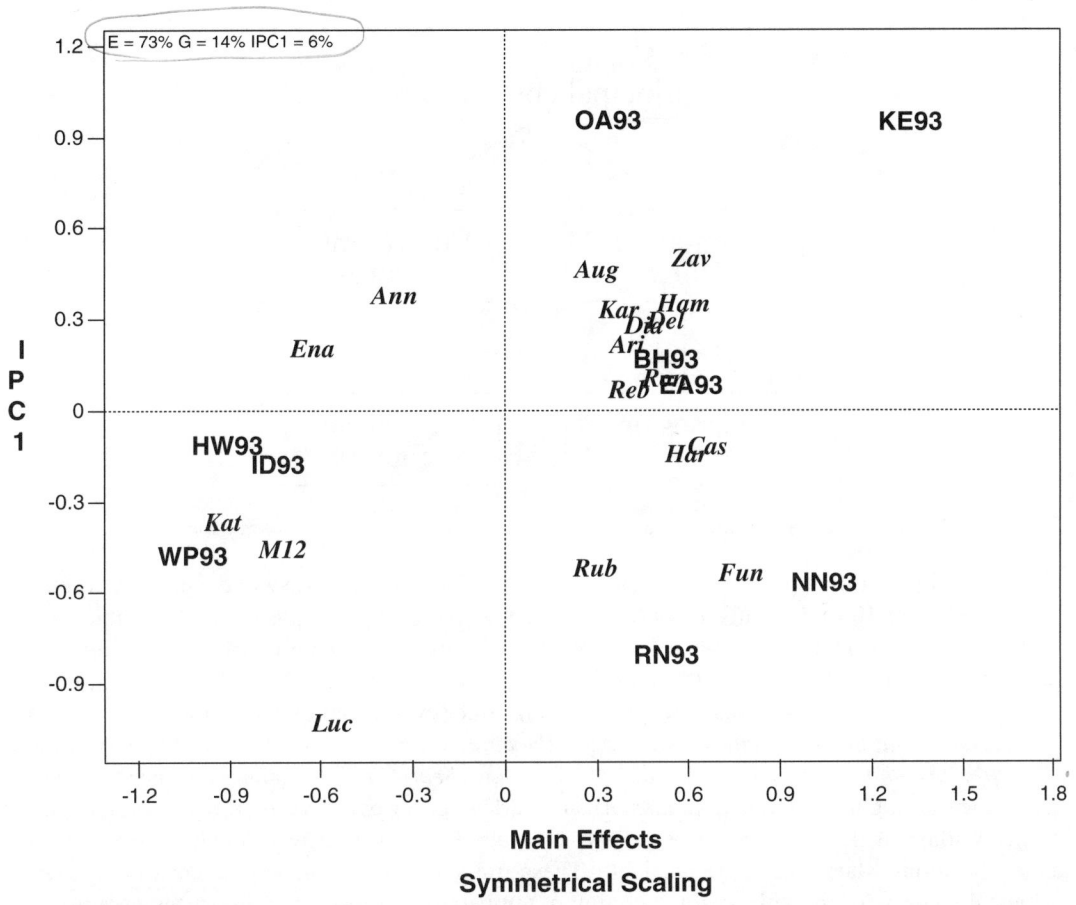

FIGURE 5.27 The biplot of main effects vs. the first interaction principal component scores; both axes are in the unit of square root of yield. This biplot is typical of AMMI (Additive Main effect and Multiplicative Interaction effect) analysis. It may suggest that cultivars *luc*, *m12*, and *kat* were superior cultivars in environments WP93, ID93, and HW93. This is actually misleading.

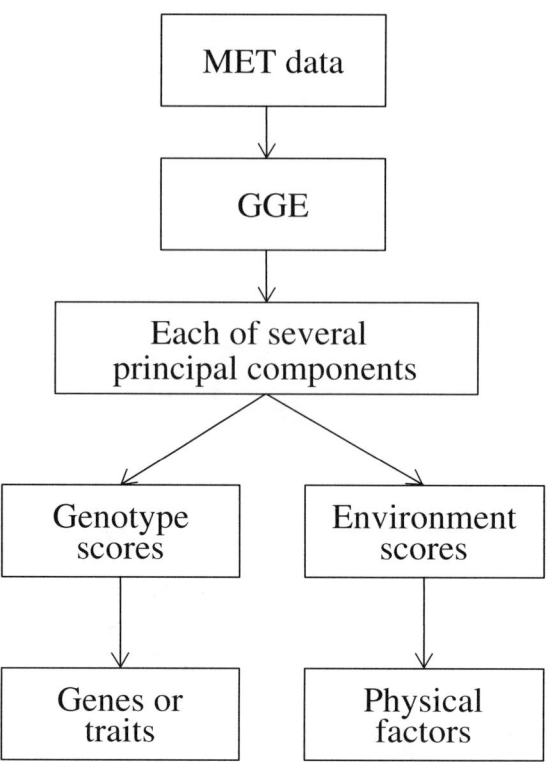

FIGURE 5.28 Flow chart indicating strategies for studying the causes of GE interaction.

We exemplify this approach using data from the 1992 to 1998 Ontario winter wheat performance trials. Each year 10 to 33 cultivars were tested at 8 to 15 locations representing the winter wheat growing region in Ontario. At each location, a randomized complete block design with 4 to 6 replicates was used. In addition to yield, several agronomic traits, i.e., date of heading, date of maturation, winter survival, plant height, lodging and several pathological traits, i.e., leaf rust (*Puccinia recondita* f. sp. *tritici*), stem rust (*Puccinia graminis* f. sp. *tritici*), powdery mildew (*Erysiphe graminis* f. sp. *tritici*), septoria leaf blotch (*Septoria tritici*), fusarium head blight (*F. gramninearum*), glume blotch (*S. nodorum*), and barley yellow dwarf virus (BYDV) were recorded at all or some of the locations. The genotypic values for each trait were obtained by averaging across locations where data were available. These traits are used as genotypic covariates. Meteorological records for monthly average minimum temperature, average maximum temperature, and total precipitation at each location in each year were used as environmental covariates. PCA on the monthly weather conditions across 84 year–location combinations revealed close associations between the monthly minimum and maximum temperatures. Consequently, only the monthly minimum temperatures were used in the analysis, which are briefly reported below.

5.7.2 Causes of GE Interaction Represented by PC1

Traits that closely associated with genotype PC1 scores and factors that closely associated with environmental PC1 scores are summarized in Table 5.3. Specifically, in 1992, greater genotypic PC1 scores were associated with shorter plant stature. Since genotypic PC1 scores were highly correlated with genotype mean yield each year (Yan and Hunt, 2001), this implied that shorter cultivars tended to yield more in 1992. On the environment side, greater environment PC1 scores were associated with cooler May to July temperatures and more precipitation in June. Collectively, the interpretation is that in 1992 shorter stature cultivars tended to yield more, particularly at

TABLE 5.3
Genotypic Trait-by-Environmental Factor Interaction Suggested by PC1

Year	Traits Associated with Greater Genotypic PC1 Scores	Environmental Factors Associated with Greater Environmental PC1 Scores
1992	Shorter	Cooler summer; more precipitation in June
1993	Less BYDV, septoria leaf blotch, and stem rust	More precipitation in October and November; cooler summer
1994	Taller; less glume blight and septoria leaf blotch	Colder winter; less precipitation in June
1995	Earlier/shorter	Warmer winter
1996	Later/winter hardier	Cooler summer; more precipitation in April and May
1997	Less septoria blotch	—
1998	Earlier/shorter	Warmer winter

locations with cooler summer temperatures and more precipitation in June. In other words, in 1992, the short stature trait interacted positively with the cooler summer factor to give higher yield.

Similarly, in 1993, greater genotypic PC1 scores were associated with better disease resistance to BYDV, Septoria leaf blotch, and stem rust; greater environmental PC1 scores were associated with more precipitation in October and November and lower summer temperature. These conditions are generally favorable to disease development. Therefore, the GEI in 1993 can be interpreted as positive interaction between the trait of better disease resistance and the factor of disease-prone conditions.

In 1994, the GEI can be interpreted as primarily due to the positive interaction between the trait of taller plant stature and the factor of colder winter and less precipitation in June. In 1995 and 1998, the GEI was due to positive interaction between the trait of short stature and early maturity and the factor of warmer winter conditions. In 1996, the interaction resulted from positive interaction between the trait of later maturity and better winter hardiness and the factor of cooler summer temperatures. In 1997, better resistance to Septoria leaf blotch was associated with greater genotypic scores or high mean yield, but no environmental factor was closely related to the environment PC1 scores.

It is interesting to note that although traits associated with genotypic PC1 scores and factors associated with environmental PC1 scores varied dramatically across years, the trait-by-factor interaction patterns were relatively consistent. For example, while shorter stature was associated with greater genotype PC1 scores and higher mean yield in 1992, 1995, and 1998, it was taller stature that was associated with greater PC2 scores. On the other hand, although more precipitation was associated with greater environmental PC1 scores in 1992, the opposite was true in 1994. Nevertheless, the underlying causes of GEI revealed by PC1 were the same. Shorter cultivars were favored in cooler and wetter summers and taller cultivar were favored in colder winters. In five of the seven years, interactions between plant height and earliness on the genotype side, and winter and summer temperature and rainfall on the environment side, were found to be responsible for the genotype-by-location interaction of winter wheat yield in Ontario. Generally, earlier maturity interacted positively with warmer winters and hotter or drier summers, whereas later maturity interacted positively with colder winters and cooler summers; and taller stature interacted positively with colder winters and hotter summers, whereas shorter stature interacted positively with warmer winters and cooler summers. Different combinations of plant height and maturity resulted in different GEI patterns. Early and tall cultivars are favored for hotter and drier summers; early and short cultivars are favored in warmer winters; late and tall cultivars are favored in colder winters; and late and short cultivars are favored in cooler summers. Based on analysis of winter wheat yield trials in Western Canada, Thomas et al. (1993) reported positive associations between plant height and yield following cold winters but negative association between them following warm winters.

TABLE 5.4
Genotypic Trait-by-Environmental Factor Interaction Suggested by PC2

Year	Traits Associated with Greater Genotypic PC2 Scores	Environmental Factors Associated with Greater Environmental PC2 Scores
1992	Taller; less stem rust	Less precipitation in June
1993	Winter hardier/taller/later	Colder winter; more precipitation in May
1994	More leaf rust	High June precipitation and low temperature in July
1995	More lodging	Cooler summer; colder winter
1996	Later	More precipitation in December
1997	Taller	Lower temperature in January
1998	Later/taller	Colder winter

5.7.3 Causes of GE Interaction Represented by PC2

Traits that closely associated with genotype PC2 scores and factors that closely associated with environmental PC2 scores are summarized in Table 5.4. In 1992, taller cultivars and those with better stem rust resistance interacted positively with locations with less precipitation in June. This is in essence the same type of trait-by-factor interaction depicted by PC1 for the same year (Table 5.3). In 1995, cultivars that experienced more lodging were favored in colder winters and cooler summers; this is the same type of interaction as we saw for PC1 for the same year. In 1996, taller cultivars were favored in more precipitation in December. In 1998, late and tall cultivars were favored in colder winters, a different way to indicate the same trait-by-factor interaction described by PC1.

Results are more interesting from years 1993 and 1997. There is no interaction between plant height or earliness and environmental factors for these two years for PC1, but we do see such interaction for PC2. In 1993, cultivars that were tall, late, or had better winter survival ratings were favored in colder winters — this is the same type of trait-by-factor interaction that we saw in 1992, 1994, 1995, 1996, and 1998 for PC1. In 1997, tall cultivars were favored in lower temperatures in January, a trait-by-factor interaction that is very familiar to us. Therefore, the interaction between plant height or maturity and winter and summer conditions was part of the basis for genotype-by-location interaction for winter wheat yield in Ontario in all years.

5.7.4 Some Comments on the Approach

The GGE model (Equation 4.5) explains what is commonly called G in terms of GEI. G is by definition a constant value for a given genotype across the tested environments, whereas the genotypic PC1 score represents a tendency of the genotypes to respond to the environmental factors represented by the environmental PC1 scores. The yield of the genotype relative to PC1 is not the same in all environments; rather, it is directly proportional to the location PC1 scores. Thus, the GGE model emphasizes the fact that the so-called G not only has a genotypic basis, but also is dependent on the environmental conditions. In other words, the G is influenced by GEI. Viewing G in terms of GEI has one potential advantage: examination of PC1 scores not only identifies genotypes with better overall performance but also suggests environmental conditions that facilitate identification of these genotypes.

As with most variety trials, genotypes and locations varied each year in the Ontario winter wheat performance trials. This, in addition to the large yearly weather variation, led to different GEI patterns across years. Nevertheless, the trait-by-factor interaction patterns were quite consistent. Particularly from PC1, the same interaction between plant height or maturity and winter or summer conditions was seen for all years except 1993 and 1997, for which the same interaction was seen from PC2. Thus, PC1 and PC2 complementarily indicated that interactions between genotypic effects, such as maturity and plant height, and environmental factors, such as winter and summer temperatures, were the major causes of GEI for winter wheat yield in Ontario.

We want to emphasize two points. First, the so-called G should be regarded as GEI from a greater scale of time and space; the GGE model is thus highly justified. Second, understanding of GEI is achievable if genotypic and environmental covariates are collected and used in performance trials.

Section III

GGE Biplot Software and Applications to Other Types of Two-Way Data

6 GGE Biplot Software — The Solution for GGE Biplot Analyses

SUMMARY

This chapter describes genotype and genotype-by-environment (GGE) biplot software, a Windows application for conducting GGE biplot analysis. GGEbiplot provides fully automatic GGE biplot analysis; it accomplishes in a few seconds a job that would take weeks or months using conventional tools. It reveals patterns and addresses breeding problems that no other method can, it turns irksome data analysis into an enjoyable interactive experience, it incorporates modules that facilitate data manipulation, and it generates graphics ready for scientific publication and professional presentation.

6.1 THE NEED FOR GGE BIPLOT SOFTWARE

After reading through the previous chapters, particularly Chapter 5, you probably have already realized that the GGE biplot methodology is indeed the most effective, most powerful, and most elegant way we have ever had for visualizing multi-environment trial (MET) data and genotype-by-environment interactions (GEI), and for addressing issues that breeders and production agronomists must deal with. Since the first publication of the methodology (Yan et al., 2000), Yan has received enthusiastic support from around the world. Some described Yan et al. (2000) as the best paper they had ever read on the use of biplots. Impressive, exceptional, intriguing, or even revolutionary are the words most frequently used in the comments. The recognition of the usefulness of the GGE biplot methodology by breeders, quantitative geneticists, and other researchers is overwhelming. Conducting GGE biplot analysis using conventional tools is, however, complex and cumbersome, even for well-trained biometricians.

Upon request, Yan sent the SAS program that he used in his initial analyses to several readers, but it was only of limited help. The frontier of GGE biplot applications is expanding rapidly. Although it was originally used for graphical analyses of MET data, it was found equally effective in visualizing other types of two-way data, including genotype-by-trait data (Yan and Rajcan, 2002), diallel cross data (Yan and Hunt, 2002), and genotype-by-genetic marker data (Yan and Falk, unpublished). For different types of data, additional utilities can be formulated. For example, in a comprehensive genotype-by-trait dataset, biplots enhance, among other things, understanding of the interrelationships among the traits, which, in turn, leads to a holistic understanding of the system (Chapter 7). For genotype-by-trait plus marker data, the GGE biplot methodology provides a superior means for quantitative trait loci (QTL) identification (Chapter 8). For diallel cross data, the information based on a GGE biplot is many folds greater than that from the conventional method (Chapter 9). For host plant-by-pathogen strain data, the GGE biplot helps in understanding relationships among host genotypes, relationships among pathogen strains, and interaction between host genotypes and pathogen strains (e.g., horizontal vs. vertical resistance) (Chapter 10). Other agriculturally related areas for which the GGE biplot methodology is potentially useful include QTL-by-environment interaction and test site-by-physical factor data analysis in conjunction with geographical information system (GIS).

Thus, there is a real need for user-friendly software to conduct GGE biplot analyses. The good news is that a Windows application called "GGEbiplot" is now available. An earlier version of the program was reported in Yan (2001), and a more advanced version will be described in this chapter.

6.2 THE TERMINOLOGY OF ENTRIES AND TESTERS

When we discuss GGE biplot analysis of MET data, genotypes or cultivars and environments or locations represent two-way data matrices. Now that the GGE biplot analysis is extended to visual analysis of other types of two-way data, additional general terms are needed. We generalize to equate an MET dataset to an entry-by-tester dataset. All factorial experiments can be generalized as a two-factor approach, in which one factor is used as entry and the other factor as tester, and a set of entry levels called entries is tested against a set of tester levels called testers. Therefore, in an MET dataset, genotype is the entry factor, and environment is the tester factor. Different genotypes are levels of the entry factor and different environments are levels of the tester factor. In a genotype-by-trait table, genotypes are entries and traits are testers. In a genotype-by-molecular marker table, genotypes are entries and molecular markers are testers. In a diallel cross data, each parent is both an entry and a tester. Conventionally, entries are presented as rows, and testers are presented as columns in a two-way table.

In certain types of two-way data, the researcher determines the factor that would be designated as entry; the other factor would be treated as tester. For example, in a host genotype-by-pathogen strain data, if the main purpose is to study the resistance of the host genotypes, genotypes should be regarded as entries and pathogen strains as testers; if the main purpose is to study the virulence of the pathogen strains, then the strains should be used as entries and the genotypes as testers. GGEbiplot has a very useful function called entry/tester switch roles, which makes it very convenient to visualize a two-way dataset in both ways. This is described under data manipulation.

6.3 PREPARING DATA FILE FOR GGEBIPLOT

When GGEbiplot is started, the snapshot on the screen looks like that shown in Figure 6.1. It has three buttons: Start, Help, and Exit. Clicking Exit will close the program; clicking Help will bring up a help file in the *html* format. Clicking Start will bring up an open file dialog as shown in Figure 6.2, which asks you to select a data file. Select a data file, for example, "w93r.dat," which is the data we used to illustrate GGE biplot of MET data in Chapter 5. Click OK and Figure 6.3 appears, which asks the user to indicate the data file format. When the correct data format is selected, a biplot based on the data will appear (refer to Figure 5.2).

GGE biplot reads data from comma delimited text files. Comma delimited text files can be generated using Microsoft Excel. For the data to be readable by GGEbiplot, however, the extension name ".cvs" may need to be changed to ".dat." GGEbiplot can read two types of data format: an Observation format and a Matrix format.

6.3.1 THE OBSERVATION DATA FORMAT

A sample data for the Observation format, which is part of the "w93r.dat," is presented in Table 6.1. The rules for this format are:

- Each row contains an observation or data point, which is made of four values — tester name, block or replication name, entry name, and measured trait value.
- The first or header row contains four items — tester, block, entry, and trait, which are separated by commas. No comma should be at the end of the row. The headers can have any names you choose, but the number of columns must be the same. The computer will not read this row; it is for the user's reference.

GGE Biplot Software — The Solution for GGE Biplot Analyses

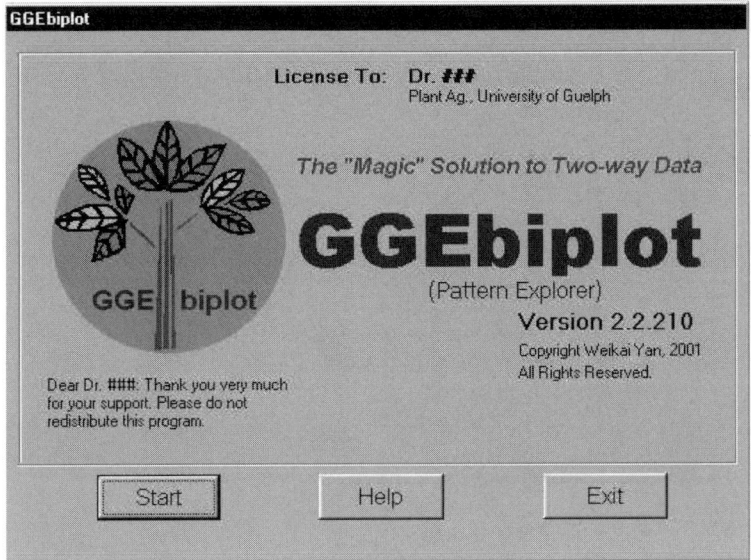

FIGURE 6.1 Snapshot of the GGEbiplot software: screen showing the Start, Help, and Exit buttons.

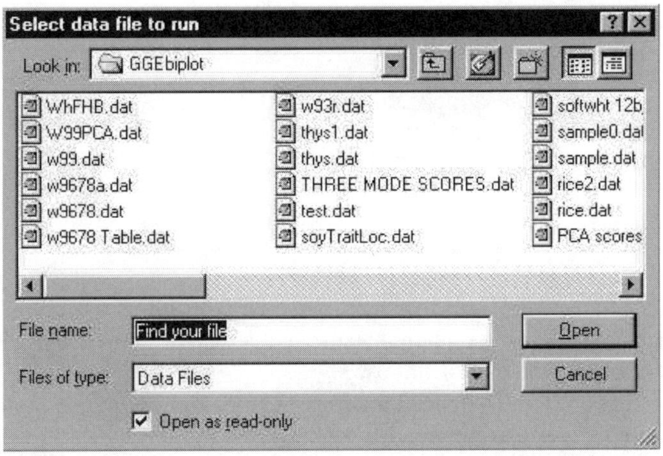

FIGURE 6.2 Snapshot of the GGEbiplot software: "Select data file to run" dialog.

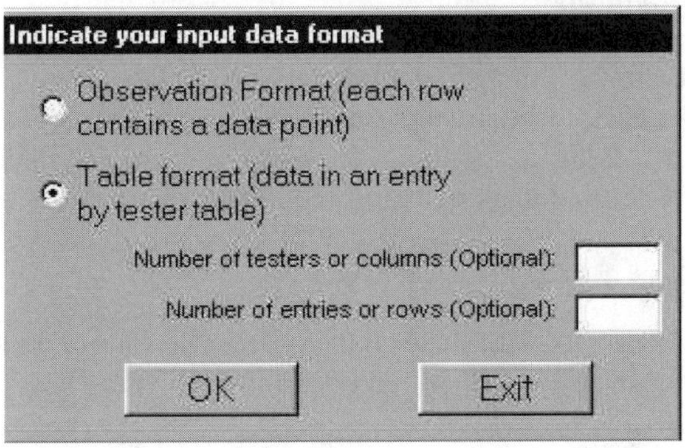

FIGURE 6.3 Snapshot of the data format selection dialog. The data can be in Observation format or in Table format. For Table format, the user may be asked to provide the number of columns and rows of two-way table.

TABLE 6.1
Example of the "Observation" Data Format

```
Tester, Block, Entry, Yield
BH93,1,AUG,3.931
BH93,1,HAR,4.323
BH93,1,ENA,2.731
BH93,1,ANN,3.984
BH93,1,HAM,3.949
BH93,1,REB,3.305
BH93,1,ZAV,2.688
BH93,1,KAR,3.236
BH93,1,RON,2.74
BH93,1,ARI,3.279
BH93,1,DEL,4.166
BH93,1,CAS,4.74
BH93,1,KAT,1.905
BH93,1,RUB,2.87
BH93,1,FUN,3.201
BH93,1,LUC,3.001
BH93,1,M12,3.07
BH93,1,DIA,4.627
BH93,2,AUG,5.14
BH93,2,HAR,4.645
BH93,2,ENA,3.444
BH93,2,ANN,4.566
BH93,2,HAM,5.001
...
```

- Other rows contain data. Each row has four columns, which are delimited by commas — tester name, block or replication name, entry name, and measured trait value. No comma should be at the end of the rows.
- The data do not have to be balanced in terms of entry-by-tester combination. The number of blocks or replicates can differ with the tester.
- Missing cells do not need to be indicated. If you do, indicate it with a period "." or "–99." The row with a missing cell can also be completely deleted. Missing cells, if any, will be replaced with the respective tester means and the user will be notified once missing cells are detected.
- The upper limit for the number of observations or data points is set at 30,000. The program is designed to accommodate a maximum of 500 entries or testers, and 650 entries plus testers, although the limit can easily be increased upon user's request.
- This format must be used to input replicated data.

6.3.2 THE MATRIX DATA FORMAT

A sample dataset for the Matrix format, which contains the mean yield of each cultivar at each location from the 1993 Ontario winter wheat trials, is presented in Table 6.2. The simple rules to follow are:

- The data are to be presented as a two-way table or matrix.
- The first or header row contains the header for entries, followed by the names of the testers, delimited by commas.

TABLE 6.2
Example of the "Matrix" Data Format

```
Names,BH93,EA93,HW93,ID93,KE93,NN93,OA93,RN93,WP93,end
ann,4.460,4.150,2.849,3.084,5.940,4.450,4.351,4.039,2.672
ari,4.417,4.771,2.912,3.506,5.699,5.152,4.956,4.386,2.938
aug,4.669,4.578,3.098,3.460,6.070,5.025,4.730,3.900,2.621
cas,4.732,4.745,3.375,3.904,6.224,5.340,4.226,4.893,3.451
del,4.390,4.603,3.511,3.848,5.773,5.421,5.147,4.098,2.832
dia,5.178,4.475,2.990,3.774,6.583,5.045,3.985,4.271,2.776
ena,3.375,4.175,2.741,3.157,5.342,4.267,4.162,4.063,2.032
fun,4.852,4.664,4.425,3.952,5.536,5.832,4.168,5.060,3.574
ham,5.038,4.741,3.508,3.437,5.960,4.859,4.977,4.514,2.859
har,5.195,4.662,3.596,3.759,5.937,5.345,3.895,4.450,3.300
kar,4.293,4.530,2.760,3.422,6.142,5.250,4.856,4.137,3.149
kat,3.151,3.040,2.388,2.350,4.229,4.257,3.384,4.071,2.103
luc,4.104,3.878,2.302,3.718,4.555,5.149,2.596,4.956,2.886
m12,3.340,3.854,2.419,2.783,4.629,5.090,3.281,3.918,2.561
reb,4.375,4.701,3.655,3.592,6.189,5.141,3.933,4.208,2.925
ron,4.940,4.698,2.950,3.898,6.063,5.326,4.302,4.299,3.031
rub,3.786,4.969,3.379,3.353,4.774,5.304,4.322,4.858,3.382
zav,4.238,4.654,3.607,3.914,6.641,4.830,5.014,4.363,3.111
```

- Each of the subsequent rows, delimited by commas, contains data for an entry, consisting of the name of the entry and the values for each tester relative to the entry.
- Missing cells may be left blank or indicated by "." or "–99."

Since "w93r.dat" is in the Observation format, we select the observation format (Figure 6.3). Upon clicking OK, data will be read, processed, and a biplot displayed as Figure 5.2.

If your data are in a table or matrix format, the Table format should be selected (Figure 6.3). For proper reading of the data, you need to provide the number of columns of the dataset. But if you put "END" (or "end," case insensitive) at the end of the header row, as shown in Table 6.2, there is no need to provide the number of columns or testers. The input box for the number of entries or rows is optional. You may ignore it, or put a number that is equal to or smaller than the real number of entries in the dataset. The program will stop reading data when this number is reached. You will get an error message if a number that is greater than the actual number of entries is used.

6.4 ORGANIZATION OF GGEBIPLOT SOFTWARE

As pointed out in Chapter 4 and demonstrated in Chapter 5, generating a GGE biplot is just the beginning, rather than the end, of biplot analysis. The GGEbiplot software fully automates GGE biplot analysis. In addition to generating a GGE biplot based on the full original data, it allows visualization of the biplot from countless perspectives. Moreover, it incorporates functions such as model selection, biplot selection, scaling (i.e., singular value partition) method selection, data manipulation, image format, and numeric output, making GGEbiplot a comprehensive tool for graphical analysis of two-way data. It produces images that are of the quality acceptable for publication in professional journals and for professional presentations. More and more scientists in North America and elsewhere in the world are now using GGEbiplot as a research tool in plant breeding, as well as a teaching aid for quantitative genetics. All biplots presented in Chapters 4 and 5 are direct outputs of GGEbiplot. For datasets from normal regional performance trials, each biplot took about one second to generate. All utilities of a GGE biplot discussed in Chapter 5, and many others not discussed so far, are just a mouse-click away.

The GGEbiplot is a program that can be installed in Windows 95 and later versions. It takes about 5 MB hard disc space and requires 5 MB RAM to run. The functions in GGEbiplot are organized under the primary menu bars *File* (Figure 6.4), *View* (Figure 6.5), *Visualization* (Figure 6.6), *Find QTL* (Figure 6.7), *Format* (Figure 6.8), *Models* (Figure 6.9), *Data Manipulation* (Figure 6.10), *Biplots* (Figure 6.11), *Scaling* (Figure 6.12), *Accessories* (Figure 6.13), and *Help*. Most of the functions are self-explanatory. Therefore, only those that are not directly visible are briefly described next.

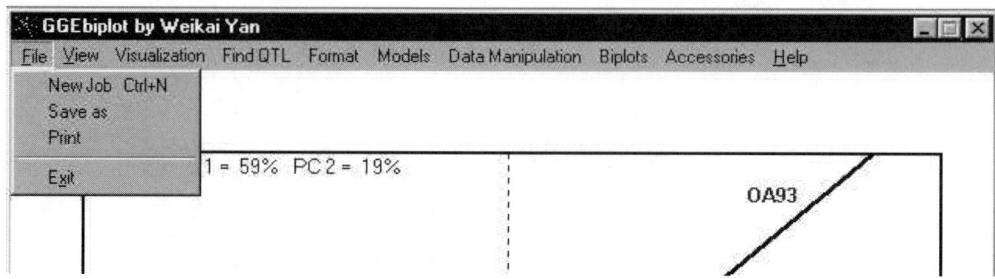

FIGURE 6.4 Snapshot showing the options of the File menu.

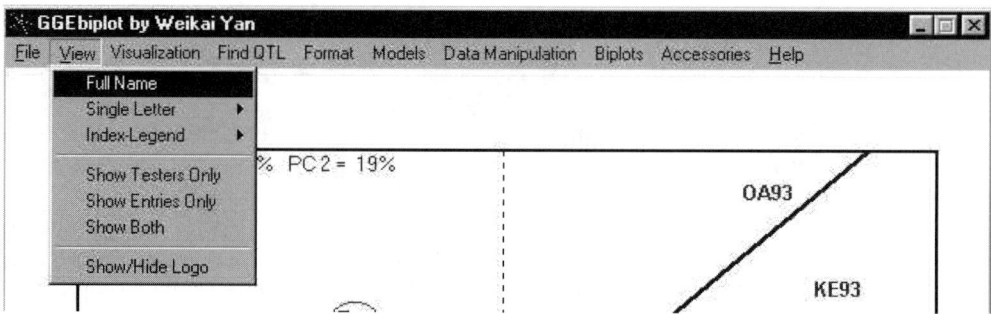

FIGURE 6.5 Snapshot showing the options of the View menu.

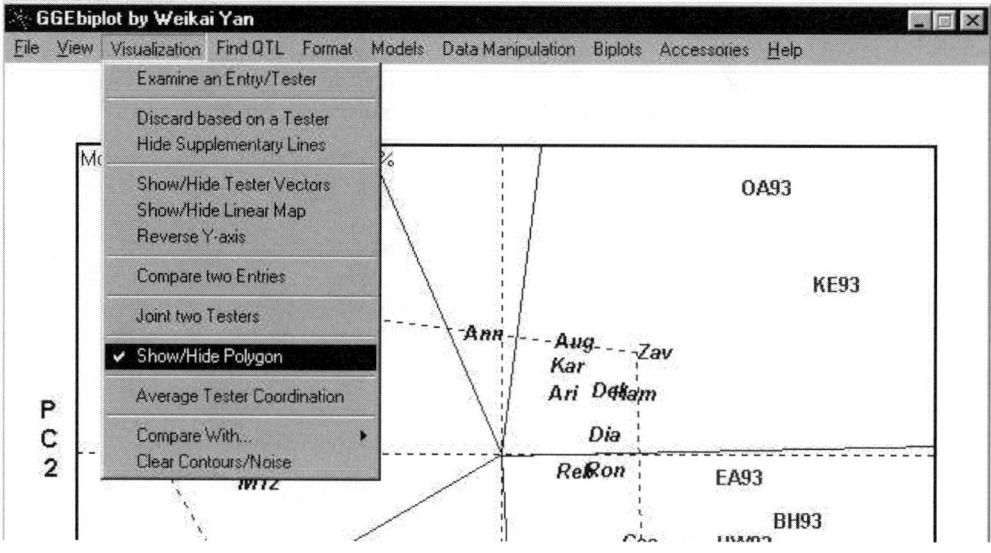

FIGURE 6.6 Snapshot showing the options of the Visualization menu.

GGE Biplot Software — The Solution for GGE Biplot Analyses

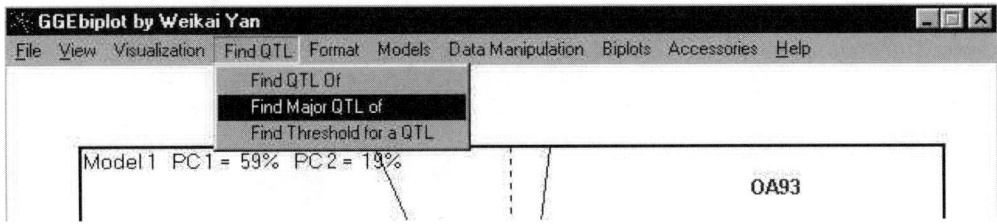

FIGURE 6.7 Snapshot showing the options of the Find QTL menu.

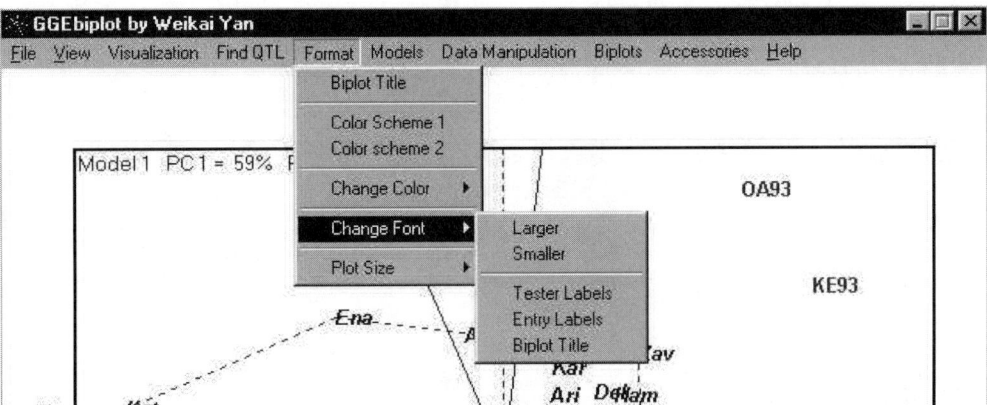

FIGURE 6.8 Snapshot showing the options of the Format menu.

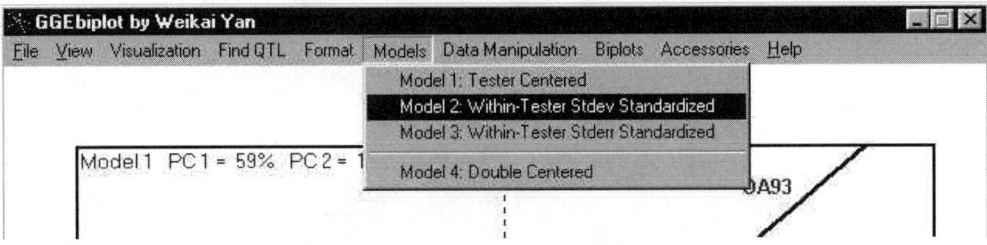

FIGURE 6.9 Snapshot showing the options of the Model selection menu.

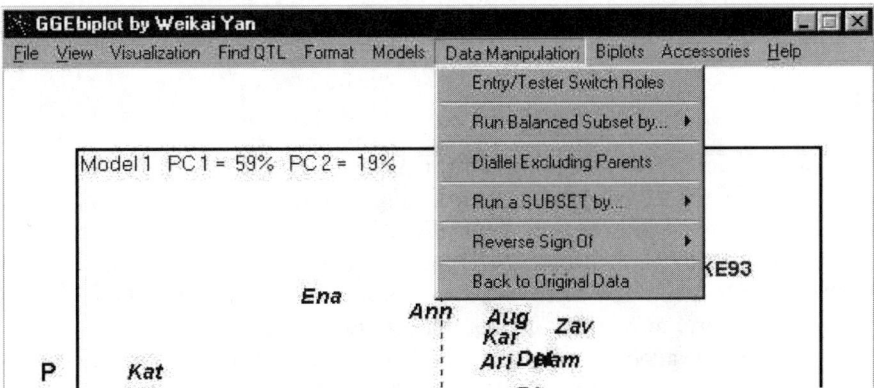

FIGURE 6.10 Snapshot showing the options of the Data Manipulation menu.

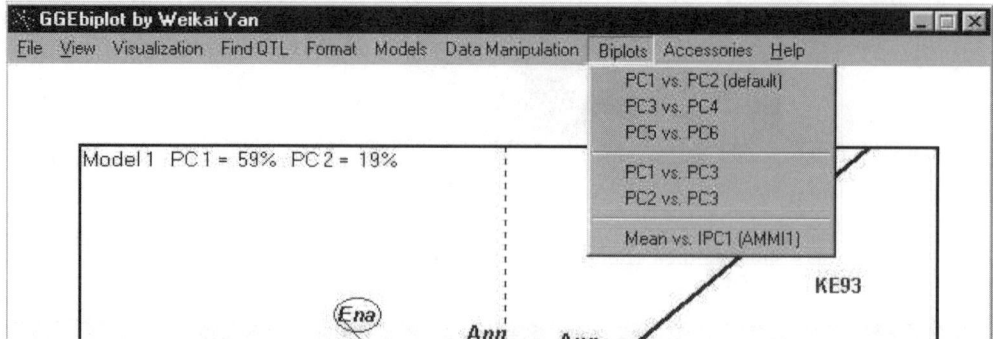

FIGURE 6.11 Snapshot showing the options of the Biplot selection menu.

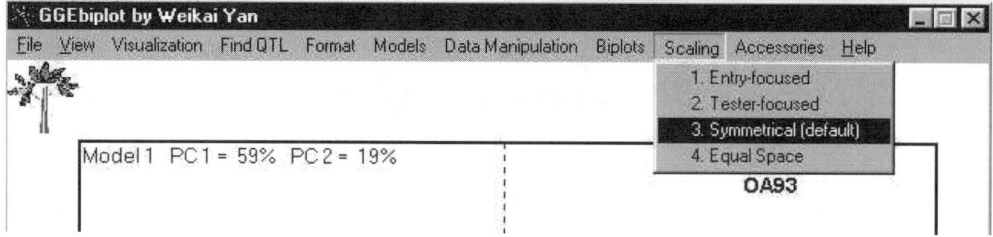

FIGURE 6.12 Snapshot showing the options of the singular value Scaling selection menu.

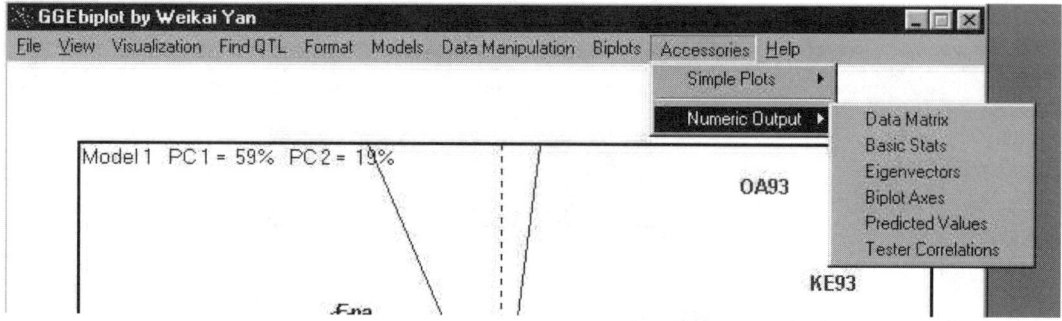

FIGURE 6.13 Snapshot showing the options of the Accessories menu. The sub-menu shows the options of Numeric output.

6.4.1 FILE

- New Job — starts the analysis of another dataset. When this function is selected, the user is reminded that a log file for the previous analysis is saved.
- Input PCA Scores — devised to generate and visualize biplots using principal component analysis (PCA) scores generated by other programs such as SAS. This function may never be used because GGEbiplot is a stand-alone program. The user, however, may find this function useful to plot any dataset that has two columns.
- Save As — not available in this version. When selected, it will remind you that the current biplot image can be printed to a pdf file using Adobe Writer. This is the recommended way for producing high-resolution biplot image files.

GGE Biplot Software — The Solution for GGE Biplot Analyses

- Print — prints the current image to any printer connected to the user's computer. This includes Adobe Writer and any fax device.
- Exit — exits the program.

6.4.2 View

- Full Name — shows the full names of both entries (e.g., genotypes) and testers (e.g., environments)
- Single Letter
  - Entry — shows each entry with a letter "c"
  - Tester — shows each tester with a letter "E"
- Index legend
  - Entry — shows entries as numerical indices, with a legend window opened with full names
  - Tester — shows testers as numerical indices, with a legend window opened with full names
- Show Testers Only
- Show Entries Only
- Show Both — shows both entries and testers
- Show/Hide Logo

6.4.3 Visualization

- Examine an Entry/Tester — allows the user to select an entry or a tester (Figure 6.14). When an entry is selected and the "Examine" button is clicked, its relative values with different testers will be displayed (Figure 5.5). When a tester is selected and the "Examine" button is clicked, performance of different entries with the selected tester will be displayed (Figure 5.3).
- Culling Based on a Tester — allows for entries (e.g., genotypes) to be discarded based on any selected tester (environment, trait, etc.) at the user-specified selection intensity. Some applications of this function are presented in Chapter 7 (Figures 7.19 to 7.22). This is a good tool to study and apply independent culling in plant breeding.
- Hide Supplementary Lines — will remove the supplementary lines on the biplot.
- Show/Hide Tester Vectors — draws lines that connect the biplot origin and each marker of the testers to facilitate visualization of the interrelationships among testers (Figure 5.23 and more in Chapter 7). When this function is selected, the tester-focused singular value partitioning method is automatically applied (defined in Chapter 4).

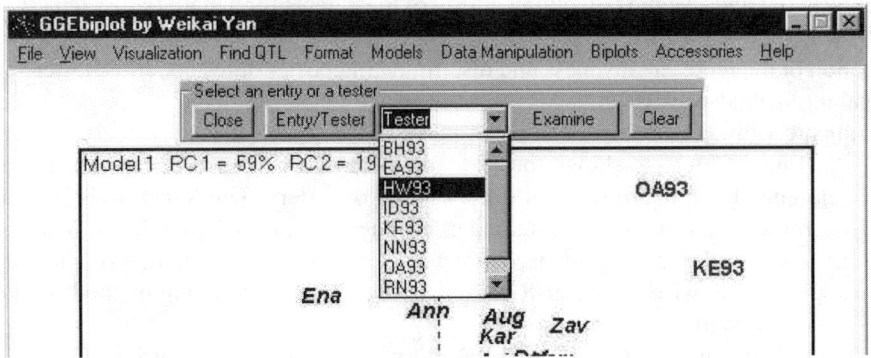

FIGURE 6.14 Snapshot showing the Examine panel.

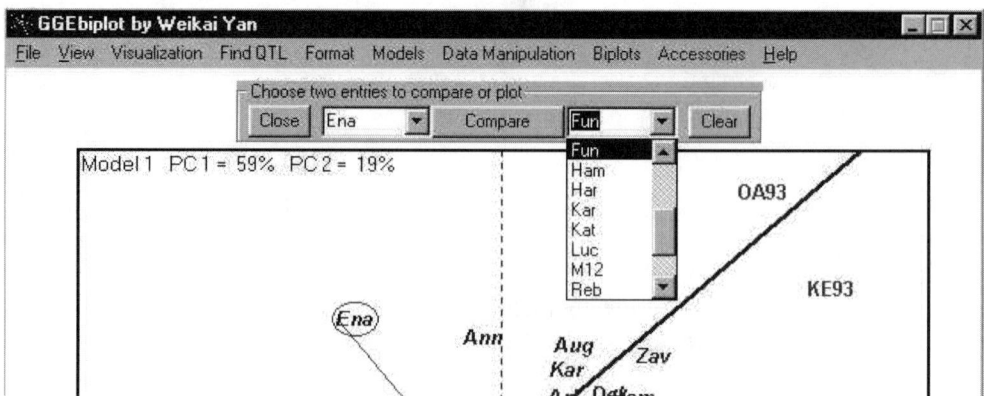

FIGURE 6.15 Snapshot showing the Compare panel.

- Show/Hide Linear Map — displays the angles among the vectors of the testers in a linear map, which mimics the linear map of molecular markers on a single chromosome (see illustrations in Chapter 8). When this function is selected, the tester-focused singular value-partitioning method is automatically applied (Chapter 4).
- Reverse Y-Axis — reverses the Y-axis of the biplot. It is useful to generate a biplot and associated linear map of the testers to be compared with available biplots or linear maps (Chapter 8). It takes a second to complete but saves the user hours or days.
- Compare Two Entries — allows visual comparison of two entries with regard to each of the testers. When this function is selected, a selection panel will appear, which allows you to select two entries to be compared (Figure 6.15). See Figure 5.7 for detailed description and interpretation.
- Joint Two Testers — allows visual evaluation of the entries based on two testers (Figure 7.11). When this function is clicked, a selection panel similar to that shown in Figure 6.15 will appear, which allows one to select two testers to be used for joint assessment of entries. This function is used to rank entries based on two testers.
- Show/Hide Polygon — the polygon view is an important characteristic of the GGE biplot methodology. It is the most efficient and elegant way to visualize the overall pattern of a two-way dataset. It is particularly useful in displaying the "which-won-where" patterns of MET data, as discussed in previous, as well as subsequent, chapters.
- Average Tester Coordination — very important function, which defines an average tester and draws an average tester coordination (ATC) on the biplot. This facilitates visualization of the mean performance and stability of a cultivar for an MET dataset (Chapter 5), and general and specific combining ability of the parents for a diallel-cross data (Chapter 8), when the entry-focused scaling method is used. At the same time, it allows visualization of the representativeness and discriminating power of testers, if the tester-focused scaling method is used.
- Compare with…
  - An Entry/Tester — allows the user to select an entry or a tester and use it as a reference to compare with all other entries or testers. This function is particularly useful when cultivars are compared as packages of traits (Figure 7.14). When entries are compared, the entry-focused singular value partitioning method is automatically applied; otherwise the tester-focused singular value partitioning method is automatically invoked.
  - The Ideal Entry — defines an ideal entry and compares all entries with it. For an MET dataset, the ideal entry is defined as one that has the highest performance in all environments and is, therefore, absolutely stable. This generates a ranking of the

cultivars in terms of both mean performance and stability (Figure 5.23). When this function is selected, the entry-focused singular value partitioning method is automatically applied.
- The Ideal Tester — defines an ideal tester and compares all testers with it. For an MET dataset, the ideal tester is defined as the most discriminating and absolutely representative tester. This generates a ranking of the test environments in terms of both criteria (Figure 5.26). When this function is selected, the tester-focused singular value partitioning method is automatically applied.
- Clear Contours/Background Noise

6.4.4 Find QTL

- Find QTL of — when clicked, a selection panel appears, which allows the user to select a tester or environment, trait, or genetic marker. When a valid tester is selected or typed, the user will be asked to input a threshold level for a marker to be considered as a quantitive trait loci (QTL), or a trait to be considered as closely associated with the target trait. The user needs to decide what percentage of the target trait must be explained by a marker or a trait for it to be regarded as a QTL or an associated trait. The user is expected to enter a value between 0 and 100. The program will then calculate the determination coefficients of all testers with the target tester and remove all testers that do not meet the criterion. A biplot and a linear map will be displayed with all testers retained. QTL is identified based on the biplot pattern. This function is designed to identify QTL of a trait, but it can also be used to identify traits associated with a marker — if you are interested in studying the pleiotropic effect of a gene locus. It can also be used to identify traits closely related to a target trait for indirect selection for the target trait.
- Auto Find QTL of — removes testers that are not well explained by the biplot of the first principal component (PC1) vs. the second principal component (PC2) due to relatively loose association with the target tester. A tester that is not well explained by the biplot is indicated by a relatively short vector. This function ensures that all testers that are retained have vectors longer than one-half of the longest vector. The minimum association with the target tester of all testers retained will be indicated in the biplot title label at the bottom of the biplot.
- Find Threshold for a QTL — estimates the determination coefficient of two randomly derived arrays, which can be used as a reference on the threshold of a marker to be considered as a QTL of a target trait.

6.4.5 Format

Functions provided in this cluster help generate biplots suitable for publication in professional journals or for making professional presentations.

- Biplot Title — allows modification of the biplot title
- Color Scheme 1 — a preset color scheme with white background
- Color Scheme 2 — a preset color scheme with more colorful settings
- Change Color
  - Background
  - Lines
  - Entry Labels
  - Tester Labels
  - Biplot Title

- Change Font
  - Larger
  - Smaller
  - Background
  - Lines
  - Entry Labels
  - Tester Labels
  - Biplot Title
- Biplot Size
  - Larger
  - Smaller

6.4.6 MODELS

- Model 0: 0–1 Data — for datasets containing only 0 and 1. The 1–0 data are directly subjected to singular value decomposition.
- Model 1: Tester-Centered — generates a GGE biplot based on tester-centered data (Equation 4.5).
- Model 2: Within-Tester STDEV Standardized — generates a GGE biplot based on tester-centered and within-tester standard deviation-scaled data (Equation 4.14).
- Model 3: Within-Tester STDERR Standardized — generates a GGE biplot based on tester-centered and within-tester standard error-scaled data (Equation 4.16). Performing this analysis requires input of replicated data.
- Model 4: Double-Centered — generates a biplot based on PC1 and PC2 of double-centered data (Equation 5.1). It displays only the entry-tester interactions.

6.4.7 DATA MANIPULATION

- Entry/Tester Switch Roles — generates a biplot based on transposed data so that entries and testers switch roles. This simple function doubles the functionality of the program. It also makes data input more flexible.
- Run Balanced Subset by...
  - Removing Entries — generates a biplot after entries with missing cells are removed.
  - Removing Testers — generates a biplot after testers with missing cells are removed.
- Diallel without parents — for analysis of diallel data only. It removes the observed values of the parents per se and replaces them with their respective row (or column) means.
- Run Any Subset by...
  - Removing Entries — generates a biplot after certain selected entries are removed. When selected, a panel appears that allows any entry to be removed or recovered (Figure 6.16).
  - Removing Testers — generates a biplot after certain selected testers are removed. When selected, a panel appears that allows any tester to be removed or recovered (Figure 6.16).

These two functions make the program many times more powerful.

- Reverse Sign of...
  - All Testers — allows visualization of entries with the smallest, rather than the largest, values for each tester.
  - One Tester — needed for selection based on traits that are so measured that a greater value means less desirable. Many traits, such as maturity, disease scores, and lodging scores, belong to this category.

GGE Biplot Software — The Solution for GGE Biplot Analyses   115

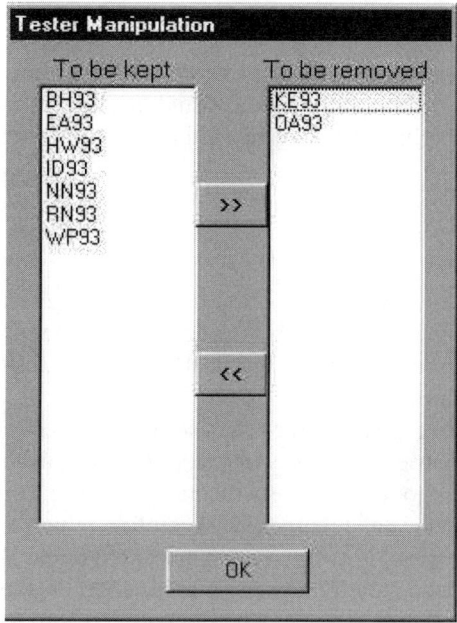

FIGURE 6.16 Snapshot showing the acquisition of a subset from panel.

- Back to Original Data — resets the biplot based on the original data.
- Back to Previous Subset — resets the biplot to a previously used subset data that is considerably different from the original data.

6.4.8 BIPLOTS

- PC1 vs. PC2 (default)
- PC3 vs. PC4
- PC5 vs. PC6
- PC1 vs. PC3
- PC2 vs. PC3
- Mean vs. IPC1 — displays the AMMI biplot (Section 5.6 and Figure 5.27). An improvement over a typical AMMI biplot is that the main effects are in the same unit as the IPC1 axis. That is, both axes use the square root of the original unit, so that the shape of the biplot will be determined by the data, rather than by the user. This biplot also displays the relative magnitude of the genotype main effect, environment main effect, and the IPC1.

6.4.9 SCALING (SINGULAR VALUE PARTITIONING)

- Tester-Focused (Default) — the singular value is entirely partitioned into testers so that the testers have the unit of the original data. This option is used when the focus is to study the interrelationships among testers (Section 4.3).
- Entry-Focused — the singular value is entirely partitioned into entries so that the entries have the unit of the original data. This option is used when the focus is to compare entries.
- Symmetrical — the singular value is symmetrically partitioned into entries and testers. This option is equally good for studying entry-by-tester interactions but is not ideal for studying interrelationships either among testers or among entries. One feature of this

scaling, which may be useful in some cases, is that both the entries and the testers for both axes have the same unit, i.e., the square root of the original unit.
- Equal Space — the singular value is so partitioned that entries and testers take equal biplot space. This option is equally good for studying the entry-by-tester interactions but is not ideal for studying interrelationships either among testers or among entries.

6.4.10 Accessories

- Simple Plots
    - Two Entries — plots two entries for all testers using tester-centered data. This is useful to compare two entries with respect to each tester (Figure 5.8). This is equivalent to the compare two entries function. The difference is that it uses measured data rather than biplot-modeled data.
    - Two Testers — plots two testers for all entries using tester-centered data. This is useful for comparing all entries based on two testers and is equivalent to the joint two testers function. Unlike the joint two testers function, however, it uses measured data rather than biplot-modeled data. Figure 6.17 exemplifies this function by comparing the performance of cultivars in two environments: OA93 and RN93. The following conclusions can be drawn from Figure 6.17: 1) in OA93 (x-axis), obviously cultivar *luc*, *m12*, and *kat* were below average, whereas cultivars *aug*, *kar*, *ari*, *ham*, *zav*, and *del* were above average. Other cultivars were about average; 2) in RN93, cultivars *fun*, *luc*, *cas*, and *rub* were above average, whereas *m12* and *aug*, among others, were below average; 3) cultivars *ron*, *ena*, *ann*, *aug*, *kar*, *ari*, *ham*, *zav*, and *del* were

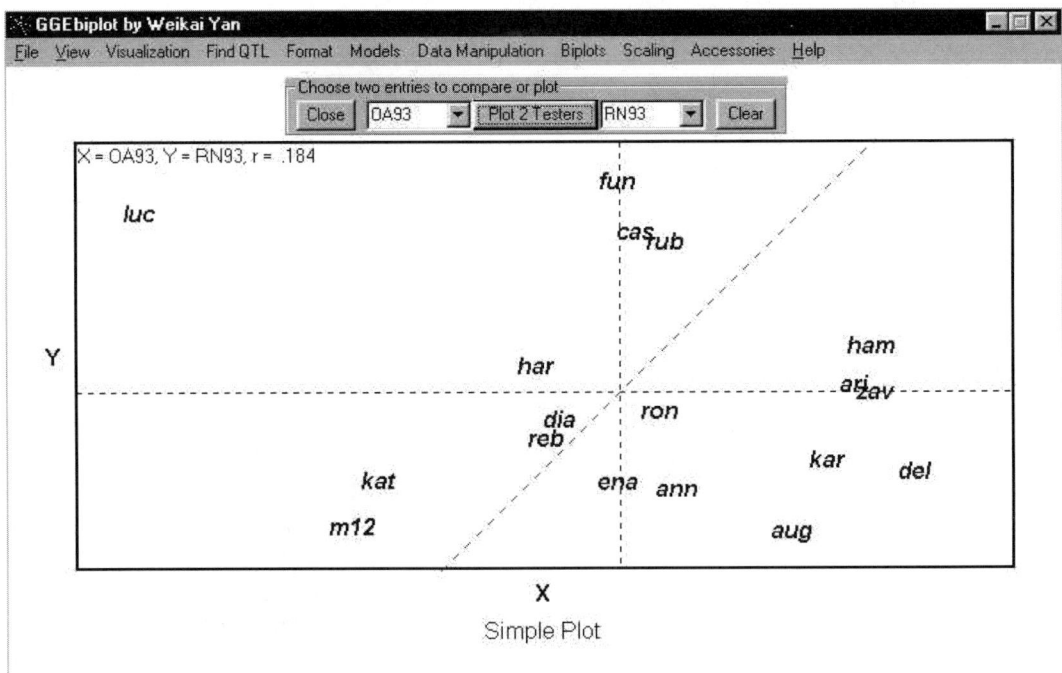

FIGURE 6.17 Snapshot showing a simple scatter plot comparing environment-centered yield of the cultivars in environment OA93 (abscissa) vs. in environment RN93 (ordinate). The guidelines indicate the mean yield in each environment. The slope line indicates equality in the two environments. The correlation between the two environments was 0.184.

relatively better in OA93 (below the equality line), whereas others were relatively better in RN93 (above the equality line); and 4) the two environments had a correlation of only 0.184.

- Numeric Output
  - Data Matrix — prints input data in the format of a table into a log file.
  - Basic Stats — prints the sum of squares and percentage of entry, tester, and entry-tester interaction into a log file.
  - Eigenvectors — prints the eigenvectors of the entries and testers for the first six principal components into a log file.
  - Axis Values — prints the x and y values that are used to generate the current biplot.
  - Predicted Values — prints the predicted values of each entry in combination with each tester based on the current biplot into a log file. The main effects of testers are added to the predictions by the biplot.
  - Tester Correlations — prints the correlation matrix among testers into a log file.

6.4.11 Help

- Displays the help file in the *html* format.

6.4.12 Log File

When running, the program generates a log file with the same name as the data file but with the extension of ".gge" in the same folder from which the current dataset was read. It records the current data file name and path, the model used, the averaged two-way table, the means of the environments, the proportions of GGE variation explained by the first six principal components, and the type of the current biplot. User-requested printouts from the Numerical Output functions will also be written to this file.

6.5 FUNCTIONS FOR A GENOTYPE-BY-ENVIRONMENT DATASET

For an MET dataset, the GGEbiplot allows the following to be performed with a click of the mouse (Chapter 5):

- Rank cultivars based on performance in a given environment
- Rank environments based on the relative performance of a given cultivar
- Rank cultivars based on their performance in any two environments
- Compare two cultivars for performance in each of the environments
- Compare cultivars with a user-specified check
- Compare cultivars with the ideal cultivar
- Show the best cultivars for each environment (i.e., show "which-won-where")
- Cull genotypes based on performance in each of the environments
- Show mean performance and stability of cultivars
- Rank cultivars based on both mean performance and stability
- Show interrelationships among environments or identify redundant environments
- Group environments based on the general biplot pattern
- Group environments based on winning cultivars for mega-environment division and GEI exploitation
- Show discriminating ability and representativeness of environments
- Rank environments based on both discriminating ability and representativeness
- Compare environments with the ideal environment

- Visualize data using different models
- Visualize data using different scaling methods
- Run a subset after removing some genotypes
- Run a subset after removing some environments
- Generate and run a balanced subset after removing genotypes
- Generate and run a balanced subset after removing environments
- Save a biplot image for use at a later time
- Generate a simple plot between any two environments
- Generate a simple plot between any two genotypes
- Print a biplot for presentation or publication
- Print input data in the format of a table into a log file
- Print sum of squares and percentage of genotype, environment, and genotype–environment interaction into a log file
- Print eigenvectors of genotypes and environments for the first six principal components into a log file
- Print axis values of the current biplot into a log file
- Print correlation matrix among environments into a log file
- Print correlation matrix among environments
- Print predicted values based on the current biplot

6.6 FUNCTIONS FOR A GENOTYPE-BY-TRAIT DATASET

Chapter 7 demonstrates the application of GGEbiplot in analyzing genotype-by-trait data. Generally, the following objectives can be achieved:

- Rank cultivars based on a trait
- Show relative levels of a cultivar for different traits
- Compare two cultivars for each of the traits
- Compare cultivars as packages of traits
- Identify the best cultivars for each trait
- Identify the poorest cultivars for each trait
- Cull cultivars based on individual traits
- Rank cultivars based on two traits
- Show the best cultivars for each trait
- Show the poorest cultivars for each trait
- Show interrelationships among traits — understanding the system and identifying challenges
- Find traits closely associated with a target trait — indirect selection
- The last 15 functions listed for genotype-by-environment data, i.e., those that follow "Visualize data using different scaling methods"

6.7 FUNCTIONS FOR A QTL-MAPPING DATASET

Application of GGEbiplot in QTL mapping is illustrated in Chapter 8. The utilities include:

- Generate a linear map among all or a subset of genetic markers
- Identify QTL for a trait at a user-specified association level
- Identify traits that are influenced by a single locus — pleiotropic effects
- Visualize genetic constitution of each individual plant or line relative to the markers
- The last 15 functions listed for genotype-by-environment data, i.e., those that follow "Visualize data using different scaling methods"

6.8 FUNCTIONS FOR A DIALLEL-CROSS DATASET

Application of GGE biplot to diallel-cross data analysis is discussed in Chapter 9. The following results can be achieved:

- Rank parents or entries based on combining ability with a tester
- Show general combining ability (GCA) of a parent (viewed as an entry)
- Show specific combining ability (SCA) of a parent (*not* a cross)
- Show the best mating partner or entry for each parent (viewed as a tester)
- Show heterotic groups
- Show the best crosses
- Show the best testers
- Formulate hypotheses relative to genetics of parents
- The last 15 functions listed for genotype-by-environment data, i.e., those that follow "Visualize data using different scaling methods"

6.9 FUNCTIONS FOR A GENOTYPE-BY-STRAIN DATASET

For a host genotype-by-pathogen strain dataset, the following questions can be graphically addressed (Chapter 10):

- Is the resistance vertical or horizontal? Determines whether or not there is race-specific resistance.
- If there is no race-specific resistance, which genotype is most resistant? Which strain is most virulent?
- If there is race-specific resistance, which genotype is resistant to which strain? Which strain is virulent to which genotype?
- What is the relationship among genotypes in response to the strains?
- What is the relationship among strains with regard to virulence?
- The last 15 functions listed for genotype-by-environment data are also applicable to genotype-by-strain data, i.e., those that follow "Visualize data using different scaling methods"

6.10 APPLICATION OF GGEBIPLOT TO OTHER TYPES OF TWO-WAY DATA

In general, GGEbiplot can be applied to any numeric data that can be put into a two-way table. In addition to the aforementioned, two-way data that are relevant to plant breeding and agricultural production include: wheat genotype-by-maize genotype interaction data generated from wheat–maize hybridization for doubled-haploid production, genotype-by-treatment (planting dates, planting density, fertilizer application, etc.), and herbicide-by-insecticide interaction data. Gene/QTL-by-environment interaction data, gene expression data generated from micro-arrays, etc., are also potential areas for GGEbiplot application.

6.11 GGEBIPLOT CONTINUES TO EVOLVE

Our understanding of the GGE biplot methodology is constantly deepening, and in reflecting this, the GGEbiplot software is constantly evolving. During the preparation and publishing of this book, the following additional important functions have become available in the GGEbiplot software. For up-to-date information on GGEbiplot, the reader is encouraged to visit the following website: http://ggebiplot.com.

6.11.1 INTERACTIVE STEPWISE REGRESSION

This module allows: 1) regression using any of the testers as dependent variable and any number of the other testers as independent variables, 2) user-monitored stepwise regression to remove non-significant variables, and 3) display of results both graphically in a biplot and in a table format in the log file. Joint application of these functions can help solve the problems of over-parameterization and collinearity and in generating robust predictive models based on agricultural and life science data.

6.11.2 INTERACTIVE QQE BIPLOT

This module generates a QQE biplot that displays both marker main effect and marker-by-environment interaction for a given trait. Here 'Q' stands for QTL, and the term 'QQE biplot' is analogous to a GGE biplot. The QQE biplot is a novel method for QTL identification based on phenotypic data of a trait measured in multiple environments.

6.11.3 THREE-WAY DATA INPUT AND VISUALIZATION

This module allows reading a genotype-by-environment-by-trait three-way dataset, with or without replication, in a single data file and offers the following options: 1) runs GGE analysis for any of the traits at the user's will; 2) runs genotype-by-trait biplot analysis based on genotypic values; 3) runs genotype-by-trait biplot analysis based on environmental values; and 4) run genotype-by-trait biplot analysis based on phenotypic values. This will save the researcher a lot of time in data preparation and enable him/her to gain another dimension in the understanding of multi-environment trial data.

6.11.4 INTERACTIVE STATISTICS

This module generates an ANOVA table, and indicates in the biplot and outputs to the log file probability of difference for both G and GGE when two entries are compared in a biplot. This function will help the researcher better appreciate the fact that many entries that do not differ in the main effect, do differ in GGE effect. In other words, many different entries may have been regarded as not different if viewed only from the perspective of main effects. Therefore, a GGE biplot can reveal a lot more information about the entries than that based on just the main effects.

7 Cultivar Evaluation Based on Multiple Traits

SUMMARY

This chapter demonstrates the numerous utilities of genotype and genotype-by-environment (GGE) biplot in visual analysis of genotype-by-trait data. These include: 1) evaluating cultivars based on a single trait; 2) evaluating cultivars based on two traits; 3) comparing cultivars as packages of traits; 4) visualizing merits of a cultivar; 5) visualizing defects of a cultivar; 6) visualizing interrelationships among traits; 7) identifying traits that are closely associated with, and therefore can be used in indirect selection for, a target trait; 8) independent culling based on single traits; 9) investigating different selection strategies; and 10) achieving systems understanding of the crop under study. Wheat end-use quality datasets are used to illustrate these points. Two concepts of plant breeding, i.e., independent culling and systems understanding of crop improvement, are discussed after illustrating some of the functions of the GGEbiplot program. This chapter is not only of importance for those interested in visual analysis of genotype-by-trait data, but it also is particularly important for those interested in breeding for wheat end-use quality, methodology of plant breeding, and systems understanding of crop improvement.

7.1 WHY MULTIPLE TRAITS?

Cultivar evaluation based on multiple traits is an important aspect of cultivar evaluation for four major reasons:

- An ideal cultivar must meet multiple requirements. For example, although yield is the universal breeding objective, end-use quality also must be met for a high-yielding genotype to be accepted as a cultivar. Other traits, such as maturity, standability, resistance to various diseases, etc., also are essential for a successful cultivar.
- Most important breeding objectives are complex traits consisting of multiple components. For example, yield can be decomposed into several yield components and quality can be decomposed into many quality components. More important, it is the sub-components, rather than the complex traits per se, that are subjected to selection when identifying parents for hybridization and individuals in early generations.
- When a target trait consists of several components or is associated with multiple traits, it is necessary to determine which components are most useful for indirect selection.
- A cultivar is a biological system rather than a simple collection of independent traits. Effective breeding requires systems understanding of cultivar improvement, including understanding of the essential components of the system and the interrelationship among them.

Although the GGE biplot methodology was originally proposed for analyzing multi-environment trials data for a given trait, it is equally applicable to all types of two-way data that assume an entry-by-tester structure, such as a genotype-by-trait two-way dataset. The only difference is that in genotype-by-trait data, different traits have different units, and the units need to be removed through standardization before meaningful biplot analyses can be made. Therefore, Equation 4.14 for mean or nonreplicated data:

$$(\hat{Y}_{ij} - \mu - \beta_j)/d_j = g_{i1}e_{1j} + g_{i2}e_{2j} + \varepsilon_{ij}$$

and Equation 4.16 for replicated data:

$$(\hat{Y}_{ij} - \mu - \beta_j)/s_j = g_{i1}e_{1j} + g_{i2}e_{2j} + \varepsilon_{ij}$$

are appropriate models. All parameters are defined in the same way as in Chapter 4 except that the word "environment" should be replaced by "trait." Multiple trait data of hard red spring wheat and soft white winter wheat will be used to demonstrate the utilities of the GGE biplot methodology.

7.2 CULTIVAR EVALUATION BASED ON MULTIPLE TRAITS

Data of 8 yield and quality traits of 21 hard red spring wheat (HRSW) cultivars grown in North Dakota, taken from the 2000 Regional Quality Report for the U.S. HRSW are presented in Table 7.1. The data for most cultivars were based on multi-locations in North Dakota from 1998 to 2000, but for some newer cultivars (*Alsen, Amidon, Argent, Kulm,* and *Mcneal*), the data were only from a few locations in 2000. Since our purpose is to demonstrate the biplot analysis rather than provide cultivar recommendations, we assume that the data are perfectly balanced and comparable.

When data are appropriately read and model 2, i.e., singular value decomposition of within-trait standard deviation-standardized data is chosen, GGEbiplot generates a genotype-by-trait biplot in about a second (Figure 7.1). If replicated data are available, model 3, i.e., singular decomposition of within-trait standard error-standardized data, may be better. The 8 traits are displayed in upper-case, and the 21 cultivars are in italics. This biplot explains 67% of the total variation. Numerous ways of viewing this type of biplot were discussed in Chapter 6. We present and discuss, however, only those that are most relevant to this particular dataset.

7.2.1 WHICH IS GOOD AT WHAT

When the *Show/Hide Polygon* function is clicked, the polygon view of the biplot will be displayed (Figure 7.2). This view helps identify cultivars with the highest values for one or more traits. The scores of three traits — yield, water absorption, and farinogram —fell in the *Mcneal* sector, suggesting that cultivar *Mcneal* had highest or near-highest values for these three traits. Cultivar *Alsen* was second to *Mcneal* for these traits. Similarly, cultivars *2375* and *Ingot* were the highest in flour extraction rate. Cultivars *Ernest* and *Kulm* had the highest test weight and grain- and flour-protein content. *Argent* had the highest loaf volume. Since the biplot did not explain all the variation, these predictions may not exactly reflect the observed numbers. Nonetheless, cultivars that are among the top with regard to a trait can be identified with confidence.

7.2.2 WHICH IS BAD AT WHAT

Upon clicking the function *Reverse Sign Of All Testers,* followed by clicking *Show/Hide Polygon*, the biplot will be transformed to resemble Figure 7.3. Since the signs of the traits are reversed, the vertex cultivars in Figure 7.3 are those that had the lowest values for one or more traits. Thus, *Mcneal* had the lowest flour extraction, *Ernest* and *Kulm* had the lowest yield, and *Mcvey* had the lowest protein content, test weight, loaf volume, etc.

Joint examination of Figures 7.2 and 7.3 reveals the merits and defects of a cultivar. Such comprehensive knowledge about cultivars is important for cultivar recommendations to farmers and for parent selection for hybridization.

TABLE 7.1
Eight Traits of 21 Hard Red Spring Wheat (HRSW) Cultivars

Cultivar	Yield[1] Bu/acre	Test Weight Lb/bu	Grain Protein %	Flour Protein %	Flour Extraction %	Farinogram (Scale 1-8)	Absorption %	Loaf Volume cm³
Alsen	67.0	60.6	15.1	14.2	67.0	6.7	66.0	1067
Amidon	52.6	59.4	15.4	14.3	68.4	6.3	66.0	985
Argent	54.3	60.4	15.7	14.5	66.9	8.0	67.3	1032
Butte86	63.6	59.2	15.2	14.3	66.9	4.7	66.0	1035
Ernest	51.4	60.1	15.7	14.6	67.7	7.7	63.8	1098
Forge	58.3	61.0	14.2	13.1	69.2	6.2	64.4	968
Grandin	52.9	58.6	15.1	14.2	67.9	6.2	64.7	1064
Gunner	52.8	61.1	15.7	14.5	68.2	6.5	66.8	1024
Hj98	60.8	58.7	14.1	13.1	67.9	7.5	64.1	1005
Ingot	54.8	61.8	14.7	13.6	69.8	5.2	64.7	998
Keene	54.7	61.2	14.7	13.6	68.3	6.7	66.9	996
Kulm	53.8	61.5	15.5	14.9	68.5	7.0	65.7	1070
Mcneal	59.6	59.7	14.7	13.8	65.7	8.0	70.1	1060
Mcvey	60.0	56.4	13.3	12.5	69.0	5.3	64.2	906
Oxen	59.3	58.9	14.3	13.1	69.5	7.0	65.2	1025
Parshall	53.8	61.3	15.4	14.3	67.7	6.8	66.5	1086
Reeder	61.4	60.2	15.1	14.0	67.5	6.0	65.8	986
Russ	57.3	59.5	14.8	13.8	67.9	6.3	67.0	1013
Trenton	54.1	59.9	15.4	14.4	68.5	6.8	65.0	1094
2375	55.9	58.7	14.9	13.8	69.9	5.0	63.9	975
2398	57.7	58.7	14.6	13.8	69.2	7.7	64.2	1020

[1] Yield data are not comparable for some newer cultivars since their yields were measured in fewer trials.

Source: U.S. Hard Red Spring Wheat 2000 Regional Quality Report.

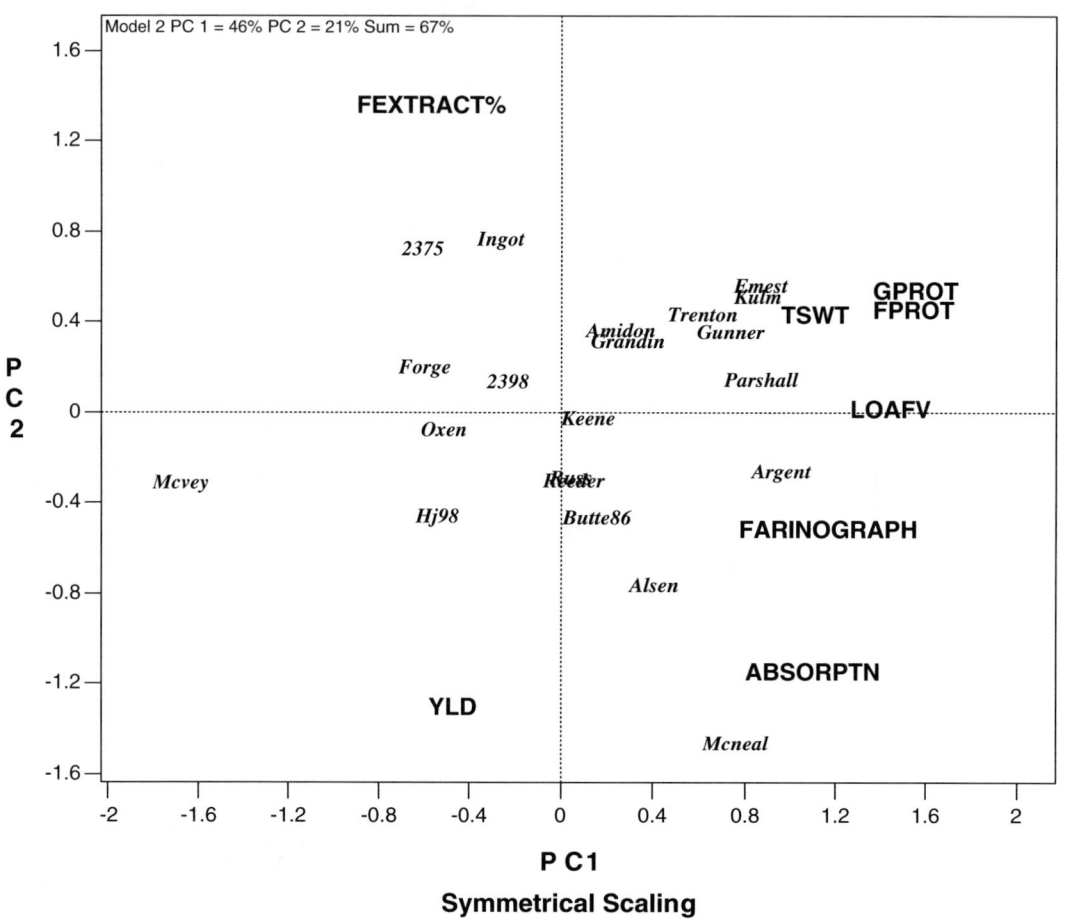

FIGURE 7.1 A genotype-by-trait biplot of 21 hard red spring wheat (HRSW) cultivars vs. 8 yield and quality traits. Genotypes or entries are in italics and traits or testers are in regular uppercase. (From 2000 U.S. HRSW report.)

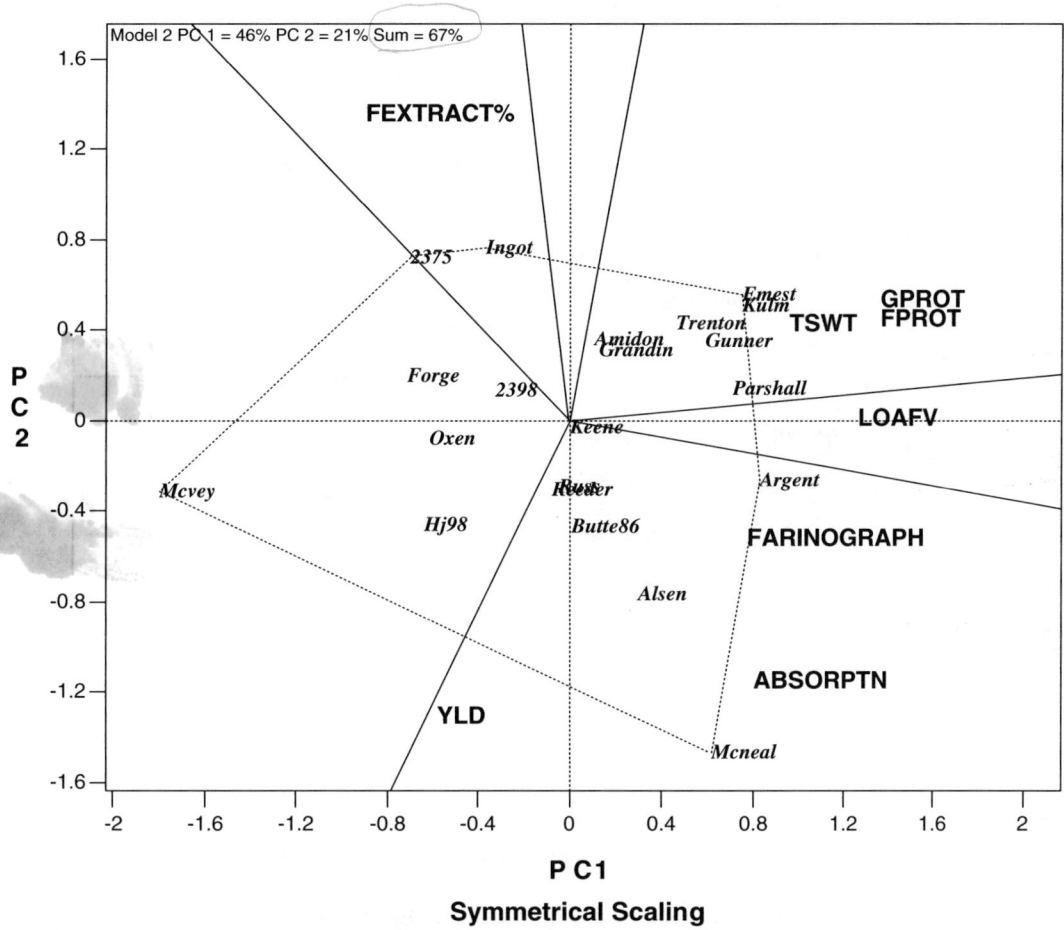

FIGURE 7.2 Polygon view of the HRSW genotype-by-trait biplot, showing which cultivar had the highest values for which traits.

126 GGE Biplot Analysis: A Graphical Tool for Breeders, Geneticists, and Agronomists

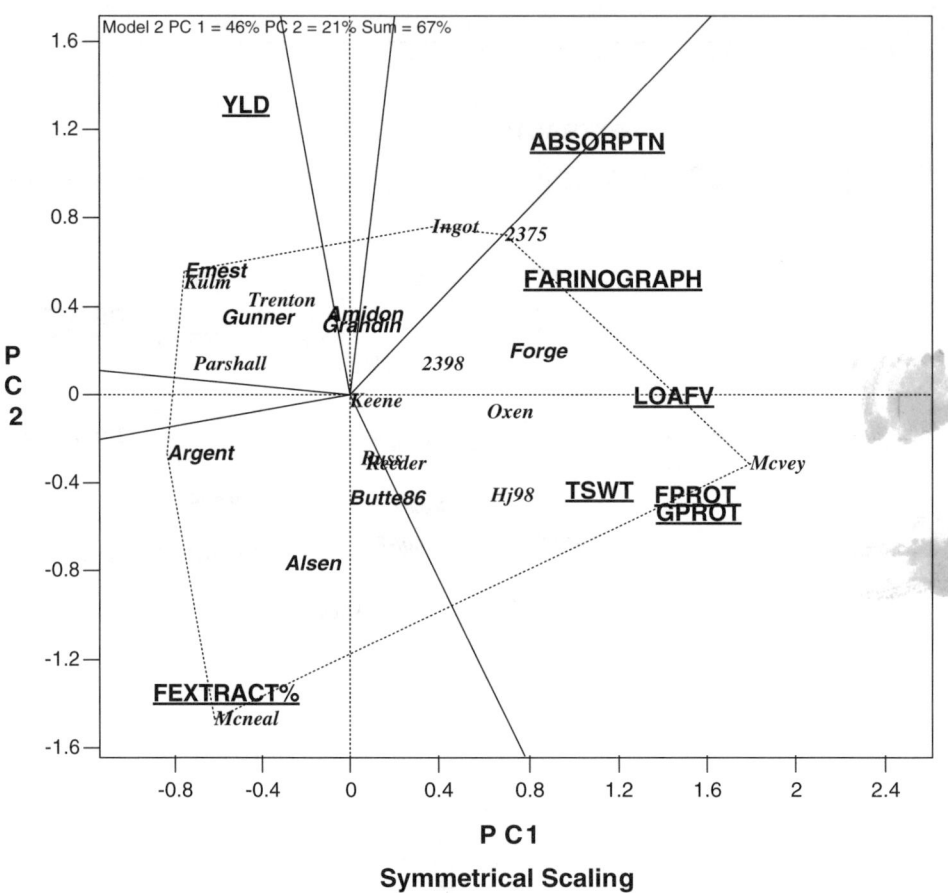

FIGURE 7.3 Polygon view of the HRSW genotype-by-trait biplot after the signs are reversed, showing which cultivar had the smallest values for which traits.

7.2.3 Comparison between Two Cultivars

To compare two cultivars, use the *Compare Two Entries* function. A panel will appear, which allows selection of any two cultivars for comparison. Here, we choose to compare *Argent* and *Grandin*, with the knowledge that *Argent* is a hard white spring-wheat cultivar derived from HRSW *Grandin*. When the *Compare* button is clicked, a line is drawn to connect the two chosen cultivars, and another line is drawn that is perpendicular to the first line and passes through the biplot origin (Figure 7.4). This latter line separates the two cultivars to be compared as well as all the traits. Figure 7.4 indicates that all traits except flour extraction are on the same side of the perpendicular line as *Argent*, meaning that *Argent* was better (i.e., had larger values) than *Grandin* for all traits except flour extraction.

Figure 7.5 compares cultivars *2375* and *Alsen*. The former is an older cultivar that is relatively resistant to Fusarium head blight, whereas the latter is a new cultivar with superior Fusarium head blight resistance derived from the famous Chinese cultivar *Sumai 3*. Figure 7.5 indicates that *Alsen* is better than *2375* in all aspects except flour extraction rate. From these two instances, an interesting question arises whether the improvement of HRSW cultivars may have been at the expense of flour extraction.

7.2.4 Interrelationship among Traits

Upon clicking the function *Show/Hide Tester Vectors*, Figure 7.6 appears. The tester vectors are the lines that originate from the biplot origin and reach markers of the traits. Since the cosine of the angle between the vectors of any two traits approximates the correlation coefficient between them, this view of the biplot is best for visualizing the interrelationship among traits. Figure 7.6 suggests close associations among test weight and flour- and grain-proteins. All three traits are closely associated with loaf volume. Traits also positively associated with loaf volume are farinogram score and water absorption. Flour extraction seems to be negatively associated with all other traits. Yield shows a weak but positive correlation with water absorption. It is remarkable that the eight vector lines approximate well the whole correlation matrix (Table 7.2).

7.2.5 The Triangle of Grain Yield, Loaf Volume, and Flour Extraction

The trait relationships (Figure 7.6) include the negative association among three traits — grain yield, loaf volume, and flour extraction (Figure 7.7). Grain yield is most important for growers, flour extraction is most important for millers, and loaf volume is most important for bakers. There is a 120° angle between each pair of the three traits, however, suggesting negative associations among them. Although each pair of the three traits is only modestly negatively correlated, it is difficult to put all three into a single cultivar. Two traits can be reasonably combined, but at the expense of the third, as illustrated next.

7.2.6 Cultivar Evaluation Based on Individual Traits

The GGEbiplot function *Examine a Tester* allows ranking of cultivars based on a single trait (a "tester"). When a trait is selected, a line called trait axis is drawn that passes through the biplot origin and the marker of the trait. The precise position of the marker is at its left end. The arrow on the trait axis points to increasing trait value. Another line, called perpendicular line, is drawn that also passes through the origin and is perpendicular to the trait axis. Based on grain yield (Figure 7.8), *Alsen* is the best cultivar, followed by *Butte86* and *Reeder*; *Mcneal*, *Mcvey*, *Hj98*; *Forge*, *Russ*, and *Oxen*. All these cultivars had above-average yield. Cultivars below the perpendicular line yielded below average, with *Ernest* being the lowest-yielding cultivar. Based on flour extraction (Figure 7.9), the best cultivars were *2375* and *Ingot*, followed by *Mcvey*, *Forge*, *Amidon*, etc. Note that the high yielding cultivars *Mcneal*, *Alsen*, and *Butte86* are the poorest in flour extraction. Based on loaf volume (Figure 7.10), *Ernest* was the best cultivar, followed by *Parshall*, *Mcneal*, *Trenton*, *Grandin*, *Argent*, *Kulm*, and *Gunner*. All these had above-average loaf volume. The poorest cultivar for loaf volume was *Mcvey*.

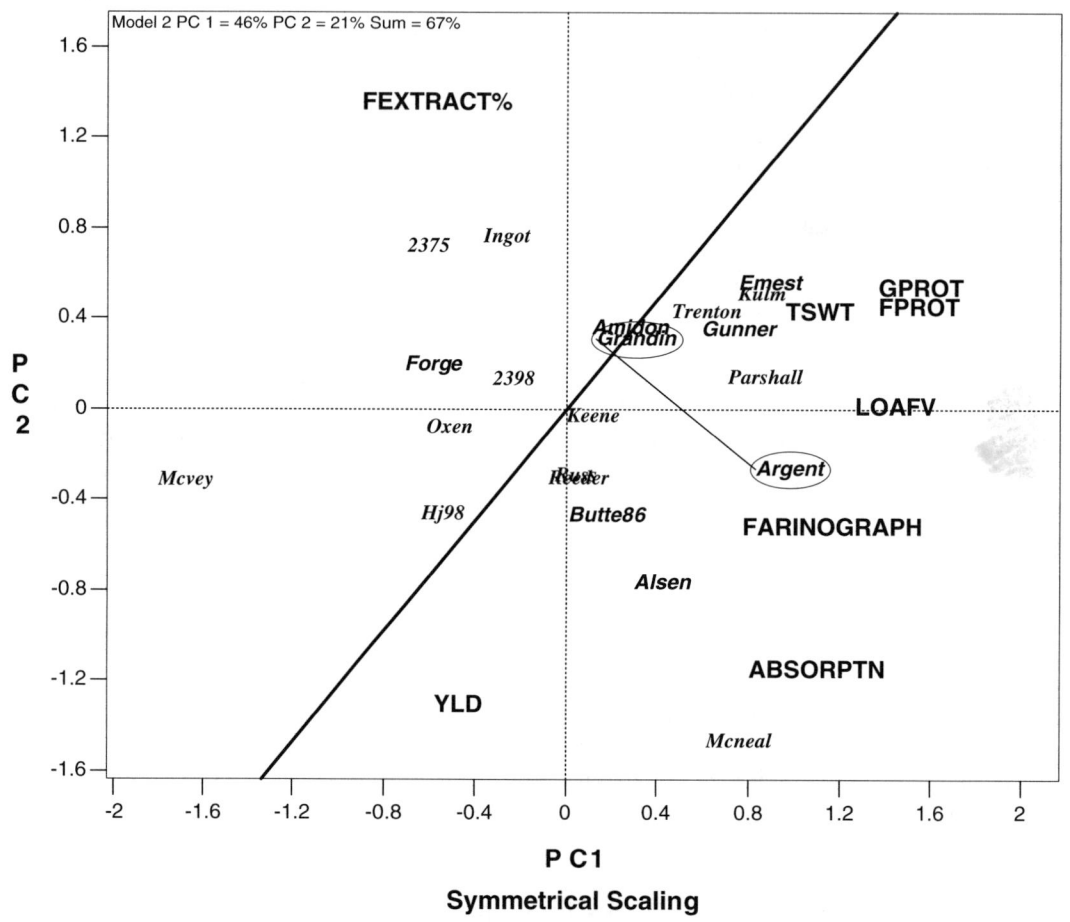

FIGURE 7.4 Visual comparison of two cultivars: *Grandin* and *Argent*, for all traits.

Cultivar Evaluation Based on Multiple Traits

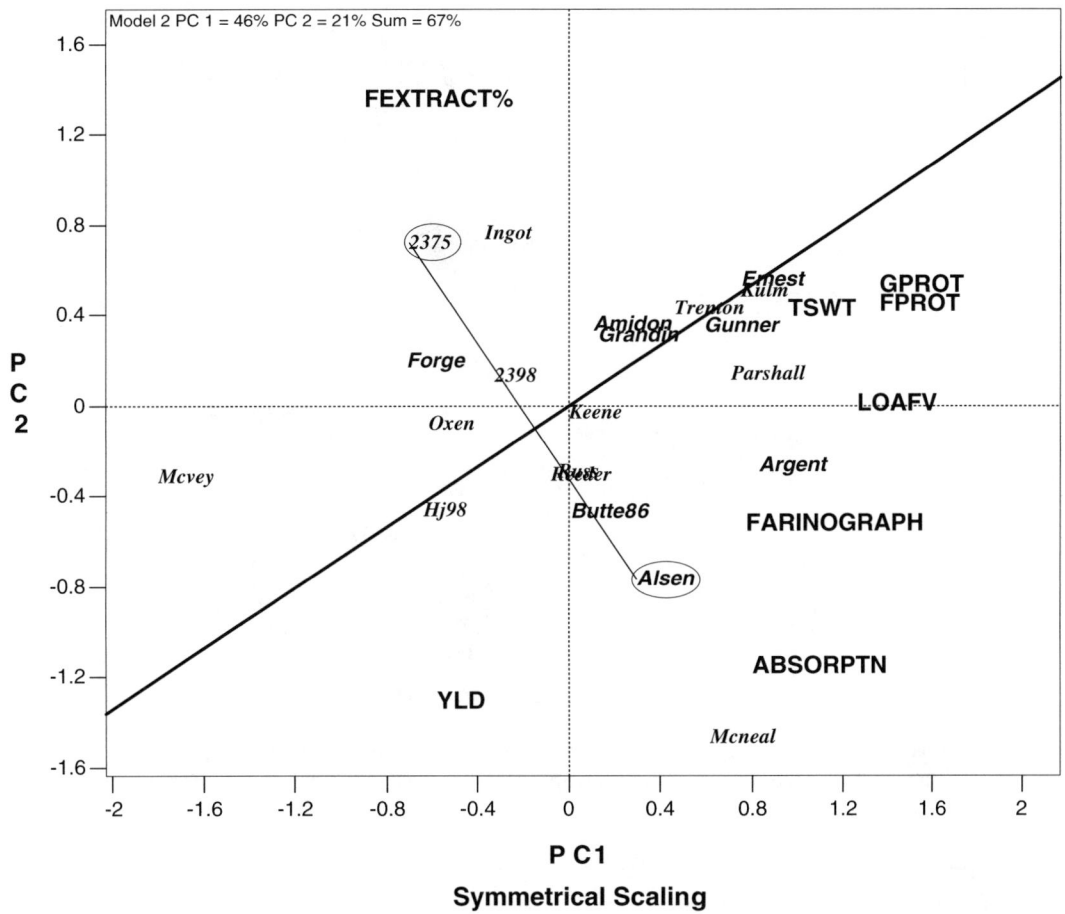

FIGURE 7.5 Visual comparison of two cultivars: *2375* and *Alsen*, for all traits.

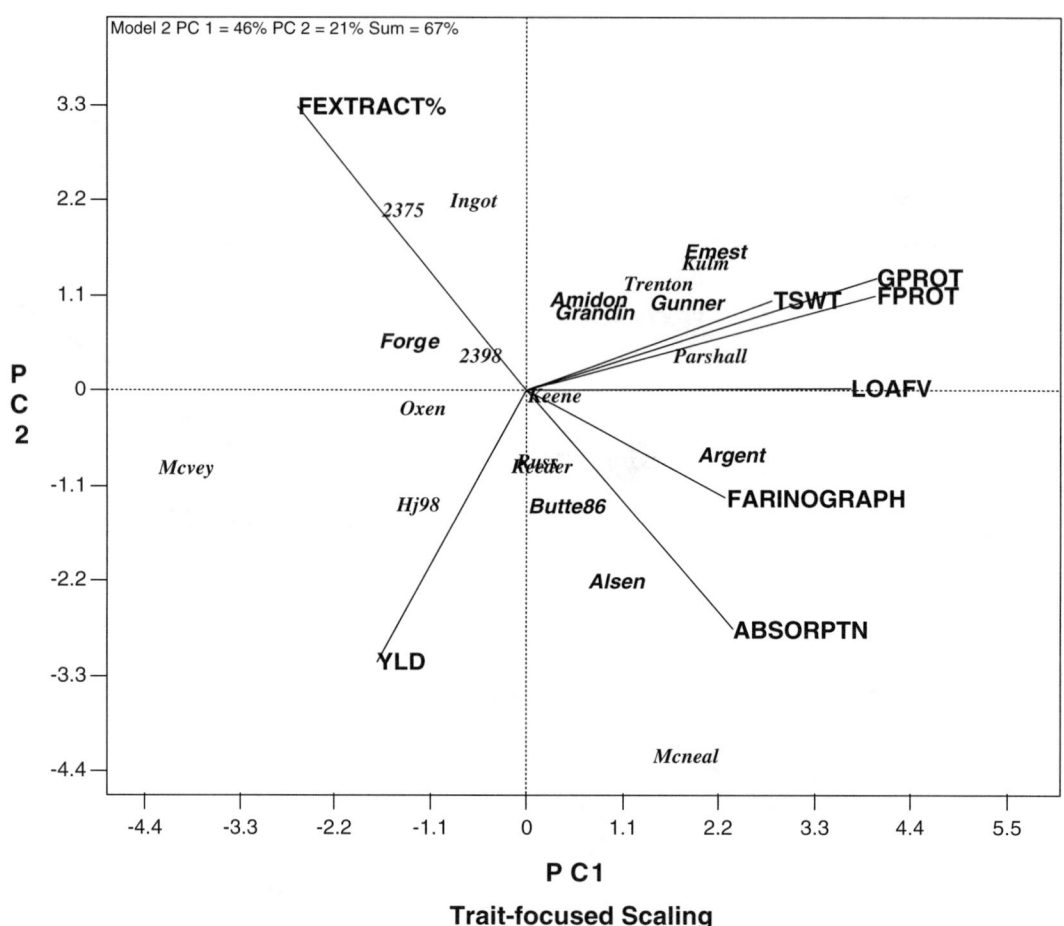

FIGURE 7.6 Vector view of the genotype-by-trait biplot, showing the interrelationships among traits. The cosine of the angle between the vectors of two traits approximates the correlation coefficients between the traits.

TABLE 7.2
Correlation Coefficients among Traits Based on Data in Table 7.1

Traits	Test Weight	Grain Protein	Flour Protein	Flour Extraction	Farinogram	Water Absorption	Loaf Volume
Grain yield	−0.276	−0.462	−0.406	−0.269	−0.185	0.066	−0.224
Test weight		0.561	0.504	−0.107	0.166	0.319	0.409
Grain protein			0.968	−0.364	0.220	0.279	0.685
Flour protein				−0.397	0.223	0.268	0.724
Flour extraction					−0.394	−0.702	−0.460
Farinogram						0.315	0.504
Water absorption							0.239

Cultivar Evaluation Based on Multiple Traits

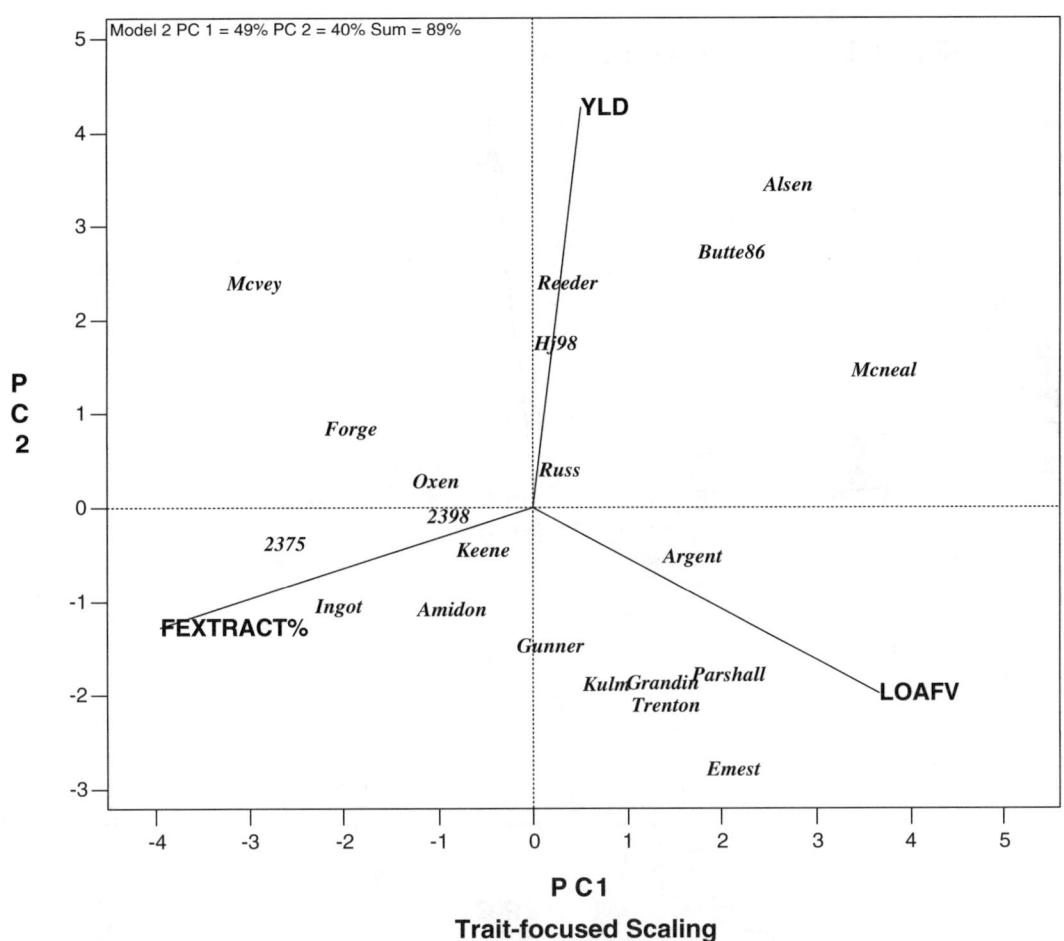

FIGURE 7.7 Vector view of the genotype by trait biplot, showing the interrelationships among grain yield, flour extraction rate, and loaf volume for 21 HRSW cultivars. The three traits are negatively correlated with one another, revealing a major challenge for HRSW breeding.

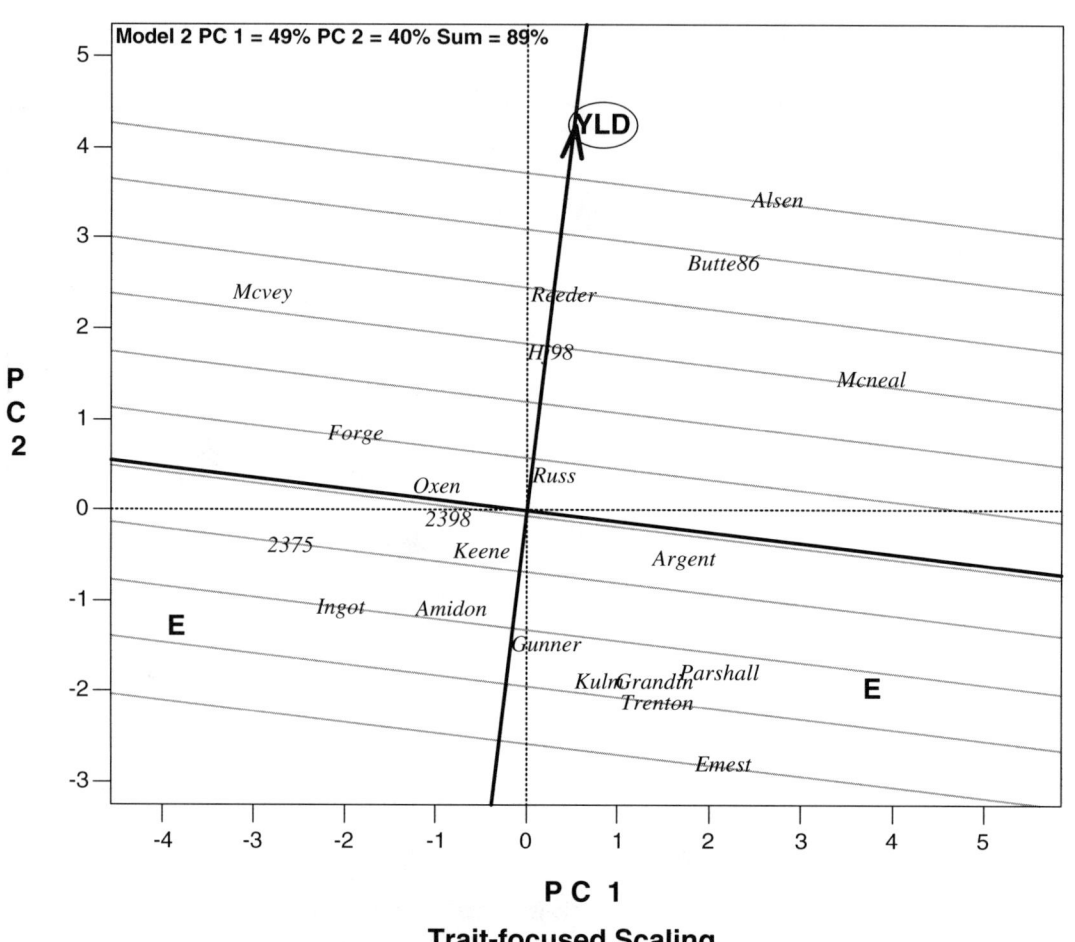

FIGURE 7.8 Cultivar ranking based on grain yield.

Cultivar Evaluation Based on Multiple Traits

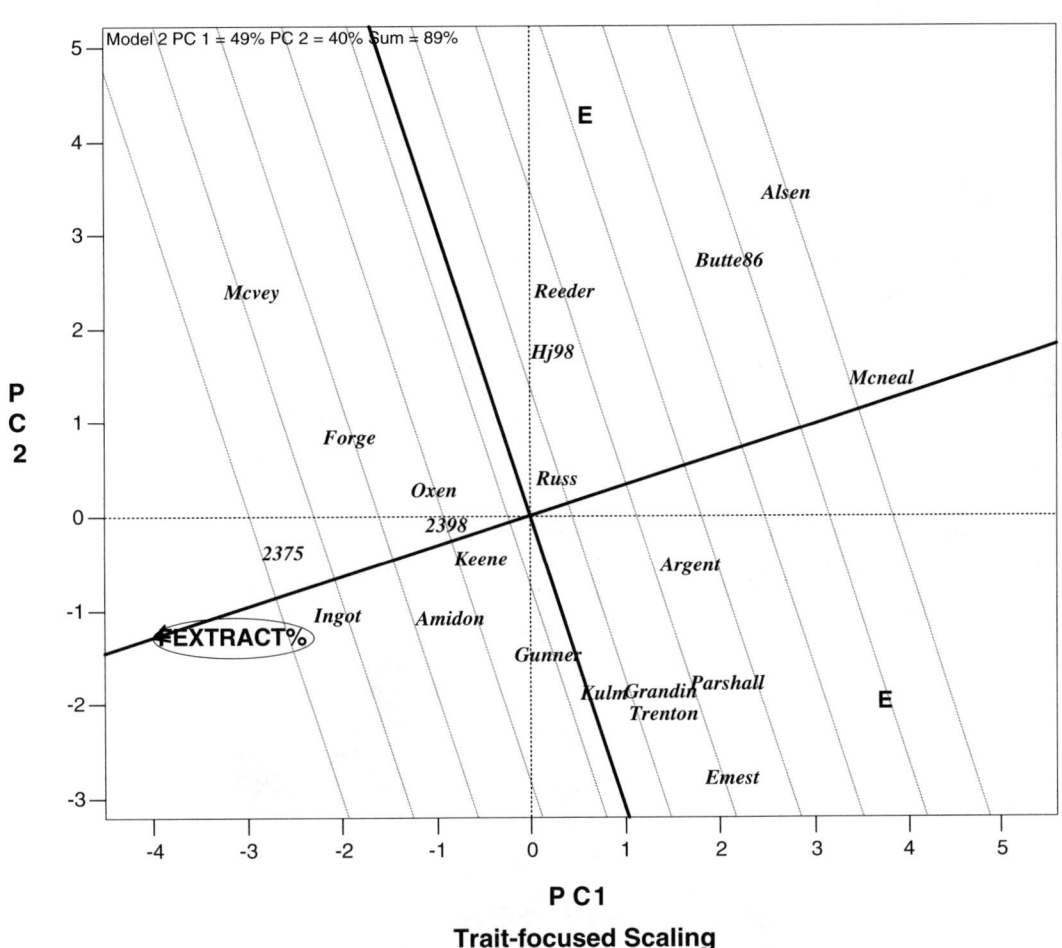

FIGURE 7.9 Cultivar ranking based on flour extraction rate.

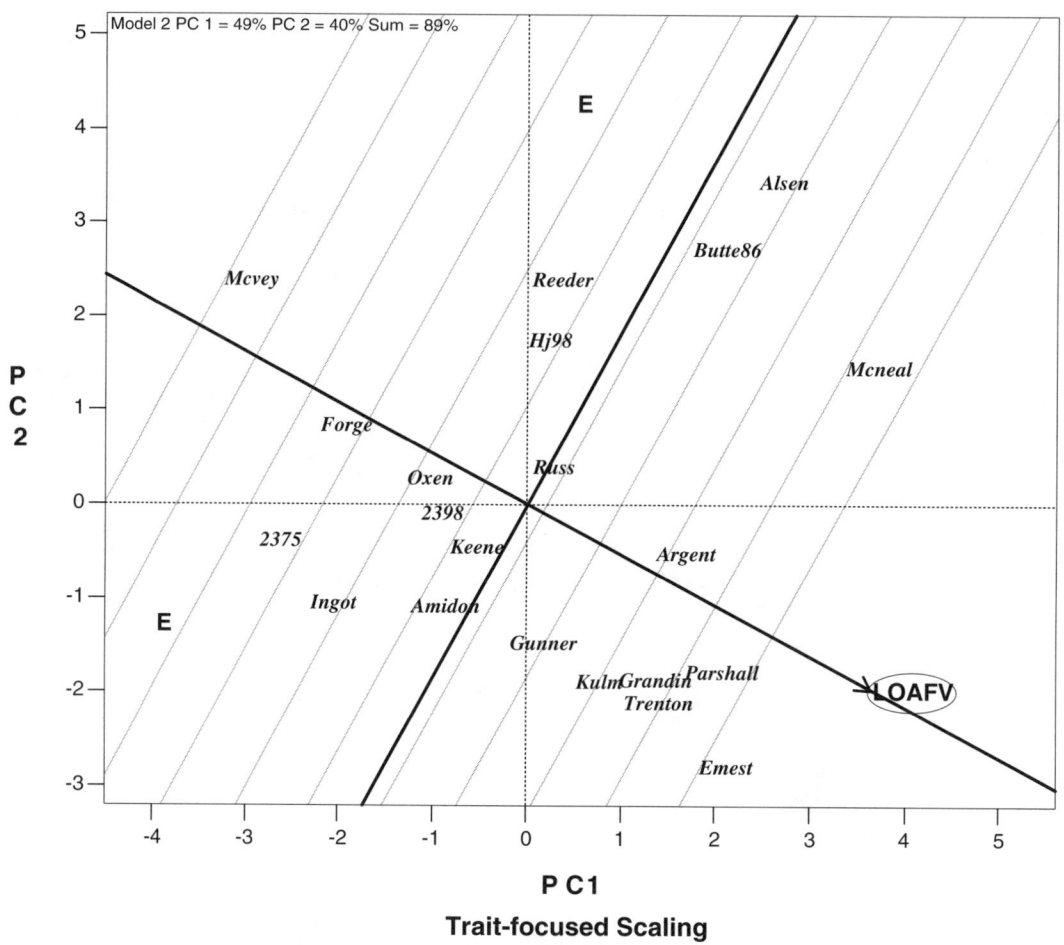

FIGURE 7.10 Cultivar ranking based on loaf volume.

7.2.7 CULTIVAR EVALUATION BASED ON TWO TRAITS

Cultivars can be visually evaluated based on two traits simultaneously. This is done by connecting the two traits, finding a median point between them, and using this point as a virtual trait. GGEbiplot contains a function *Joint Two Testers*, which allows visualization of cultivars based on any two traits. If yield and loaf volume are considered at the same time (Figure 7.11), *Mcneal*, *Alsen*, and *Butte86* were the best. Unfortunately, they were also the poorest in flour extraction. Cultivars *2375* and *Ingot* were just the opposite of these three cultivars.

7.3 IDENTIFYING TRAITS FOR INDIRECT SELECTION FOR LOAF VOLUME

Table 7.3 contains 29 quality traits for wheat samples collected in 15 U.S. HRSW sub-regions in 1999. We use this dataset to illustrate the use of biplots in identifying traits that can be used for indirect selection for a target trait. There are many quality measures related to bread wheat quality. Brief explanations for each of these measures are presented in Table 7.4.

The ultimate measure of bread wheat quality is loaf volume. Reliable determination of loaf volume requires a large quantity of seeds and can be conducted only in later generations of the breeding cycle. Therefore, selection for HRSW quality in earlier generations must rely on indirect selection for other traits that are predictive of loaf volume. For this reason, numerous indirect selection procedures and numerous indices have been formulated. These numerous indices did not make selection for HRSW quality easier, however. They make the selection more expensive and decision-making more difficult. It would be ideal to identify a few measures that are closely related to loaf volume. In addition, these measures should be determined easily and cheaply.

Figure 7.12 is a biplot that displays all 29 traits and 15 sub-regions. It explained 69% of the total variation. Much of the total variation is explained by the first two PCs, which suggests close associations among traits. The trait vectors are shown to help examine the interrelationship among traits. Most of the measures are positively or negatively associated with loaf volume. Only a few traits are apparently independent of loaf volume, which are sedimentation (SDS), extensibility, flour ash, and kernel weight, as indicated by their near–90° angles with loaf volume. Traits that are positively correlated with loaf volume are located on the right half of the biplot. These include protein content, measures related to grain sprouting (falling number, measures of alpha amylase activity) and measures of gluten strength, i.e., those generated by alveograph, extensograph, and farinogaph. Traits on the left side of the biplot are negatively associated with loaf volume. These include grain moisture, flour extraction, percentage of large kernels, farinograph-determined mixing tolerance index (MTI), and Alveograph-L (extensibility). All four sub-regions from Montana are on the right side of the biplot, indicating that wheat samples from these sub-regions have greater loaf volume and associated traits. Wheat samples from all other sub-regions have lower-than-average loaf volume.

GGEbiplot has a built-in function *Auto Find QTL*, which allows identification and display of traits that are closely associated with a target trait. Invoking this function and selecting loaf volume as the target trait, the biplot is transformed to Figure 7.13. You can see that traits less closely associated with loaf volume were removed, and all traits that are retained explained at least 21% of the loaf volume variation. The correlation matrix among these traits is presented in Table 7.5 as a reference to Figure 7.13.

Traits that are most closely associated with loaf volume are in the order of farinograph stability (Farino-stab), alveograph energy requirement (ALV-W), extensograph area and resistance, and farinograph peak time (Farino-PT), all being measures of dough strength (Figure 7.13). Since these measures are highly correlated, there is no need to measure all of them. Any of these measures should be effective for indirect selection for loaf volume.

It would be desirable to identify traits that are more easily measured. In the upper part of the biplot, grain and flour protein contents are most easily measured traits; on the lower part of the

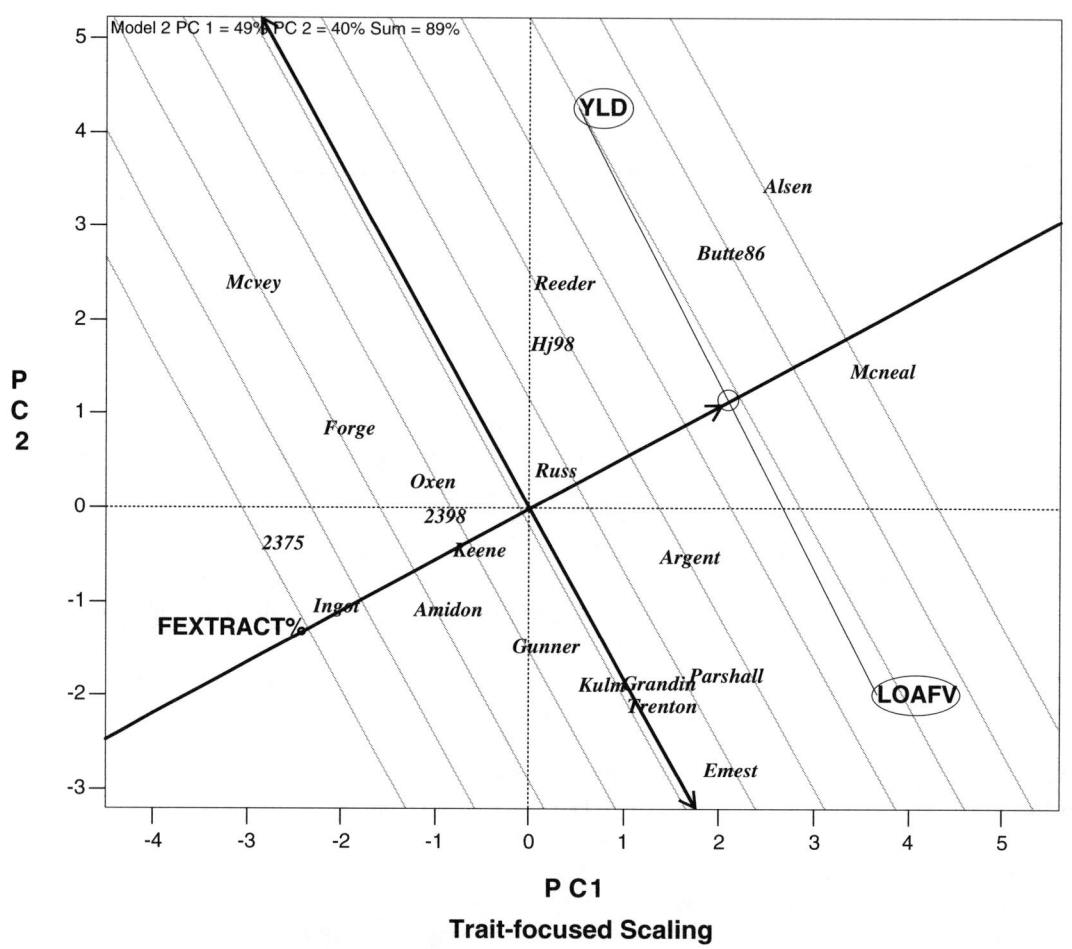

FIGURE 7.11 Cultivar ranking based on grain yield and loaf volume.

TABLE 7.3
Twenty-nine Quality Traits of Hard Red Spring Wheat (HRSW) Measured on Wheat Samples from 15 Sub-regions of Montana, North Dakota, South Dakota, and Minnesota

AREA	GMOIST	KW	KERNEL-M	KERNEL-L	GPROT	GASH	GRAIN FN	SDS	FLOUR EXTRACT	FASH	FPROT	WETGLU	FLOUR FN	VISCO65	VISCO100	FARIN-ABSN	FARIN-PT	FARIN-STAB	FARIN-MTI	EXTENSBLTY	EXTN-RESIST	EXTN-AREA	ALV-P	ALV-L	ALV-W	BAKING-ABSN	LOAFV	TSWT	VITREOUS
MN-A	12.8	31.6	23	73	13.7	1.67	311	49	70.5	0.5	12.5	34.4	320	315	925	65.2	7.5	11.5	40	22.5	7.6	123	88	113	334	63.7	1010	61.0	49
MN-B	12.2	31.1	27	67	13.8	1.69	376	45	69.6	0.4	12.5	34.7	423	820	2620	63.7	7.5	14.0	40	22.0	8.1	114	79	123	325	62.2	990	60.6	74
MT-A	9.6	30.1	38	53	16.3	1.65	355	62	67.8	0.5	15.1	40.6	449	685	3080	65.0	20.0	40.0	10	24.8	11.8	196	100	116	432	63.5	1105	59.0	98
MT-B	10.4	32.7	30	66	13.7	1.62	407	47	68.7	0.5	12.7	34.3	443	896	3010	65.5	10.0	40.0	5	20.7	10.9	154	118	86	395	64.0	1005	61.3	89
MT-C	8.8	30.2	49	42	16.9	1.74	385	48	66.9	0.5	15.6	41.2	497	915	3360	67.3	33.0	47.0	5	22.2	12.1	190	119	95	452	65.8	1115	57.6	92
MT-D	9.9	33.3	37	57	15.0	1.66	395	52	67.9	0.5	14.1	38.2	453	810	3380	66.9	13.0	24.0	10	21.7	9.9	146	101	118	427	65.4	1100	61.0	87
ND-A	12.3	31.4	39	65	14.6	1.67	304	59	69.3	0.5	13.6	39.5	315	265	980	64.3	7.0	10.5	40	22.7	6.6	111	71	137	300	62.8	1035	59.5	58
ND-B	12.4	29.1	29	66	14.6	1.62	193	53	69.3	0.5	13.3	38.1	220	60	225	64.6	5.5	7.0	55	23.5	5.9	106	69	133	279	63.1	910	58.4	41
ND-C	12.9	32.7	29	66	14.4	1.65	320	52	70.1	0.5	13.1	37.1	331	280	1065	65.1	7.0	9.0	45	21.8	7.3	115	82	110	299	63.6	985	60.1	71
ND-D	11.3	30.6	36	58	14.1	1.65	348	57	70.2	0.5	13.1	36.4	366	585	1900	65.2	7.0	11.5	40	21.8	7.1	114	90	133	379	63.7	970	60.5	77
ND-E	12.5	31.2	27	67	14.5	1.68	370	54	69.5	0.5	13.2	37.5	410	680	2520	65.9	7.0	11.0	40	22.4	7.2	116	93	127	373	64.4	1035	59.6	50
ND-F	12.3	31.7	28	66	14.1	1.69	397	53	70.1	0.4	12.8	34.6	409	750	2730	65.0	7.5	11.5	40	21.8	7.0	114	83	131	346	63.5	1010	60.9	78
SD-A	10.8	30.7	32	61	13.7	1.66	402	44	67.9	0.5	12.5	34.3	431	695	2650	64.0	7.0	12.5	35	24.2	7.3	127	78	128	326	62.5	1000	61.3	83
SD-B	12.1	31.9	33	61	14.1	1.72	382	53	69.1	0.5	12.6	33.3	414	740	2420	62.9	8.5	13.0	30	23.1	9.4	151	81	132	366	61.4	1005	60.3	65
SD-C	12.4	31.0	28	66	14.0	1.68	332	53	68.5	0.5	12.6	33.0	343	360	1280	63.6	9.0	14.0	30	24.7	9.0	142	81	121	335	62.1	1045	60.2	73

MN — Minnesota, MT — Montana, ND — North Dakota, SD — South Dakota.

Source: 2000 Regional Quality Report for U.S. Hard Red Spring Wheat.

TABLE 7.4
A List of Wheat Quality Traits Discussed in this Chapter

ALV-L — the dough extensibility measured by alveograph, which traces a curve that measures the air pressure necessary to inflate a piece of dough to the point of rupture.

ALV-P — the maximum pressure needed to deform the piece of dough during the inflation process. It is an indication of resistance or dough stability.

ALV-W — the amount of energy needed to inflate the dough to the point of rupture. It is an indication of the dough strength.

BAKING-ABSN — water absorption measured in baking experiments.

CSPREAD — cookie width.

CW/T — cookie width/thickness ratio.

EXTENSBLTY — extensibility measured by extensograph, which measures dough strength by stretching a piece of dough on a hook until it breaks.

EXTN-AREA — the area beneath the extensograph curve.

EXTN-RESIST — the resistance to extension.

FARIN-ABSN — the amount of water that can be added to the flour until the dough reaches a definite consistency. It is measured by faringraph, which traces a curve during the dough mixing process to record variations in gluten development and the breakdown of gluten proteins over time.

FARIN-MTI — The difference in Brabender Units (BU) from the top of the curve at peak mixing time to the top of the curve five minutes after the peak mixing time, with higher values indicating lower mixing stability.

FARINOGRAM — a scale of 1 to 8, with higher values indicating stronger mixing properties.

FARIN-PT — The time interval, in minutes, from the first addition of water until the curve reaches its maximum height.

FARIN-STAB — The number of minutes the top of the curve remains above the 500 unit line when the highest portion (peak) is centered on the 500 unit line.

FASH —% of flour ash.

FEXTRACT — flour extraction (%).

FN-FLOUR — falling number of flour, a measure of soundness or alpha-amylose activity of flour, with higher values indicating less sprouting.

FN-GRAIN — falling number of grain, a measure of soundness or alpha-amylose activity of grain, with higher values indicating less sprouting.

FPROT — flour protein content (%).

GASH — grain ash (%).

GMOIST — grain moisture.

GPROT — grain protein content (%)

HARDNESS — grain hardness, with higher values indicating greater softness.

KERNEL-M — percentage of medium-sized kernels.

KERNEL-L — percentage of large-sized kernels.

LOAFV — loaf volume in cm^3.

PROTDIFF — difference between grain protein and flour protein.

RVAFNLV — viscosity at the end of pasting curve measured by Rapid Visco Analyzer.

RVAPKT — peak time of pasting curve measured by Rapic Visco Analyzer.

RVAPKV — viscosity at the peak time measured by Rapid Visco Analyzer.

RVATROUGH — viscosity at breakdown measured by Rapid Visco Analyzer.

SDS — sedimentation in cm^3.

TKW — thousand-kernel weight.

TSWT — test weight.

VISCO100 — amylograph peak viscosity measured with 100 grams of flour, with higher values indicating less sprouting.

VISCO65 — amylograph peak viscosity measured with 60 grams of flour, with higher values indicating less sprouting.

VITREOUS% — percentage of vitreous kernels. It is a measure of grain hardness.

WETGLU — percentage of wet gluten.

Cultivar Evaluation Based on Multiple Traits

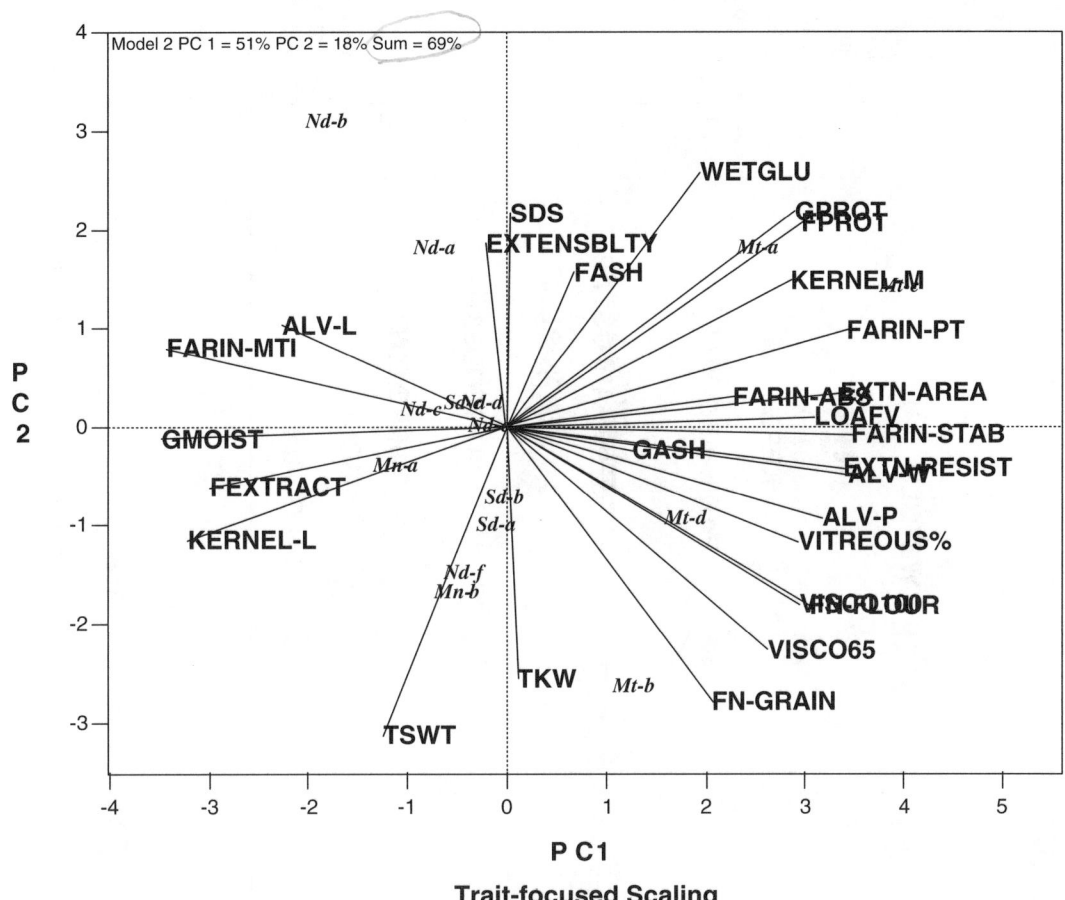

FIGURE 7.12 Vector view of a biplot based on quality traits measured in HRSW samples taken from different regions of the HRSW production area in USA, showing the interrelationships among various traits related to bread-making quality. Nd — North Dakota, Sd — South Dakota, Mn — Minnesota, and Mt — Montana.

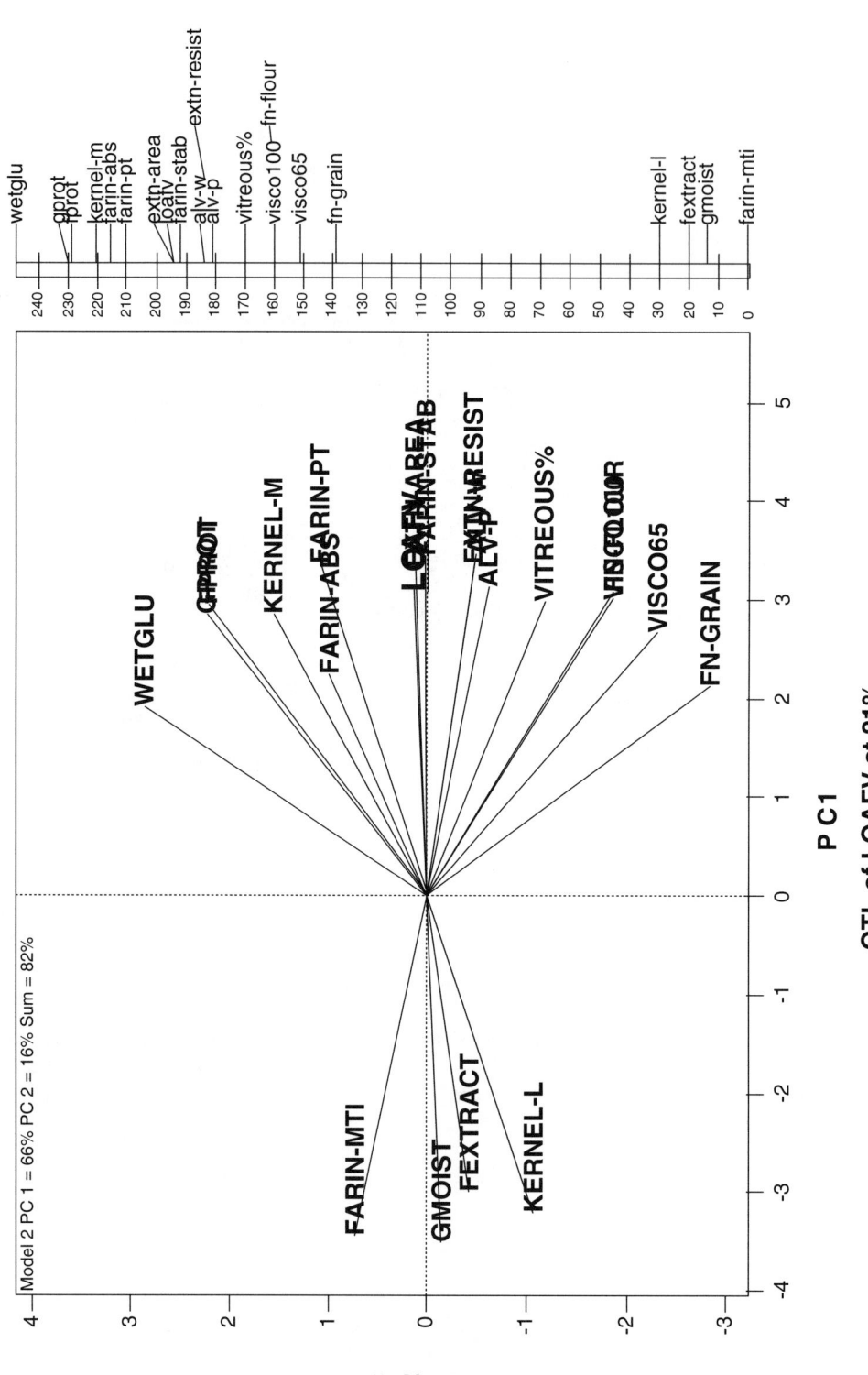

FIGURE 7.13 Vector view of the biplot showing interrelationships among traits that are more closely associated with loaf volume. This can be used to identify traits that may be suitable for indirect selection for loaf volume.

Cultivar Evaluation Based on Multiple Traits

TABLE 7.5
Correlation Coefficients among Traits (Loaf Volume and Closely Associated Traits)

Names	FARIN-MTI	ALV-W	EXTN-AREA	FARIN-PT	EXTN-RESIST	FPROT	FN-FLOUR	GPROT	FARIN-STAB	FEXTRACT	GMOIST	VISCO100	KERNEL-L	ALV-P	VITREOUS%	KERNEL-M	FN-GRAIN	VISCO65	FARIN-ABSN	WETGLU
LOAFV	-0.779	0.773	0.758	0.751	0.749	0.699	0.694	0.684	0.680	-0.675	-0.662	0.635	-0.624	0.610	0.600	0.597	0.513	0.504	0.499	0.461
FARIN-MTI		-0.876	-0.880	-0.740	-0.945	-0.576	-0.778	-0.538	-0.923	0.755	0.865	-0.758	0.646	-0.866	-0.786	-0.579	-0.601	-0.691	-0.485	-0.271
ALV-W			0.814	0.765	0.846	0.660	0.812	0.629	0.824	-0.613	-0.841	0.817	-0.741	0.873	0.697	0.602	0.609	0.750	0.621	0.373
EXTN-AREA				0.850	0.958	0.694	0.659	0.725	0.894	-0.780	-0.809	0.611	-0.744	0.721	0.695	0.615	0.393	0.519	0.321	0.365
FARIN-PT					0.809	0.865	0.607	0.891	0.851	-0.744	-0.810	0.547	-0.875	0.714	0.596	0.784	0.285	0.469	0.583	0.612
EXTN-RESIST						0.605	0.735	0.617	0.928	-0.729	-0.801	0.690	-0.671	0.817	0.744	0.547	0.507	0.636	0.368	0.268
FPROT							0.382	0.985	0.699	-0.650	-0.737	0.381	-0.840	0.535	0.446	0.83	0.024	0.223	0.645	0.903
FN-FLOUR								0.372	0.691	-0.590	-0.724	0.974	-0.601	0.711	0.791	0.452	0.911	0.955	0.367	0.103
GPROT									0.681	-0.654	-0.698	0.360	-0.852	0.491	0.436	0.817	0.003	0.209	0.587	0.870
FARIN-STAB										-0.727	-0.878	0.669	-0.703	0.884	0.737	0.622	0.429	0.609	0.538	0.440
FEXTRACT											0.835	-0.574	0.763	-0.508	-0.621	-0.69	-0.330	-0.453	-0.313	-0.411
GMOIST												-0.737	0.859	-0.762	-0.796	-0.771	-0.464	-0.639	-0.565	-0.501
VISCO100													-0.579	0.692	0.776	0.42	0.878	0.967	0.395	0.122
KERNEL-L														-0.562	-0.657	-0.919	-0.309	-0.486	-0.480	-0.625
ALV-P															0.631	0.452	0.545	0.674	0.725	0.296
VITREOUS%																0.533	0.696	0.706	0.291	0.163
KERNEL-M																	0.196	0.333	0.445	0.712
FN-GRAIN																		0.908	0.185	-0.223
VISCO65																			0.307	-0.035
FARIN-ABSN																				0.611

biplot, vitreous and flour falling number are most easily measured traits. Particularly, grain protein content and grain falling number (indicative of resistance to pre-harvest sprouting) are virtually orthogonal to each other, with an angle of about 90° between them and a correlation of 0.003 (Table 7.5). Mechanistically, the two traits are independent of each other, although both contribute to loaf volume. Therefore, joint use of these two traits as indirect selection for loaf volume should be an economic and promising option in early generations. Percentage of vitreous kernels is more closely associated with loaf volume than falling number and protein content, and it can be visually assessed. It is, therefore, a promising trait for indirect selection for loaf volume. Grain protein and flour protein also are almost identical; there is, therefore, no point in measuring both.

Grain moisture is highly negatively correlated with loaf volume, but since it is highly influenced by nongenetic factors, it has probably little use in selection for loaf volume. Wet gluten content is closely associated with protein content, but it is more difficult to measure than protein content and also less closely associated with loaf volume than protein content. It is therefore not recommended as a trait for indirect selection.

This example demonstrates that the GGEbiplot program can be used to identify traits for indirect selection for a target trait. Specifically, the selection for loaf volume for HRSW during a breeding cycle can follow four progressive steps: 1) visual selection for vitreous kernel rate and kernel size composition; 2) selection for grain protein content and grain falling number; 3) selection for dough properties using one, not all, of the dough evaluation methods, preferably farinograph; and finally 4) for loaf volume per se.

7.4 IDENTIFICATION OF REDUNDANT TRAITS

In addition to revealing traits that may be used for indirect selection, biplots like Figure 7.13 also can be used in identifying redundant traits. Scientists complain of lack of research funds, but much money is being spent on collecting information that may be redundant. For example, we already mentioned the close correlations among the measures from various dough property procedures (farinograph, mixograph, and alveograph); it seems that there is no need for doing all the tests. Another example is the close relationship between grain protein and flour protein (Figures 7.13 and 7.17). Clearly, through better understanding of the interrelationships among various traits, e.g., using biplots like Figures 7.13 and 7.17, much money can be saved without sacrificing useful information.

7.5 COMPARING CULTIVARS AS PACKAGES OF TRAITS

There are cases where many traits are measured, but it is not clear which ones are more important for cultivar evaluation. In food industry, procedures are usually customized according to an established cultivar or a mixture of established cultivars. Therefore, it is usually not known which single traits are responsible for making a cultivar or cultivars ideal to best meet the requirement of industrial processing. For example, the cultivar *Augusta* has been used as the absolute check for making cookies in Ontario, Canada, and all new soft wheat cultivars are compared with this check for suitability for cookie-making. Many traits are measured (Table 7.6), but it is not clear what an ideal cultivar should have. In such cases, what we can do, as an initial step, is to assume all measured traits are equally important and compare the cultivars as packages of traits. Further analysis may lead to specific traits that can be used for indirect selection for cookie-making suitability.

Table 7.6 contains data on 14 winter wheat genotypes for 19 traits that are thought to be relevant to cookie-making property. Among them, *Augusta* is the ideal check, and Canadian eastern winter wheat mixture 1 (*CEWW1*) is the mixture of soft winter wheat cultivars currently grown in Ontario. All other 12 genotypes are new breeding lines to be evaluated for cookie-making suitability.

TABLE 7.6
Nineteen Quality Traits of 14 Soft Winter Wheat Genotypes

GENO	TSWT	KW	HARDNESS	GPROT	GASH	FN	FEXTRACT	FASH	FPROT	PROTDIFF	CSPREAD	CW/T	ALV-P	ALV-L	ALV-W	RVAPKT	RVAPKV	RVATROUGH	RVAFNLV
AUGUSTA	73.7	31.9	73.2	9.45	1.54	289	73.0	0.47	8.11	1.34	8.40	11.79	17	137	41	6.0	170	104	184
GEN99-53	75.8	32.1	71.6	9.64	1.54	348	71.5	0.42	8.24	1.40	8.16	9.75	32	113	79	6.3	217	160	268
TW97613	78.4	34.4	74.8	10.38	1.52	332	67.2	0.38	8.91	1.47	8.56	9.78	26	120	86	6.0	220	139	246
RC98110	75.9	32.2	73.1	10.08	1.63	286	72.2	0.46	8.83	1.25	8.16	10.04	30	121	100	6.1	218	138	238
RC98111	78.4	31.3	72.3	10.64	1.56	371	72.6	0.42	8.69	1.95	8.34	10.11	32	80	58	6.3	249	172	290
RC98113	78.8	32.3	71.8	10.71	1.58	375	74.6	0.44	8.85	1.86	8.24	9.99	32	80	57	6.2	239	162	280
RC98021	80.5	28.5	73.3	11.16	1.65	323	72.8	0.44	9.81	1.35	8.32	9.64	33	113	97	6.0	231	136	244
CM753	75.7	30.8	71.1	9.78	1.49	365	71.6	0.41	8.34	1.44	8.13	10.48	35	98	83	6.3	223	167	277
WBK0290B1	80.3	35.0	70.6	10.55	1.58	355	69.4	0.42	8.59	1.95	8.32	10.74	37	93	84	6.2	190	139	224
WBL0274C1	77.0	34.3	70.7	9.24	1.49	328	74.0	0.44	8.08	1.16	8.61	11.68	29	102	56	6.0	196	129	224
WBL0476D1	77.5	30.9	72.5	9.60	1.55	337	74.5	0.42	8.21	1.39	8.27	10.18	34	59	51	6.3	176	132	223
PRC9603	74.8	32.1	72.6	9.94	1.60	311	70.1	0.48	8.22	1.71	8.48	11.30	29	73	43	5.9	197	104	194
PRC9624	76.0	33.8	72.5	10.28	1.66	275	72.7	0.47	8.96	1.32	8.35	10.77	19	141	47	5.8	156	77	155
CEWW1	82.0	36.0	72.9	9.80	1.51	325	74.0	0.45	8.50	1.40	8.50	10.51	30	123	79	6.3	198	142	233

Source: 2000 Ontario winter wheat registration trials.

The biplot based on this dataset is presented in Figure 7.14. As in all other biplots, the cultivars are in italics, and the traits are in regular uppercase. The biplot explained 56% of the total variation.

7.5.1 Comparing New Genotypes with the Standard Cultivar

The GGEbiplot has a *Compare With a Standard Entry* function. When this function is invoked and *Augusta* is selected as the standard entry (cultivar), a set of 10 concentric circles will appear, with *Augusta* at the center (Figure 7.14). These concentric circles allow comparison of all genotypes with *Augusta*. Since *Augusta* is the standard cultivar, the closer a genotype is to *Augusta*, the more ideal it is. From Figure 7.14, we can see that genotypes *PRC9603*, *PRC9624*, and *WBI0174c1* are close to *Augusta*; they are closer to *Augusta* than *CEWW1*, which represents the acceptable level for cookie-making quality. These three genotypes, thus, have better than acceptable cookie-making quality. Other genotypes, however, are further away from *Augusta* than *CEWW1*. Their cookie-making quality is, therefore, questionable. Genotypes *RC98021*, *RC98113*, *RC98111*, and *CM753* are far away from *Augusta*, and therefore, should be rejected.

7.5.2 What is Good with the Standard Cultivar?

If we knew all the good attributes of the standard cultivar *Augusta*, we would understand better what the important traits for cookie-making quality are. With this knowledge, genotypes can be evaluated based on some quality traits, with or without the presence of the standard cultivar. The polygon view of the biplot (Figure 7.15), brought up by the GGEbiplot function *Show/Hide Polygon*, indicates that *Augusta* had highest values in cookie width/thickness ratio (CW/T) and cookie spread (CSPREAD).

The GGEbiplot function *Examine an Entry* allows visualization of the relative levels of a cultivar or entry with regard to various traits. Figure 7.16 indicates that *Augusta* had above-average values in CW/T, flour ash (FASH), CSPREAD, kernel weight (KW), and dough extensibility (ALV-L). It had below-average levels for all other traits, including grain and flour protein content, dough strength (ALV-W), dough stability (ALV-P), viscosity (RVAPKV), etc. Therefore, these results suggest that good cookie-making quality includes high CW/T and low protein content and low gluten strength.

7.5.3 Traits for Indirect Selection for Cookie-Making Quality

Using the GGEbiplot function *Auto Find QTL* and selecting CW/T as the target trait, traits that have associations with CW/T at ≥14% level are identified and displayed in Figure 7.17. All traits, except flour ash and cookie spread, are negatively correlated with CW/T (Figure 7.17 and Table 7.7). Table 7.7 is presented to validate interrelationship among traits displayed in Figure 7.17 and to help build confidence in the biplot interpretation. Four traits are most closely associated with CW/T. They are: gluten strength (ALV-W), flour protein, viscosity at peak value (RVAPKV), and viscosity at the end of the pasting curve (RVAFNLV). More easily determined traits include grain and flour protein, which are closely correlated (Figure 7.17 and Table 7.7). This justifies the practice of using low protein content as a means of selecting for cookie-making quality in early generations. As was seen for hard wheat, a trait that is orthogonal to protein content is falling number, which also is negatively correlated with CW/T. The high similarity in the interrelationships among various traits for both hard wheat and soft wheat, even though the data were completely unrelated, is interesting. It is also interesting that grain protein content and falling number can be used as traits for indirect selection for both bread wheat (hard wheat) and pastry wheat (soft wheat) but in the opposite directions.

Cultivar Evaluation Based on Multiple Traits

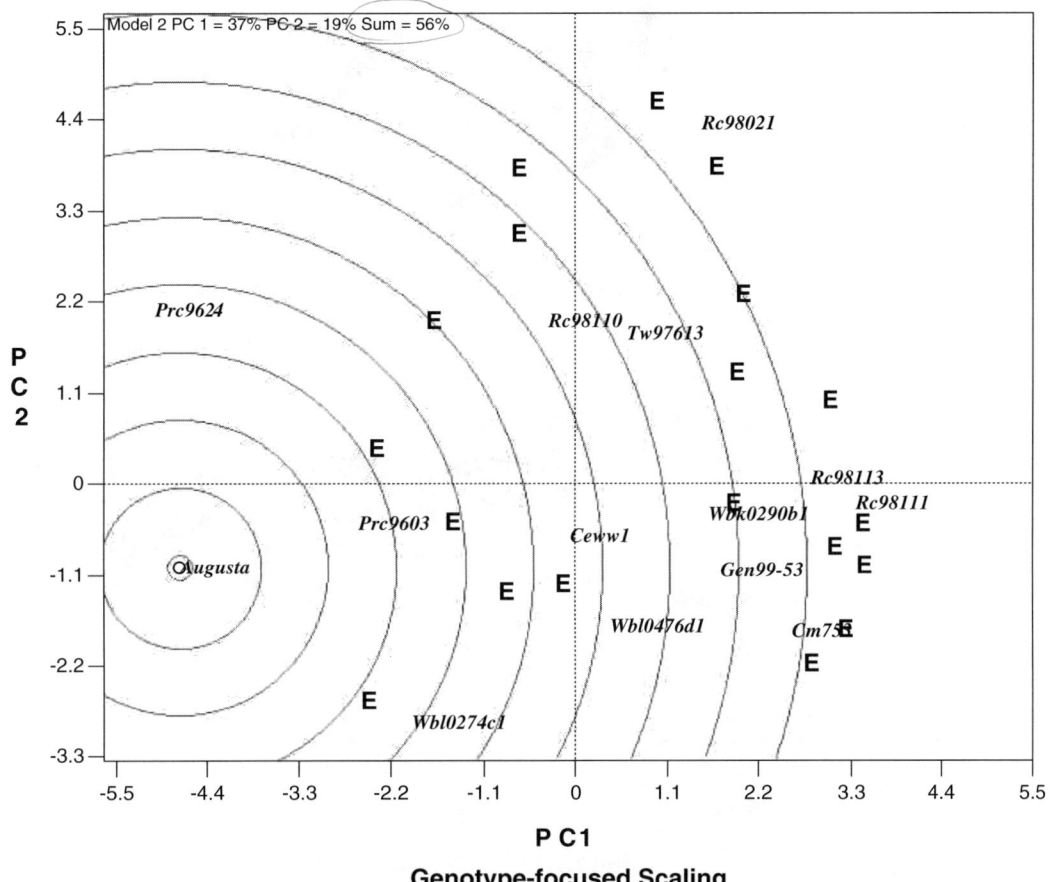

FIGURE 7.14 Comparison of all cultivars with a standard cultivar *Augusta* for similarity in 19 traits related to cookie making quality, assuming all traits are equally important.

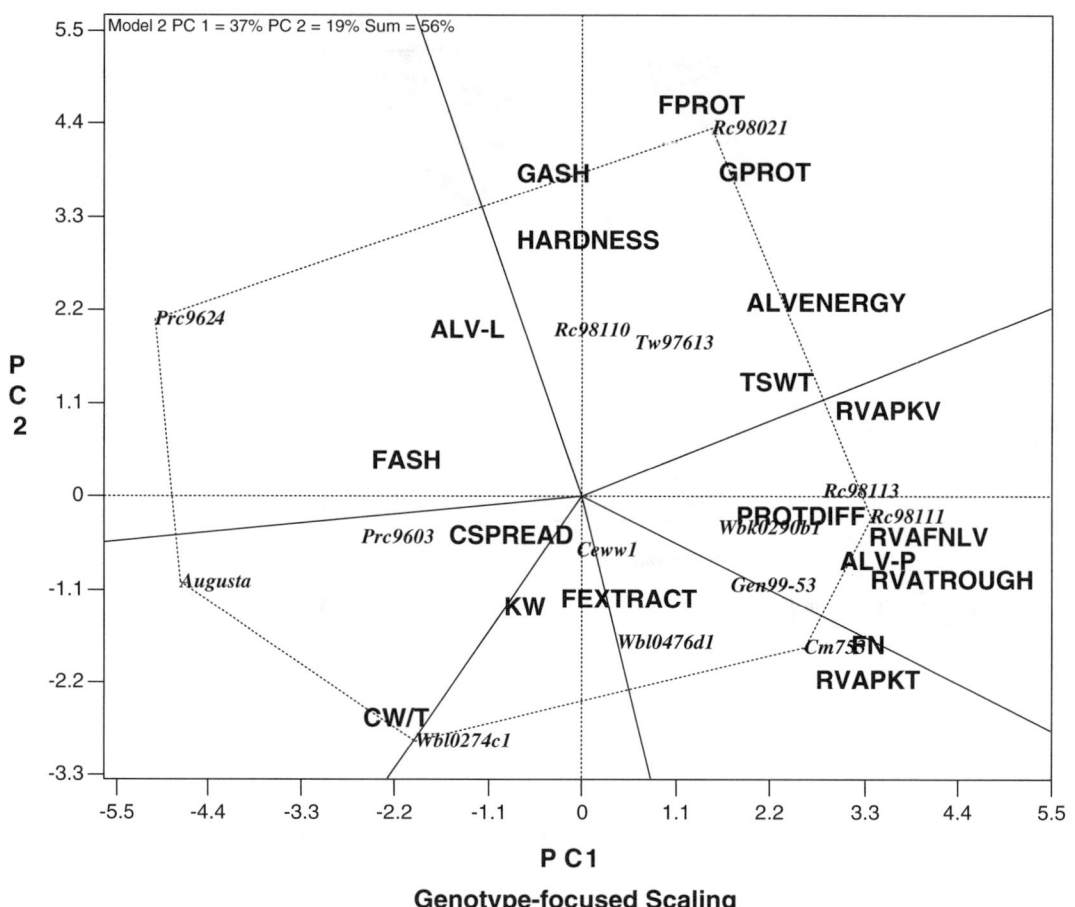

FIGURE 7.15 Polygon view of the genotype-by-trait biplot, showing which cultivars have the highest value for which traits.

Cultivar Evaluation Based on Multiple Traits

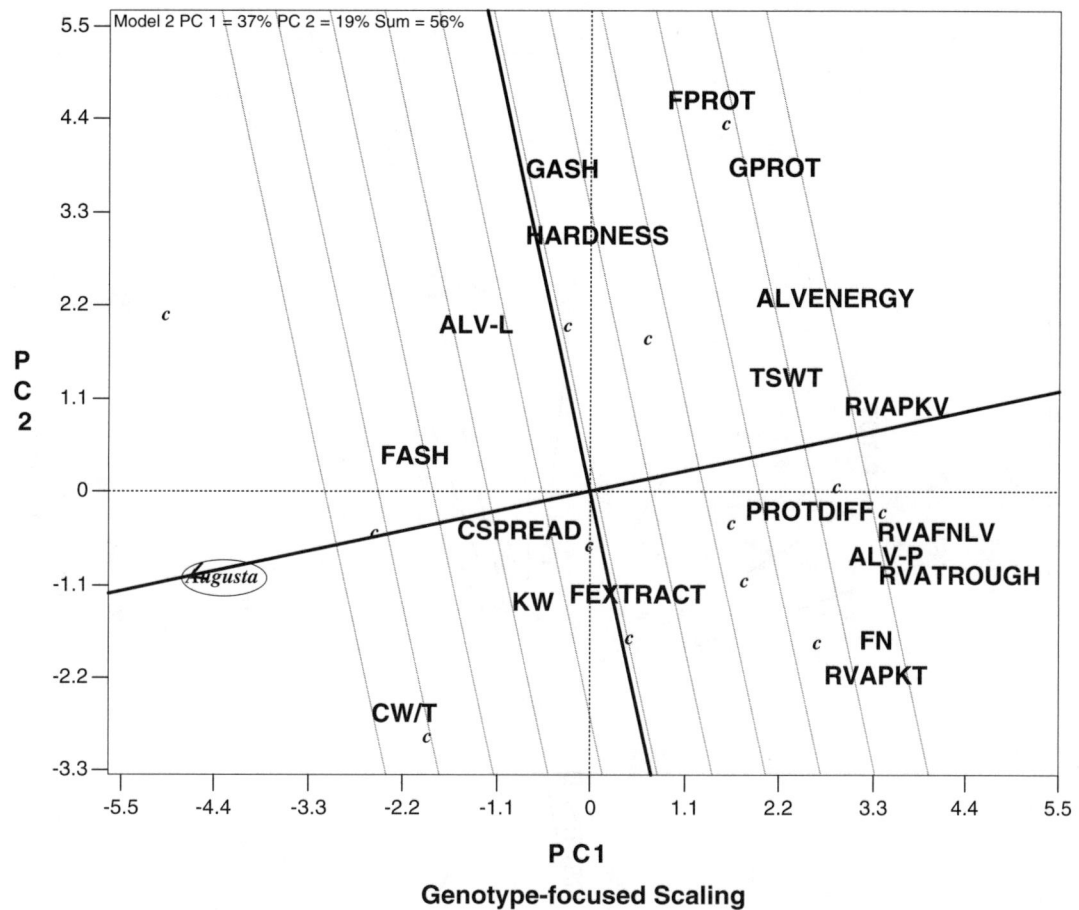

FIGURE 7.16 Visualizing the relative magnitudes of cultivar *Augusta* for various traits.

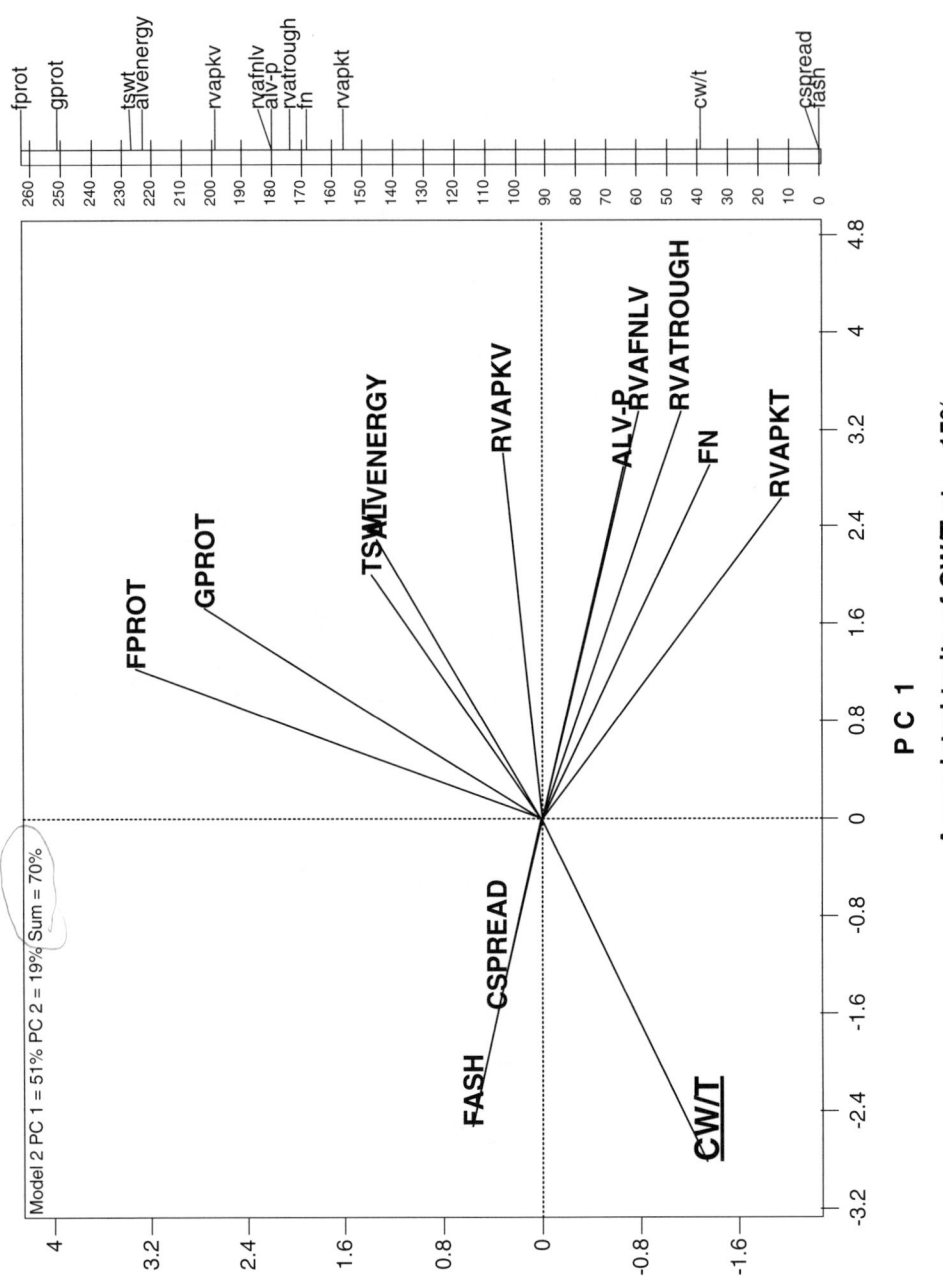

FIGURE 7.17 Vector view of the genotype-by-trait biplot, showing the interrelationships among various traits that are associated with cookie width/thickness ratio, which is an important measure of cookie-making quality.

TABLE 7.7
Correlation Coefficients among Traits (CW/T and Closely Related Traits)

Names	ALVENERGY	RVAFNLV	RVAPKV	FPROT	GPROT	RVATROUGH	FASH	CSPREAD	ALV-P	TSWT	RVAPKT	FN
CW/T	-0.632	-0.627	-0.620	-0.610	-0.588	-0.572	0.547	0.484	-0.475	-0.441	-0.413	-0.382
ALVENERGY		0.467	0.488	0.515	0.370	0.481	-0.439	-0.312	0.493	0.455	0.283	0.147
RVAFNLV			0.904	0.132	0.282	0.983	-0.638	-0.39	0.681	0.336	0.737	0.830
RVAPKV				0.378	0.504	0.830	-0.444	-0.262	0.543	0.313	0.429	0.648
FPROT					0.877	0.035	-0.078	-0.121	0.079	0.497	-0.251	-0.065
GPROT						0.200	-0.163	-0.167	0.257	0.534	-0.084	0.242
RVATROUGH							-0.652	-0.401	0.728	0.372	0.831	0.850
FASH								0.082	-0.485	-0.346	-0.506	-0.667
CSPREAD									-0.383	0.191	-0.469	-0.236
ALV-P										0.492	0.680	0.736
TSWT											0.344	0.400
RVAPKT												0.726

7.6 INVESTIGATION OF DIFFERENT SELECTION STRATEGIES

For soybean cultivars, three traits are most important: grain yield, protein content, and oil content. Figure 7.18 is a biplot based on the 1999 Ontario soybean uniform test in the 2800 crop heat unit area. The 3 traits are in regular uppercase and the 119 genotypes are in italics. The biplot explains 90% of the total variation. Trait vectors are displayed to show the interrelationship among them, which is typical of all other datasets (Yan and Rajcan, 2002). Specifically, grain yield is almost independent of protein content and oil content, whereas the latter two are negatively associated. Given this scenario, what is the most effective and economic way to select?

The GGEbiplot has a *Discard based on a Tester* function, which allows culling of genotypes based on a single trait, one at a time, at a user-specified level. This function can, therefore, be used to compare various selection strategies. Figure 7.19 is the selection result based on yield alone, whereby genotypes with yields below the median are discarded. Figures 7.20 to 7.22 are based on yield and oil content, yield and protein content, and oil content and protein content, respectively, using the same criteria. Obviously, selection based on both protein content and oil content is not a viable strategy since, due to the strong negative association between the two traits, most, if not all, genotypes would be discarded, and those that survive the selection would inevitably be low yielding (Figure 7.22). Selection based on yield and oil content eliminated all genotypes with high protein content (Figure 7.20), and selection based on yield and protein content eliminated all genotypes with high oil content (Figure 7.21). Clearly, selection based on yield alone gives results (Figure 7.19) that are equivalent to putting Figures 7.20 and 7.21 together. This exercise in selection using GGEbiplot suggests that selection based on yield alone in the early stages of selection is not only most economical but also most effective. GGEbiplot also allows genotype evaluation based on two traits as illustrated in Section 7.2 and Figure 7.11.

7.7 SYSTEMS UNDERSTANDING OF CROP IMPROVEMENT

7.7.1 Systems Understanding, Independent Culling, and the Breeder's Eye

As a wheat breeder, it has been a long-time interest of Yan to acquire a systems view of crop improvement (Yan, 1993; Yan and Wallace, 1995; Wallace and Yan, 1998), and now the GGE biplot methodology provides a quantitative approach for achieving this goal. Recall the interesting, albeit unfavorable, triangle of grain yield, flour extraction, and loaf volume of hard wheat (Figure 7.7). Although there is no strong negative correlation between any pair of the three traits, as can be inferred from the angles among them and verified from Table 7.2, it is difficult to find a cultivar that is good in all three traits. Any two traits can be reasonably combined but at the expense of the third. This relationship reminds us of a triangle consisting of earliness, winter hardiness, and head size in winter wheat (Yan and Wallace, 1995). It also reminds us of a relationship consisting of yield, protein content, and earliness in spring wheat observed in western Canada. This is in addition to the most commonly observed triangle among morphological components of yield, i.e., number of heads per unit area, number of kernels per head, and weight per kernel.

Do these triangles mean that it is difficult to improve all traits at the same time? Yes, as is often observed in practice. Do these triangles mean that it is impossible to improve all traits simultaneously? No, there are instances where a superior cultivar seems to be better than some nonadapted genotypes in most, if not all, aspects. But improving all aspects that make up the triangle requires seeking solutions outside these aspects. That is, we need additional dimensions to fully understand the system and to improve its efficiency.

Figure 7.23 presents a holistic hypothesis on the wheat system, which is a summary of nearly 20 years of observation and analysis of Yan as a wheat breeder. The system of wheat is depicted as "a piece of cake" and several traits or system components compete for a share of it. These include winter hardiness, earliness, yield components or potential, milling quality e.g., flour extraction, and protein content which is a major determinant of loaf volume (Figure 7.13).

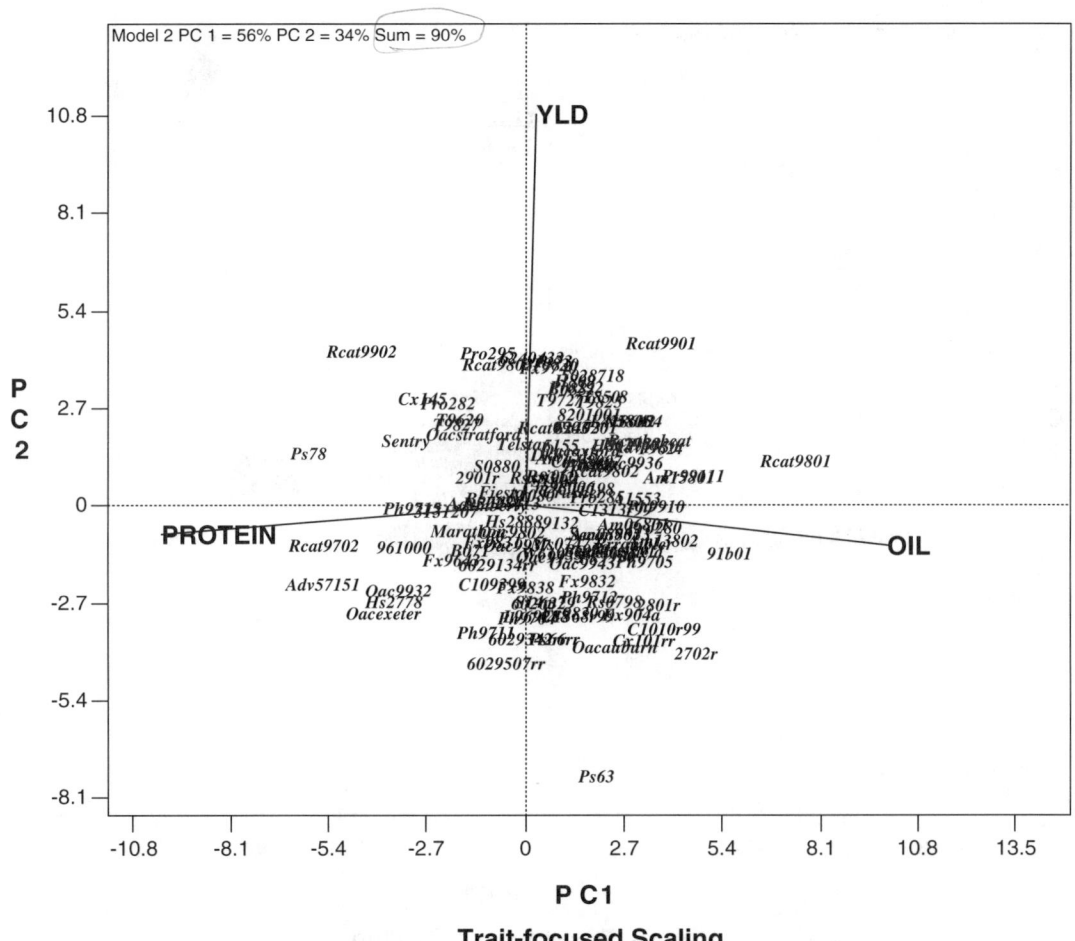

FIGURE 7.18 A genotype-by-trait biplot based on soybean data, showing the interrelationships among seed yield, oil, and protein concentration for 119 soybean genotypes. The genotypes are in italics and the traits are in regular uppercase.

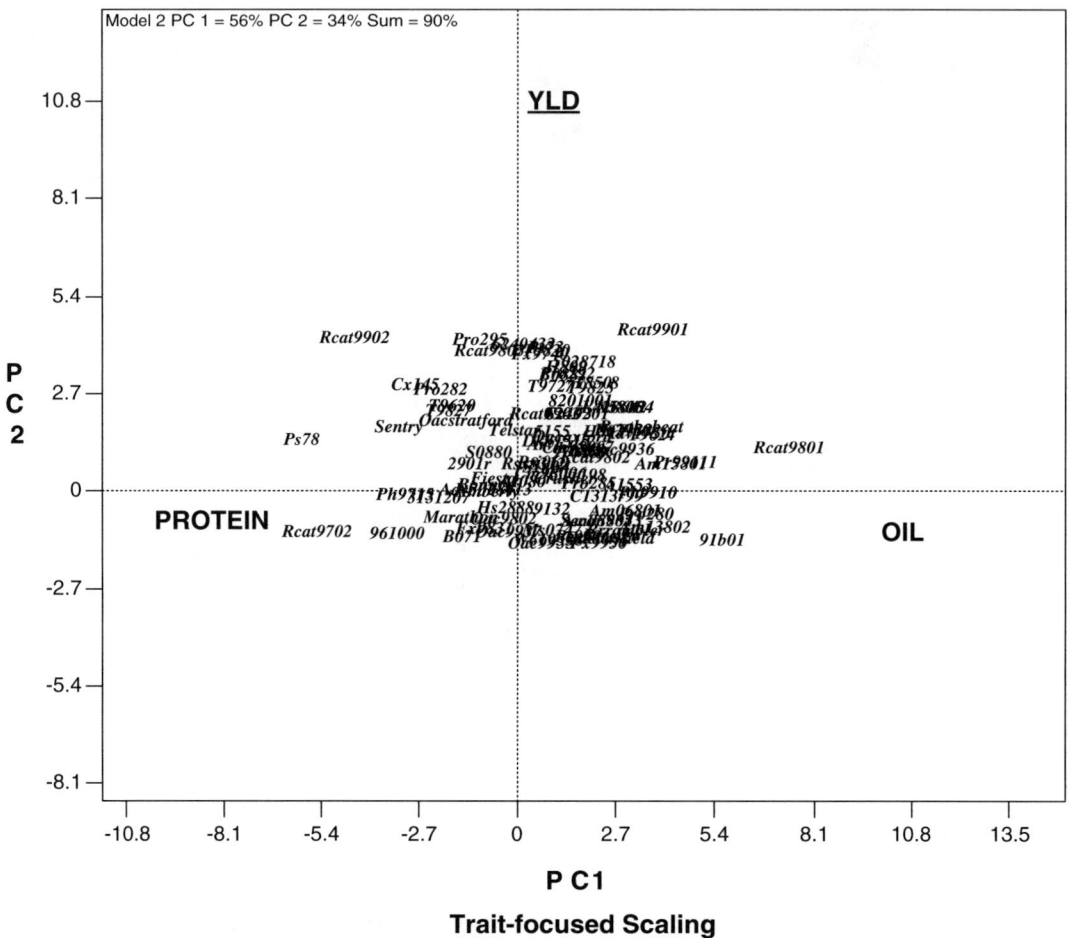

FIGURE 7.19 The genotype-by-trait biplot after genotypes with below-median yield were removed or culled.

Cultivar Evaluation Based on Multiple Traits

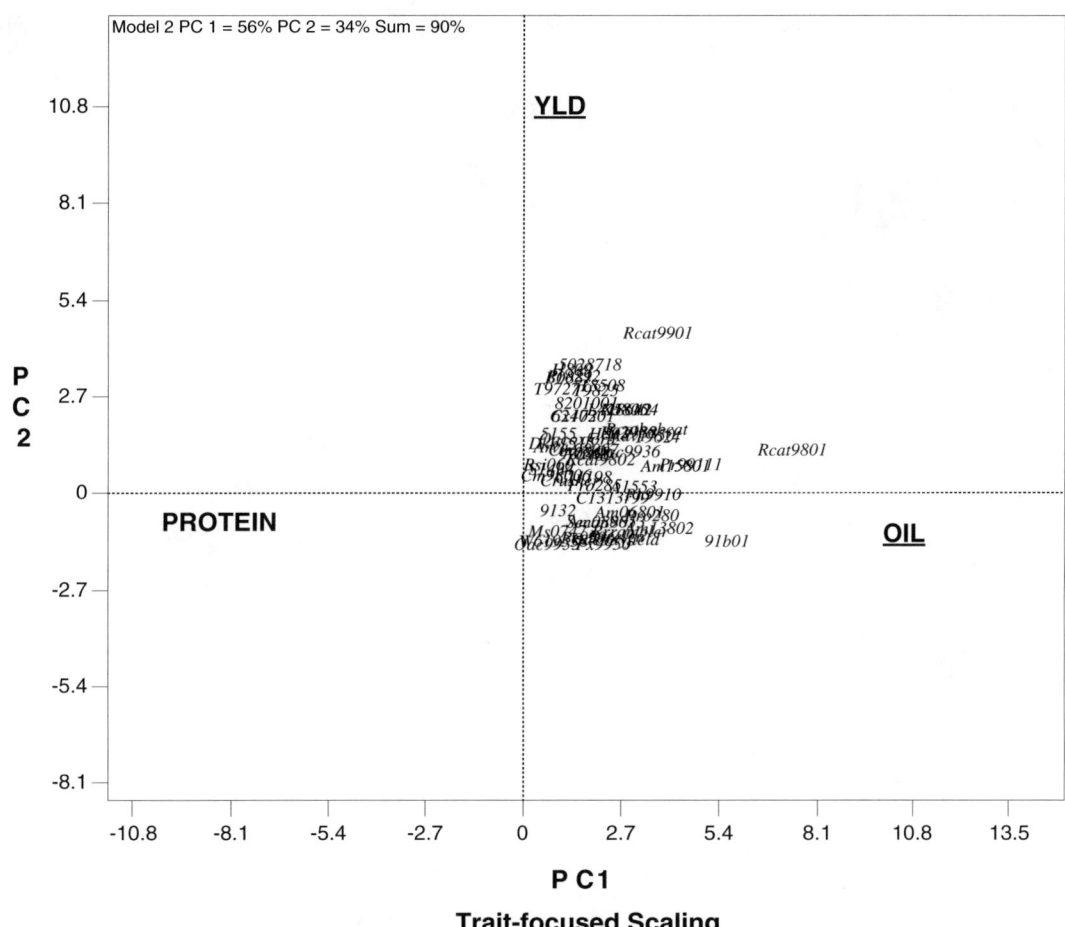

FIGURE 7.20 The genotype-by-trait biplot after genotypes with below-median grain yield and genotypes with below-median oil concentration were removed or culled.

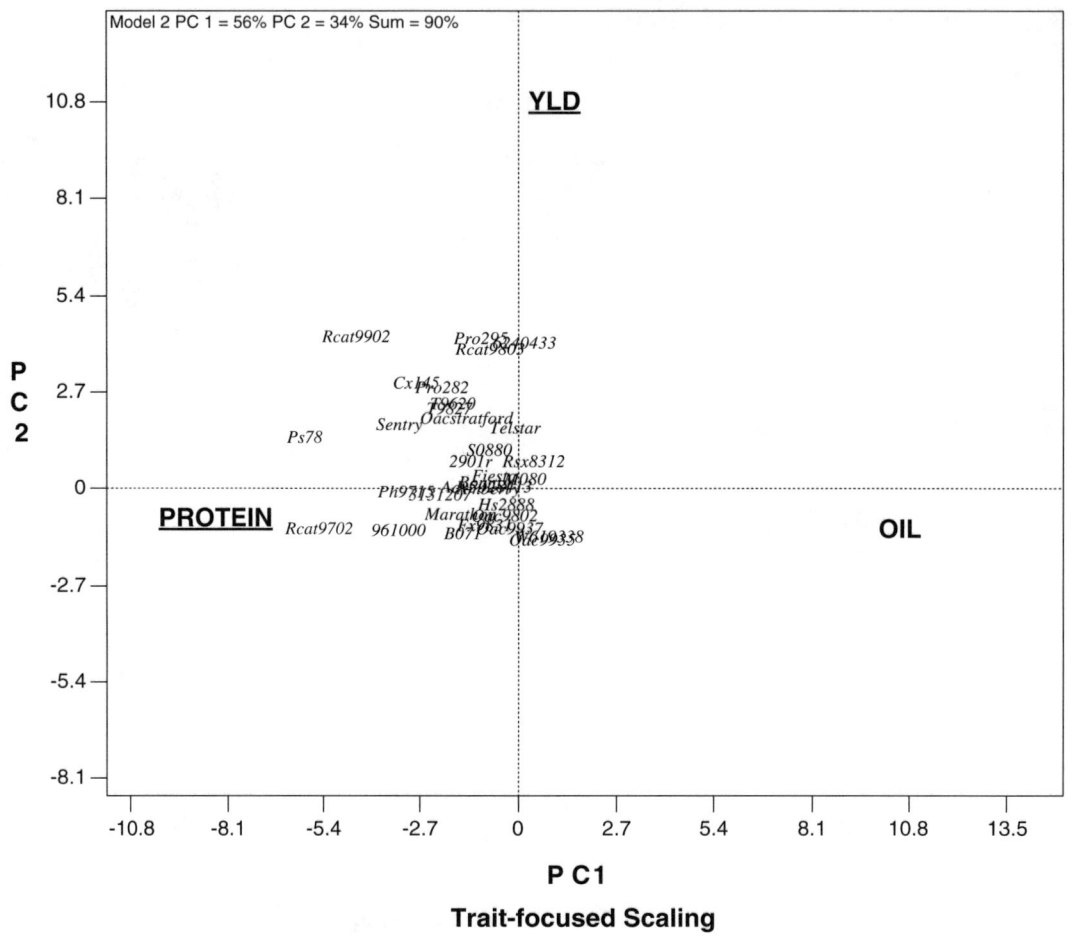

FIGURE 7.21 The genotype-by-trait biplot after genotypes with below-median grain yield and genotypes with below-median protein concentration were removed or culled.

Cultivar Evaluation Based on Multiple Traits

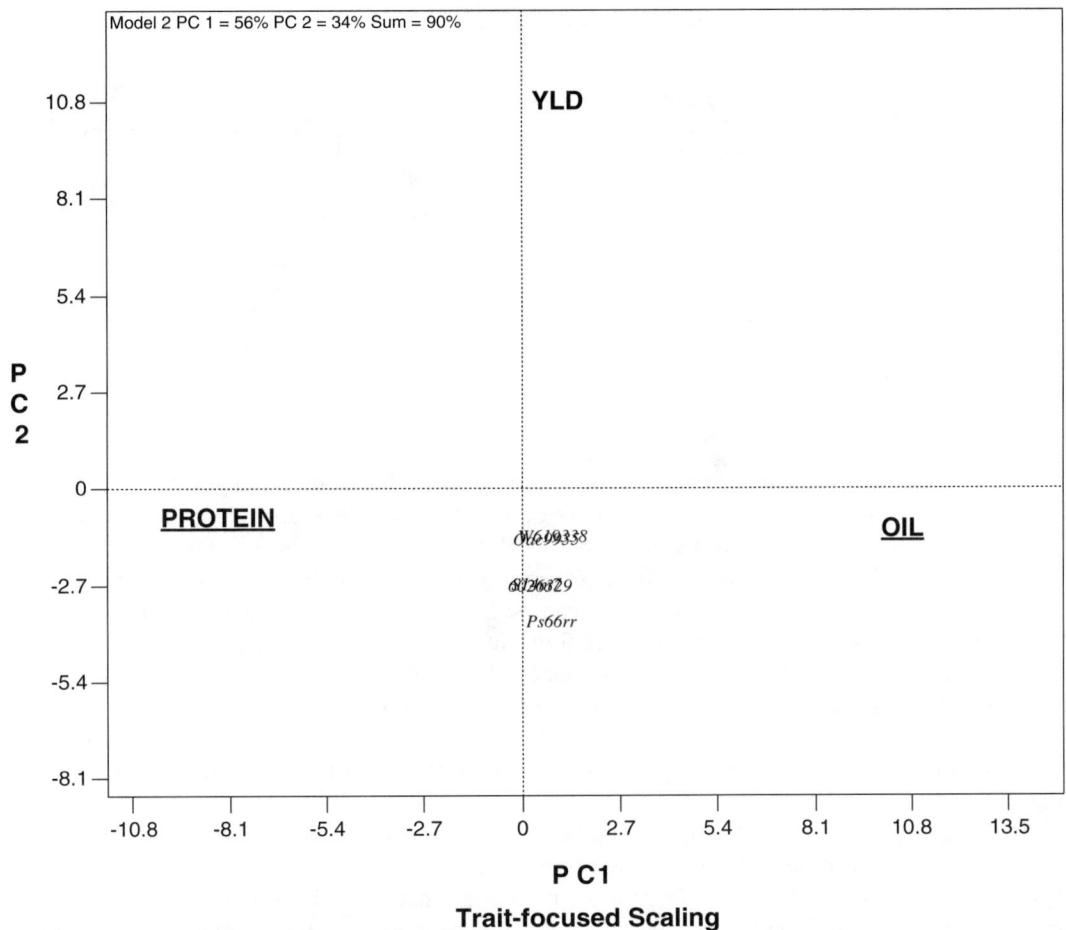

FIGURE 7.22 The genotype-by-trait biplot after genotypes with below-median oil concentration or below-median protein concentration were removed or culled.

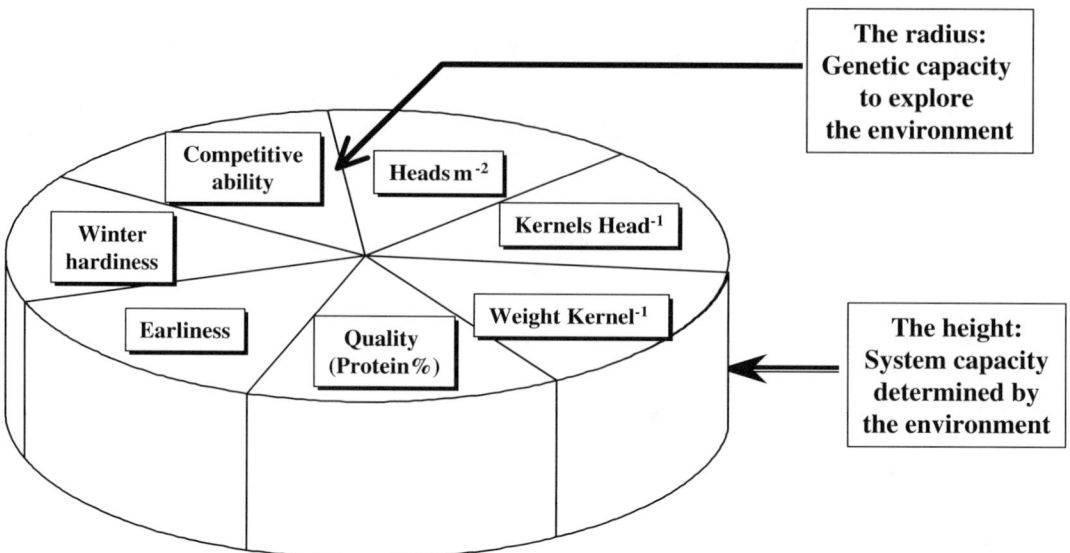

FIGURE 7.23 The hypothesis of constant system capacity, showing the interconnectedness among various components of a wheat cultivar.

Some interesting and important inferences can be made based on this hypothesis or model. First, if the size of the "cake" is fixed, it is not possible to improve all traits at the same time. Second, any trait can be improved at the expense of one or more other traits. On the other hand, reduction in the level of one trait will cause improvement in the levels of one or more other traits. Third, two or more, but not all, traits can be improved simultaneously, but this will be at the expense of other traits. Fourth, the relationship between any two traits is not fixed; it is dependent on the levels of other traits. Experienced breeders should have plenty of observations to support these inferences. Trade-offs among the system components are inevitable, as often found in studies comparing historical cultivars. A general trend in wheat is that yield is increased through more heads per unit area and more kernels per head, at the expense of kernel weight (e.g., Donmez et al., 2001).

If the size of the cake is fixed, can the breeder do anything about it? Yes. The breeder can adjust the allocation of the cake to different traits so that the system output is maximized. To do this, the breeder needs to know which traits are more important and which traits are less important, for a given temporal and spatial framework, and to allocate more resources to more important traits and less resources to less important ones.

The interconnectedness among traits suggests that to improve a single trait to an extremely high level will not necessarily lead to a superior cultivar; it is more likely to lead to an inferior genotype that has one or more traits below the acceptable levels. On the other hand, to reduce a single trait to an excessively low level will certainly lead to an inferior genotype. Therefore, we cannot select superior cultivars by selecting for a single trait; but we can cull inferior genotypes based on single traits. As a corollary, independent culling is the most rational method for selection in early stages of a breeding cycle. We use the word "select" for high selection intensity so that only a small fraction of the population is retained, and the word "cull" for low selection intensity so that the majority of the selection population is retained.

If culling were conducted on only one trait, selection efficiency would be very limited. Independent culling based on multiple traits, however, can jointly achieve high selection intensity. The following formula describes how the joint selection intensity (I), which is a measure of selection efficiency, is influenced by the number of traits, given a fixed selection

Cultivar Evaluation Based on Multiple Traits

intensity for each trait (*i*) which is also a measure of risk to mistakenly discard promising genotypes:

$$I = 1 - (1-i)^n \qquad (7.1)$$

with

$$n = \sum_{j=1}^{E} t_j \qquad (7.2)$$

where *n* represents the number of traits, *E* is the number of environments in which selection is conducted, and t_j is the number of traits selected for in environment *j*. An environment can be a year–location combination or an artificial environment, e.g., disease nursery, off-site nursery, off-season nursery, etc. Equation 7.1 is illustrated in Figure 7.24. For a given overall culling intensity (*I*), the culling intensity based on individual traits (*i*) need not be high if traits to be culled are many.

Equation 7.1 helps understand the power of "the Breeder's Eye." An experienced breeder intuitively selects for numerous traits while observing his/her material. Because the breeder is looking at many traits at the same time, joint selection intensity (I) should be fairly high even though the selection intensity for each trait (*i*) is low, which means much of the population is discarded with little risk of mistakenly discarding promising genotypes. Indeed, the Breeder's Eye is the single most valuable asset of a breeding program. It is unfortunate that many laboratory-oriented scientists supposedly facilitating crop improvement and many managers who are at the helm of hiring breeders are not aware of this.

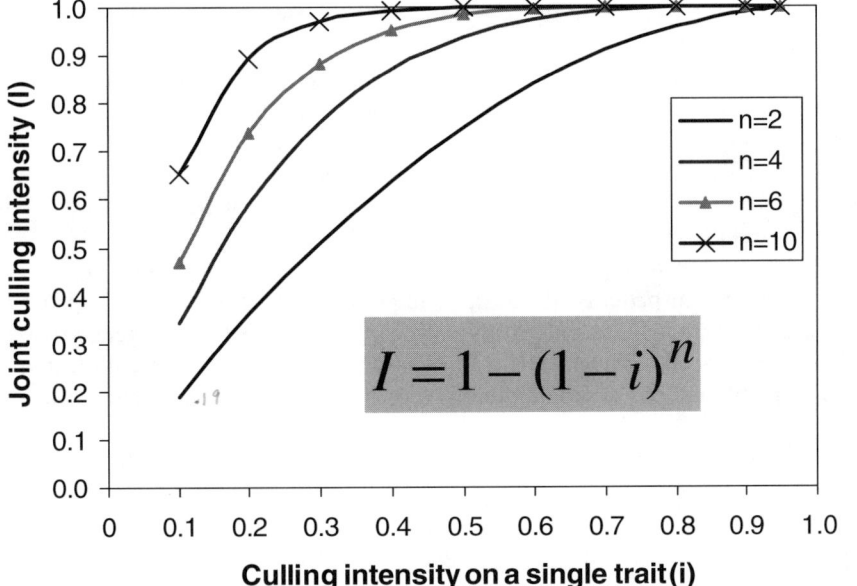

FIGURE 7.24 Relationship between the overall culling intensity (*I*), culling intensity for single traits (*i*), and the number of traits (*n*) to be selected..

7.7.2 ENLARGING THE SYSTEM CAPACITY BY REINFORCING FACTORS OUTSIDE THE SYSTEM

Improvement of all traits simultaneously is difficult; it is possible to do so, however, if the size of the cake (Figure 7.23) can be enlarged. Figure 7.23 identifies two possibilities. The first is to increase the thickness of the cake, and the second is to enlarge the radius of the cake. The thickness of the cake represents the environmental capacity. It is a common observation that most, if not all, traits can be enhanced in a better environment with irrigation, plenty of mineral nutrient supply, plenty of sunshine, effective measures to control diseases and pests, etc. This is the agronomic approach to enlarging the cake. The importance of this approach is repeatedly manifested in multi-environment trials, as the environment main effect is always a predominant source of yield variation. To enlarge the cake through improving growing conditions is, in most cases, beyond human power.

As breeders, we are more interested in enlarging the cake genetically. The cake's radius is hypothesized to represent genetic factors that can enlarge the cake. A big question, however, is, "What in the world are they?" It seems that resistance to various biotic stresses and tolerance to various abiotic stresses are the fundamental reasons to enlarge the cake and thereby improve all system components simultaneously or to improve some of the components without adversely affecting others. Disease resistance has been, and will continue to be, one of the major aspects in cultivar improvement in all crops. Tolerance to abiotic factors has been recognized as the major cause for the improvement of maize (Tollenaar, 1989; Duvick, 1996; Troyer, 1996) and other crops.

7.8 THREE-MODE PRINCIPAL COMPONENT ANALYSIS AND VISUALIZATION

With regard to biplot display of genotype-by-trait data, it is relevant to briefly mention a variant of biplot called joint plot, which is generated by three-way principal component analysis (PCA) (Kroonenberg, 1983). Detailed description of this methodology is beyond the scope of this book, but interested readers may refer to Basford et al. (1991). Recent applications of this method in genotype-by-environment-by-trait three-way analysis include de la Vega and Chapman (2001), Chapman et al. (1997), and Basford et al. (1996). In these publications, three-way PCA was conducted for three-mode genotype-by-environment-by-trait data. Genotype-by-trait biplots or joint plots were presented for each of the first two environmental principal components. The genotype-by-trait biplot for the first environment principal component (PC1) is identical to the genotype-by-trait biplots discussed in this chapter, when the trait values for each genotype were obtained by averaging across all environments. The genotype-by-trait joint plot for the second environment, principal component (PC2) shows different genotype-by-trait relations for different environments or environment groups.

A possible approach to replacing three-way PCA with two-way PCA is to first study the genotype-by-environment patterns of the most important trait, e.g., yield, using the GGE biplot methodology. If there are no clear-cut genotype-by-environment patterns, a genotype-by-trait biplot based on values across all environments should suffice. If there are clear-cut clusters of environments, a genotype-by-trait biplot should be constructed and studied for each cluster of environments.

8 QTL Identification Using GGEbiplot

SUMMARY

A North American two-row barley mapping dataset was used to demonstrate quantitative trait loci (QTL) mapping with genotype main effect and genotype–by-environment (GGE) biplot. We demonstrate that biplots can give correct clusters of markers that are known to be closely linked and that they provide similar linear arrangements of markers within clusters of markers in each of the seven barley chromosomes. We also demonstrate that GGEbiplot provides a convenient means for identifying QTL for a target trait. QTL mapping using a biplot takes a different approach from interval mapping but achieves the same results and is much simpler. Besides, the GGE biplot approach provides information about the trait and marker values of individuals in the mapping population, which directly leads to marker-based selection. The biplot pattern also allows missing marker data to be replaced with considerable accuracy. Finally, QTL-by-environment interaction and marker-based selection are discussed.

8.1 WHY BIPLOT?

QTL represents the location of a gene that affects a quantitative trait. QTL mapping relates genetic markers to a target trait for selection based on associated markers, or helps isolate the genes that are responsible for the target trait. There are well-established strategies for QTL mapping (Tinker, 1996). The simplest method to assess the association of a marker with a trait is to test if genotypes with one marker allele are statistically different from those with a different marker allele for the target trait. This can be done by a t-test or simple linear regression. A more sophisticated and widely used method is internal mapping. The principle behind interval mapping is to test a model for the presence of a QTL at many positions between two mapped marker loci. The model is fit, and the goodness of fit is tested using the maximum likelihood method. Models are evaluated by computing the likelihoods of the observed distributions with and without fitting a QTL effect. The goodness of fit is measured via LR or LOD, which is the LN or LOG of the ratio between the two likelihoods. Alternatively and more simplistically, phenotypes can be regressed on the flanking markers to test the goodness of fit. This method is called simple interval mapping since it fits and tests QTL one at a time and does not consider possible existence of QTL at other locations. A more sophisticated method, called composite interval mapping, handles interval mapping by including partial regression coefficients from markers in other regions of the genome. Tinker (1996) provides a good description of the various QTL-mapping methods.

All available QTL-mapping methods treat the individual plants or lines in the mapping population as random samples. This is justified if the sole purpose is to do QTL mapping. For breeders, however, each of the lines in the mapping population is unique, and knowing their genotype and characteristics is equally, if not more, important. A method that can identify QTL for a target trait and display the genotypic information of the individual lines is lacking.

From the previous chapters, we have learned that a GGE biplot of an entry-tester two-way table displays: 1) the interrelationship among testers, e.g., environments or traits; 2) the interrelationship among entries or genotypes; and 3) the interaction between entries and testers. Now imagine that the testers consist of genetic markers and traits. Displaying the interrelationship among markers and traits is, in essence, a matter of QTL mapping. Displaying the interactions between entries and testers reflects trait levels and marker values of the genotypes. We show in this chapter that a biplot provides an effective means for QTL mapping and for marker-based selection.

8.2 DATA SOURCE AND MODEL

8.2.1 THE DATA

The data that will be used to illustrate QTL mapping using GGEbiplot is from a North American barley genome project (Kasha et al., 1995), which was a joint effort of many barley scientists across North America. A cross was made between *Harrington*, an important malting barley cultivar in North America, and *TR306*, a line poor in malting quality but with other merits. A population of 150 random doubled-haploid (DH) lines was produced from F_1 using the *Hordeum bulbosum* method (Kasha and Kao, 1970). The population was first mapped using over 200 genetic markers (Kasha et al., 1995), and later 127 markers were chosen to generate a more robust map (Tinker et al., 1996). Agronomic trait data were obtained from the DH lines grown in 28 year–location combinations across North America during 1992 and 1993. Many agronomic and quality traits were measured from six environments. These traits were: yield, heading date (HEADINGD), maturity, plant height, kernel weight (KW), lodging score, powdery mildew scores (PM-group), seed plumpness (PLM), test weight (TSWT), plump kernel weight (PKW), protein content (PRO), soluble protein (PSL), total protein (PST), β-glucan content, viscosity, amylase activity (AMY), diastatic power, fine extraction (XF1), extraction at 70°C (X70), and difference between the two extraction methods. Dr. Nick Tinker, Agriculture Canada at Ottawa, graciously provided the dataset used here. The dataset contains 145 DH lines, 127 genetic markers, and 22 traits.

8.2.2 THE MODEL

The model for QTL mapping is exactly the same as that shown for studying genotype-by-trait data in Chapter 7, i.e., the within-tester standard deviation-standardized model (Equation 4.14):

$$(\hat{Y}_{ij} - \mu - \beta_j)/d_j = g_{i1}e_{1j} + g_{i2}e_{2j} + \varepsilon_{ij}.$$

All parameters are defined in the same way as in Chapter 4 except that trait or marker replaces the word environment. The GGEbiplot substitutes means of the trait or of the marker in question for missing values.

8.3 GROUPING OF LINKED MARKERS

GGEbiplot reads data, fits the model, and displays the biplot of principal component 1 (PC1) vs. principal component 2 (PC2) – all in a few seconds. For clarity, Figure 8.1 displays only the testers, i.e., markers + traits; the entries or DH lines are hidden. Although only 7 + 6 = 13% of the total variation is explained by this biplot, its ability of clustering markers or traits is apparent. We see two clusters of markers on chromosome 1 (preceded by "1"), two clusters of markers on chromosome 7 (preceded by "7"), one cluster of markers on chromosome 6 on the right, and one cluster of markers on chromosome 5 on the left. It is also apparent that some quality traits — PSL, PST, X70, AMY, XF1 — are clustered together in the left part of the biplot, adjacent to a group of chromosome markers. Some other traits — PLM, PRO, TSWT — fell in a cluster in the right part. HEADINGD is located close to a cluster of chromosome 7 markers. Later, we will show that there is a QTL for heading date in that chromosomal region. Traits and markers with relatively long vectors are displayed in the form of a linear map (Figure 8.1), with the scale of degrees. Markers or traits, among or within clusters, are more discernible in the linear map, but one should bear in

QTL Identification Using GGEbiplot

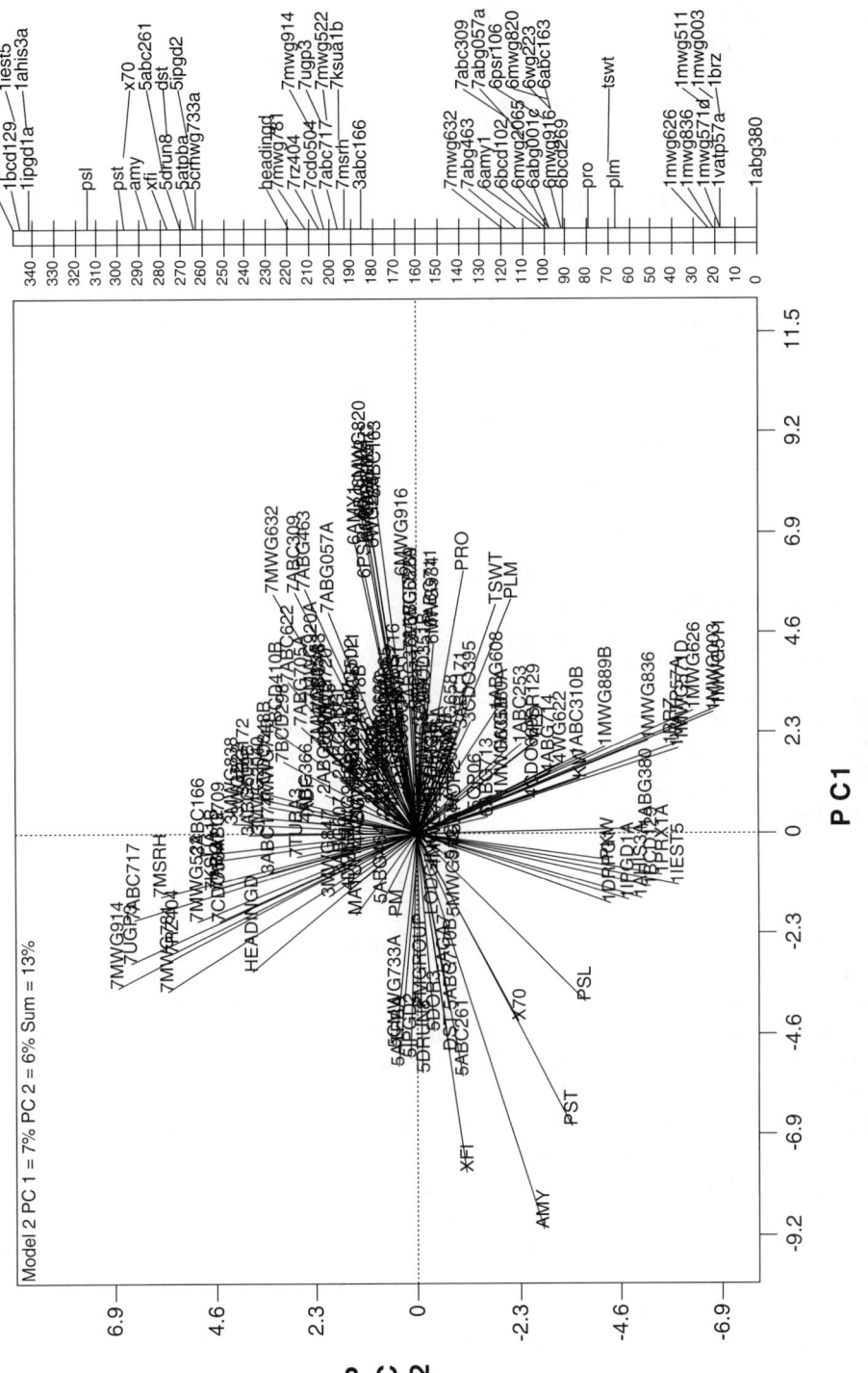

FIGURE 8.1 The biplot of PC1 vs. PC2 based on the North American barley genomics project data. Genotype labels are hidden for clarity. Markers are preceded by the chromosome number with which they are associated.

mind the simple geometric rule that angles are measured from 0 to 360°, and 360° equals 0°. For example, the two groups of markers on chromosome 1 are on the two ends of the linear bar. This indicates that they are close to, rather than remote from, each other.

Traits and markers with short vectors are not well represented in the PC1 vs. PC2 biplot, so they are not shown in the linear map. These traits or markers are likely to be represented by other principal components. Indeed, the biplot of principal component 3 (PC3) vs. principal component 4 (PC4) displays additional clusters of markers or traits (Figure 8.2). These include a cluster of markers on chromosome 3 (preceded by "3"), a cluster of markers on chromosome 6, which was also shown in Figure 8.1, a cluster of markers on chromosome 4, which is clearly associated with powdery mildew resistance (PM), PM-group, and maturity. In addition, the association of KW, PKW, and PRO with a group of markers on chromosome 7 is obvious. Later, we show that these associations are indeed indications of QTL for the respective traits. Note that Figure 8.2 explains only 11% of the total variation. Clusters of markers on chromosome 2 are not detected until principal component 5 (PC5) and principal component 6 (PC6) (Figure 8.3) are used. Also shown in Figure 8.3 are clusters of markers on chromosomes 1 and 5.

This preliminary analysis of the entire dataset suggests that GGEbiplot is a promising tool for gene and QTL mapping. This point is further elaborated later in this chapter.

8.4 GENE MAPPING USING BIPLOT

To assess the ability of GGEbiplot for gene mapping, biplot analysis was done for markers on each chromosome separately (Figures 8.4 to 8.10) and compared with the standard maps based on standard mapping methods (Figure 8.11).

The order of markers on chromosome 6 based on biplot (Figure 8.9) is almost exactly the same as that based on the standard method (Figure 8.11). The only difference is that marker WG223 was misplaced between *bcd102* and *abc106*. For all other chromosomes, except chromosome 2, the biplot correctly identified the apparently independent clusters of markers on each chromosome, and the order of markers within each cluster has good agreement between the two methods. Thus, although the biplots cannot be used for accurate gene mapping, the methodology can be applied to identify gene clusters with considerable confidence. This utility of GGEbiplot may be valuable as an initial attempt at gene mapping, since it uses only a few seconds.

The poorest match is for chromosome 2 (Figure 8.4 vs. Figure 8.11). The two maps only partially agree. Considering the reasonably good agreement for the other chromosomes, this discrepancy was unexpected. One reason may be that the marker density on this chromosome is relatively sparse as compared with other chromosomes.

The biplot approach has its own advantages, however. In addition to showing the order of markers, it shows the marker combinations of the DH lines. Take, for example, the biplot for chromosome 6 (Figure 8.9). The DH lines are arranged in the shape of an oval, with its four projections labeled 1, 2, 3, and 4. The DH lines around projection 1, with all markers on its side of the biplot origin, should have a value of 1 for all markers. That is, they should have the allele of the first parent, *Harrington*, for all markers. On the other side of the biplot origin, the DH lines around projection 3 should have the allele of the second parent, *TR306*, for all markers. These can be verified from Table 8.1. The DH lines around the oval projection 2 have their alleles from *Harrington* for markers in the lower part of the biplot, but from *TR306* for markers in the upper part of the biplot. The DH lines around the oval projection 4 should be just the opposite of those around projection 2 (Table 8.1). The genotypes of all DH lines relative to chromosome 6 can be roughly visualized. Clockwise, from projection 1 to projection 3, the lines have progressively fewer alleles from *Harrington* and more alleles from *TR306*; from projection 3 to projection 1, the lines have progressively fewer *TR306* alleles and more *Harrington* alleles. DH lines that are apparently off the oval are generated through double crossovers in this cross (Group 5 in Table 8.1).

QTL Identification Using GGEbiplot

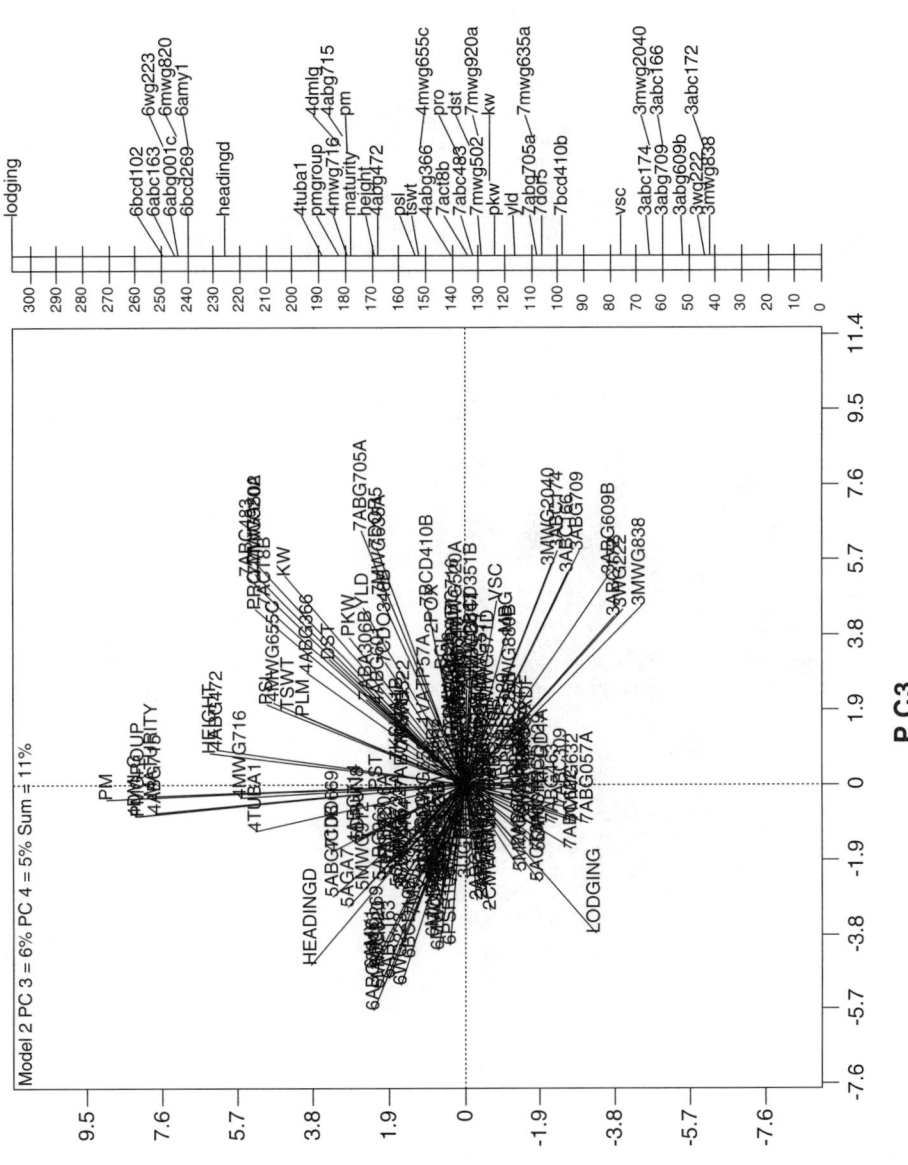

FIGURE 8.2 The biplot of PC3 vs. PC4 based on the North American barley genomics project data. Genotype labels are hidden for clarity. Markers are preceded by the chromosome number with which they are associated.

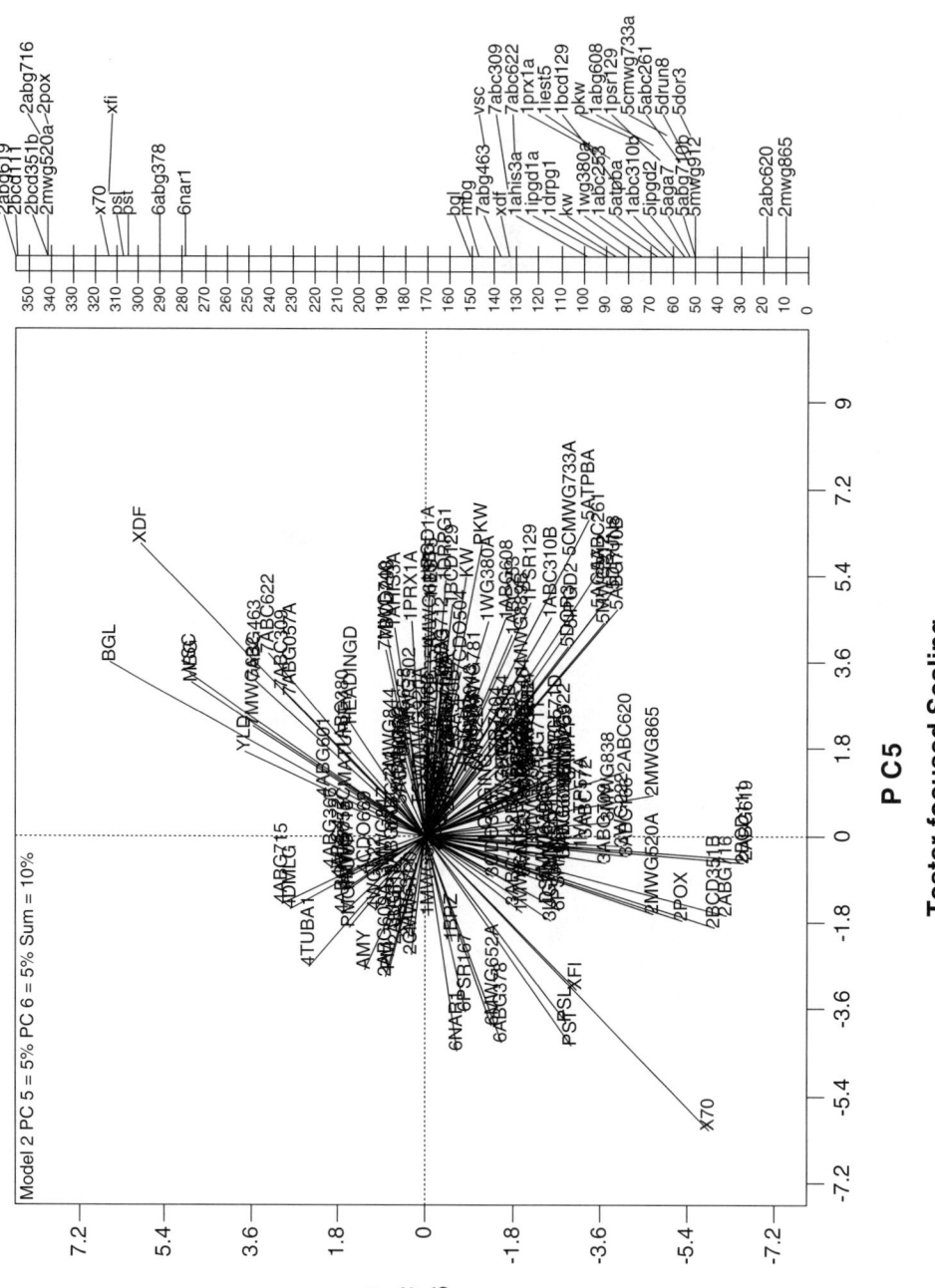

FIGURE 8.3 The biplot of PC5 vs. PC6 based on the North American barley genomics project data. Genotype labels are hidden for clarity. Markers are preceded by the chromosome number with which they are associated.

QTL Identification Using GGEbiplot

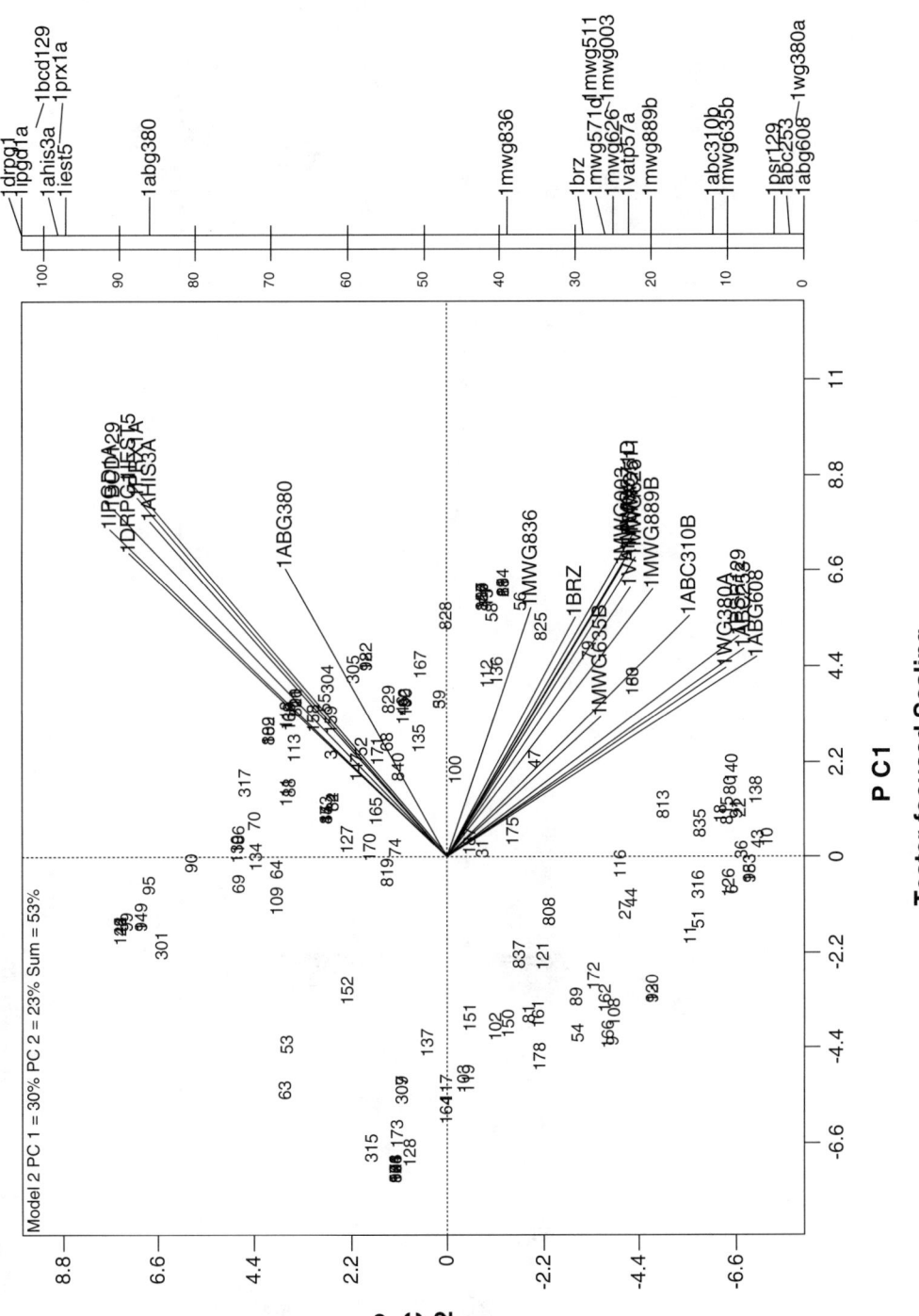

FIGURE 8.4 Biplot for markers on barley chromosome 1. Based on data of the North American barley genomics project.

166 GGE Biplot Analysis: A Graphical Tool for Breeders, Geneticists, and Agronomists

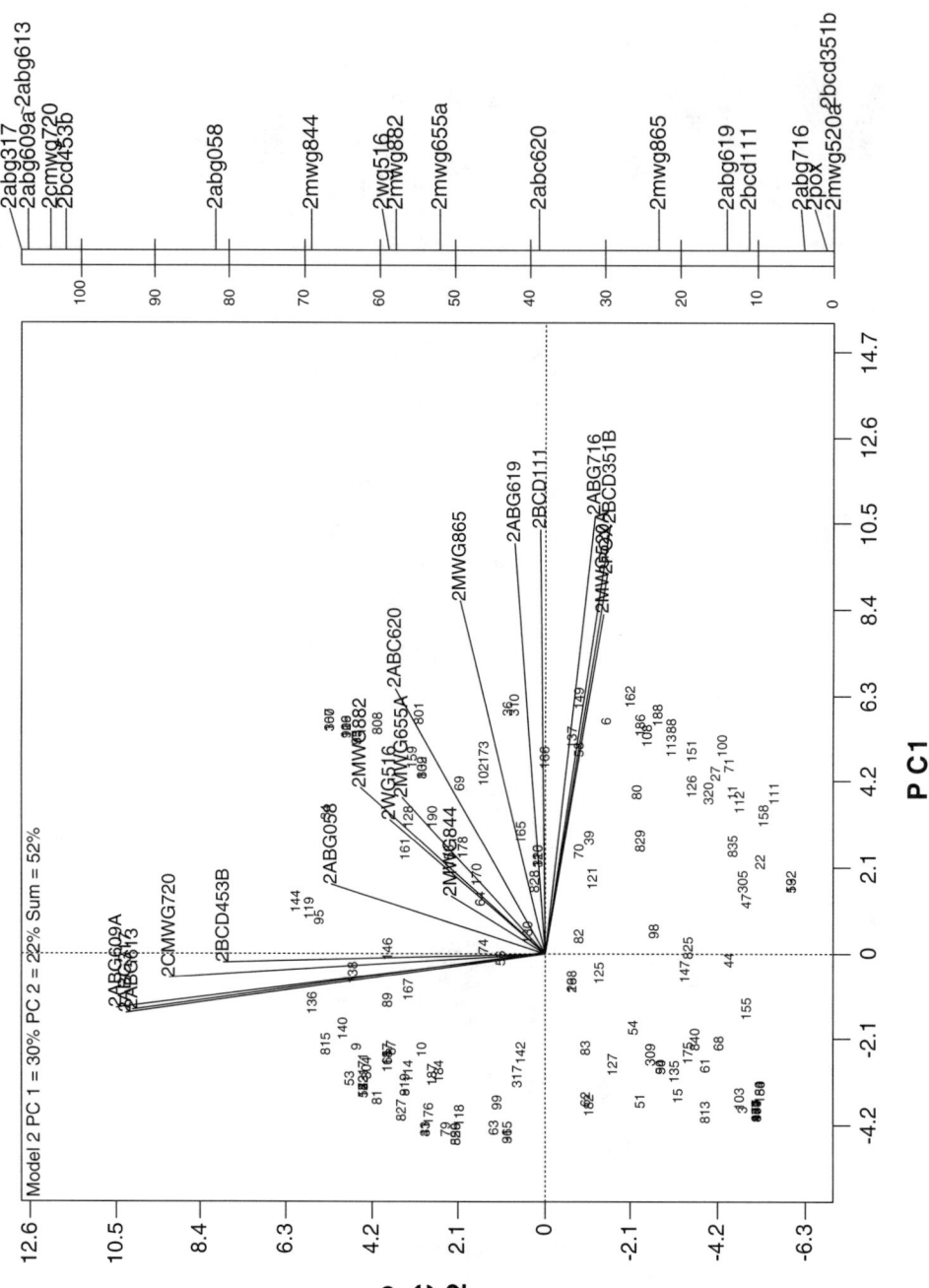

FIGURE 8.5 Biplot for markers on barley chromosome 2. Based on data of the North American barley genomics project.

QTL Identification Using GGEbiplot

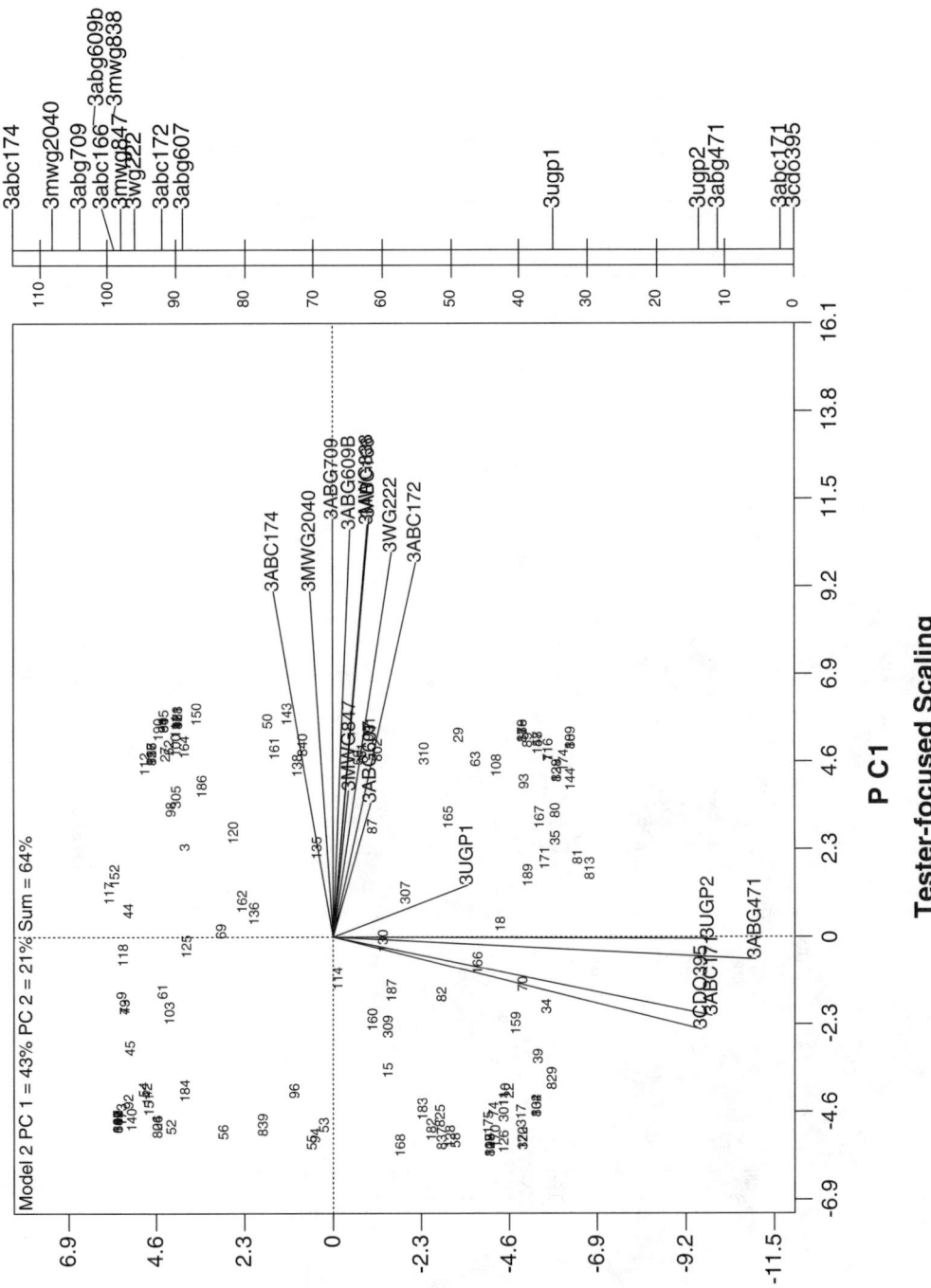

FIGURE 8.6 Biplot for markers on barley chromosome 3. Based on data of the North American barley genomics project.

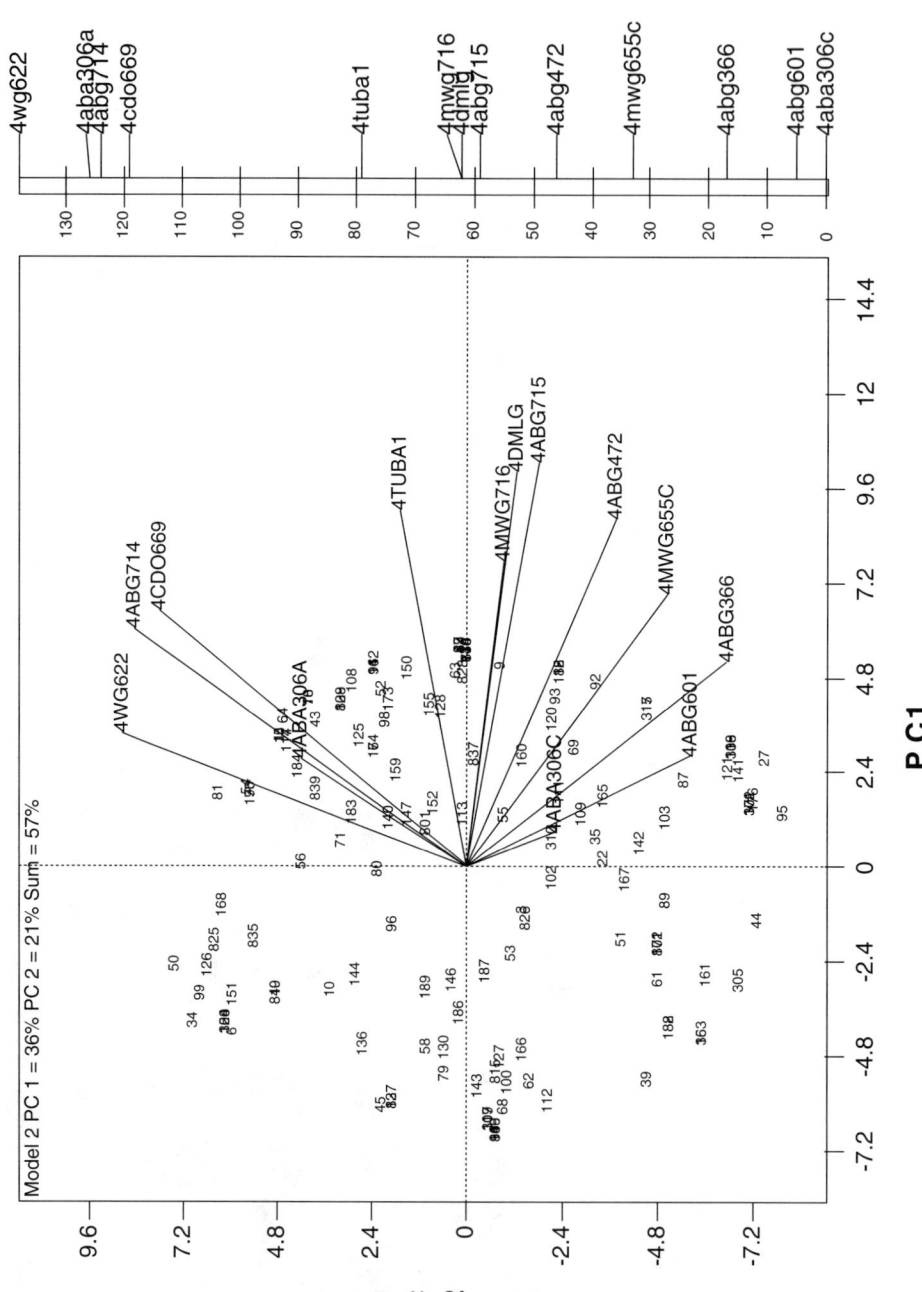

FIGURE 8.7 Biplot for markers on barley chromosome 4. Based on data of the North American barley genomics project.

QTL Identification Using GGEbiplot

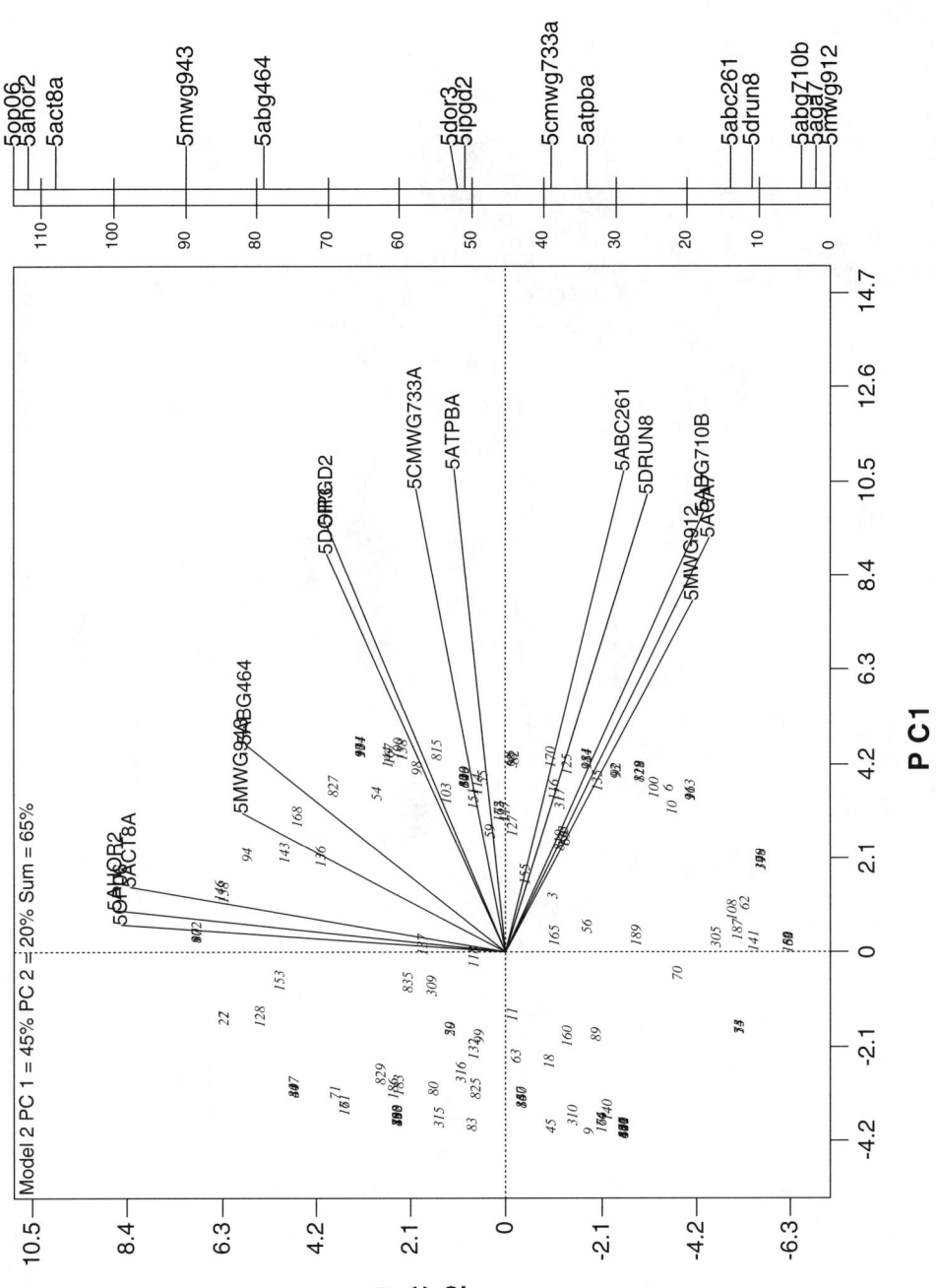

FIGURE 8.8 Biplot for markers on barley chromosome 5. Based on data of the North American barley genomics project.

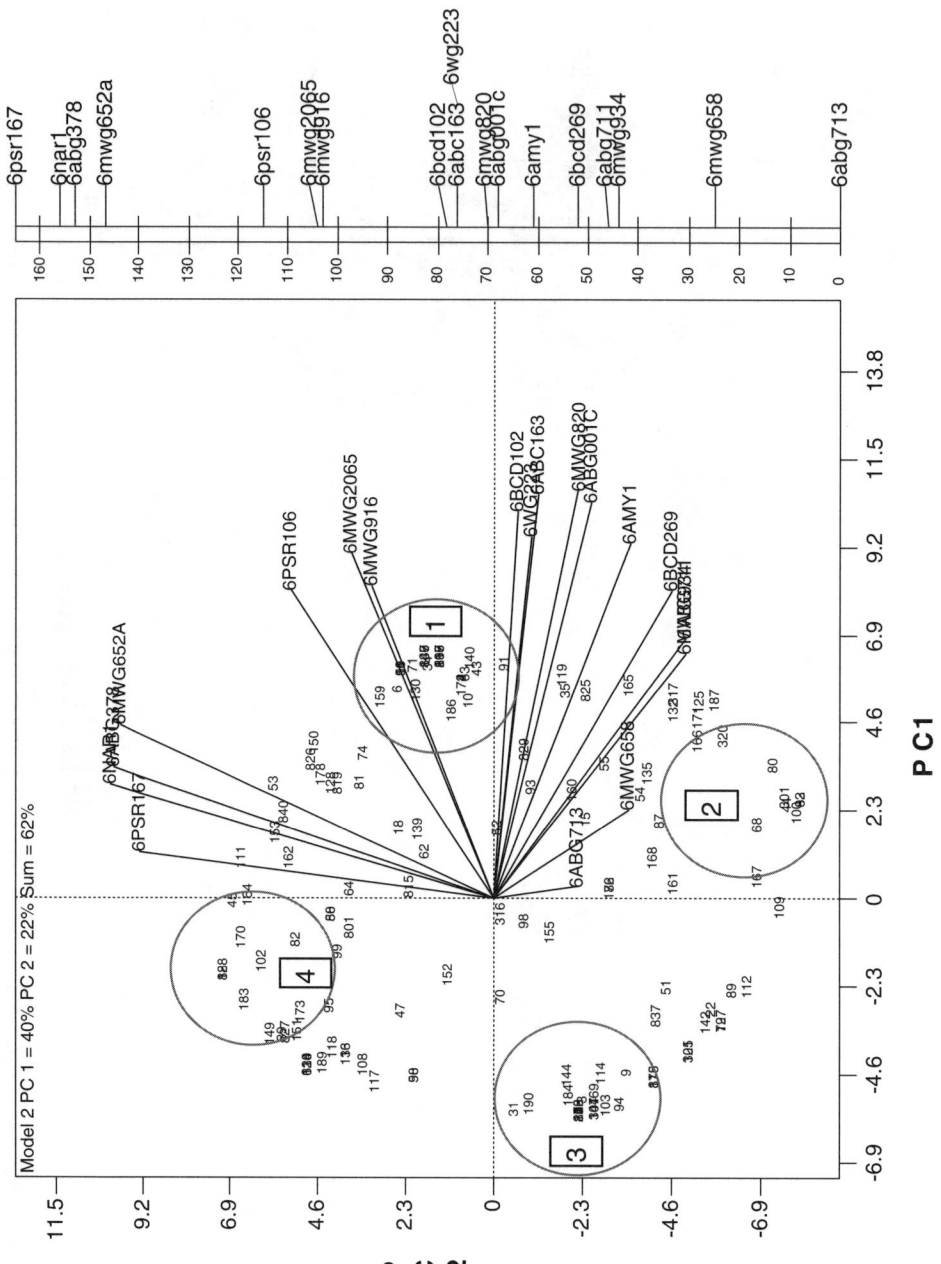

FIGURE 8.9 Biplot for markers on barley chromosome 6. Based on data of the North American barley genomics project.

QTL Identification Using GGEbiplot

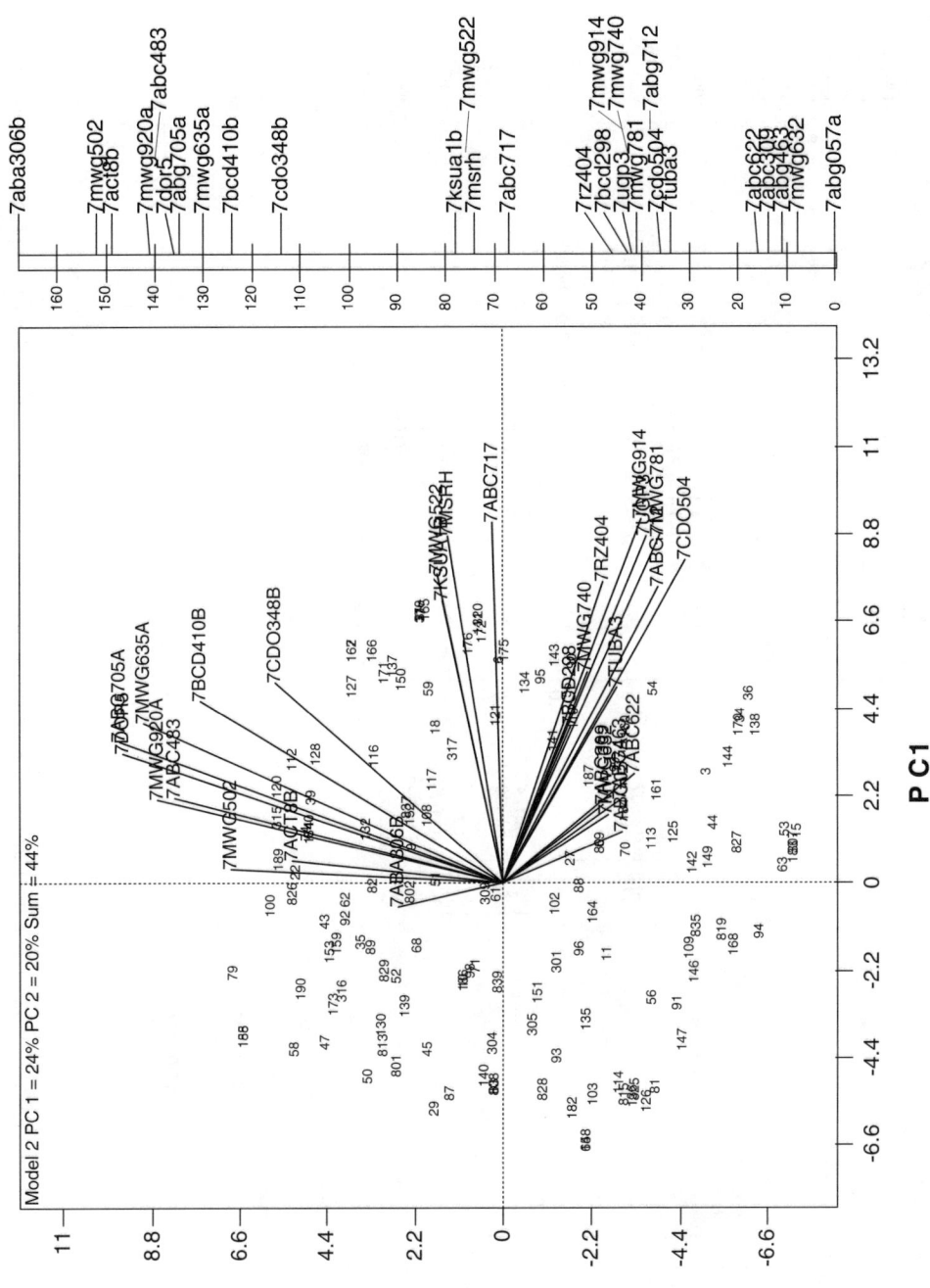

FIGURE 8.10 Biplot for markers on barley chromosome 7. Based on data of the North American barley genomics project.

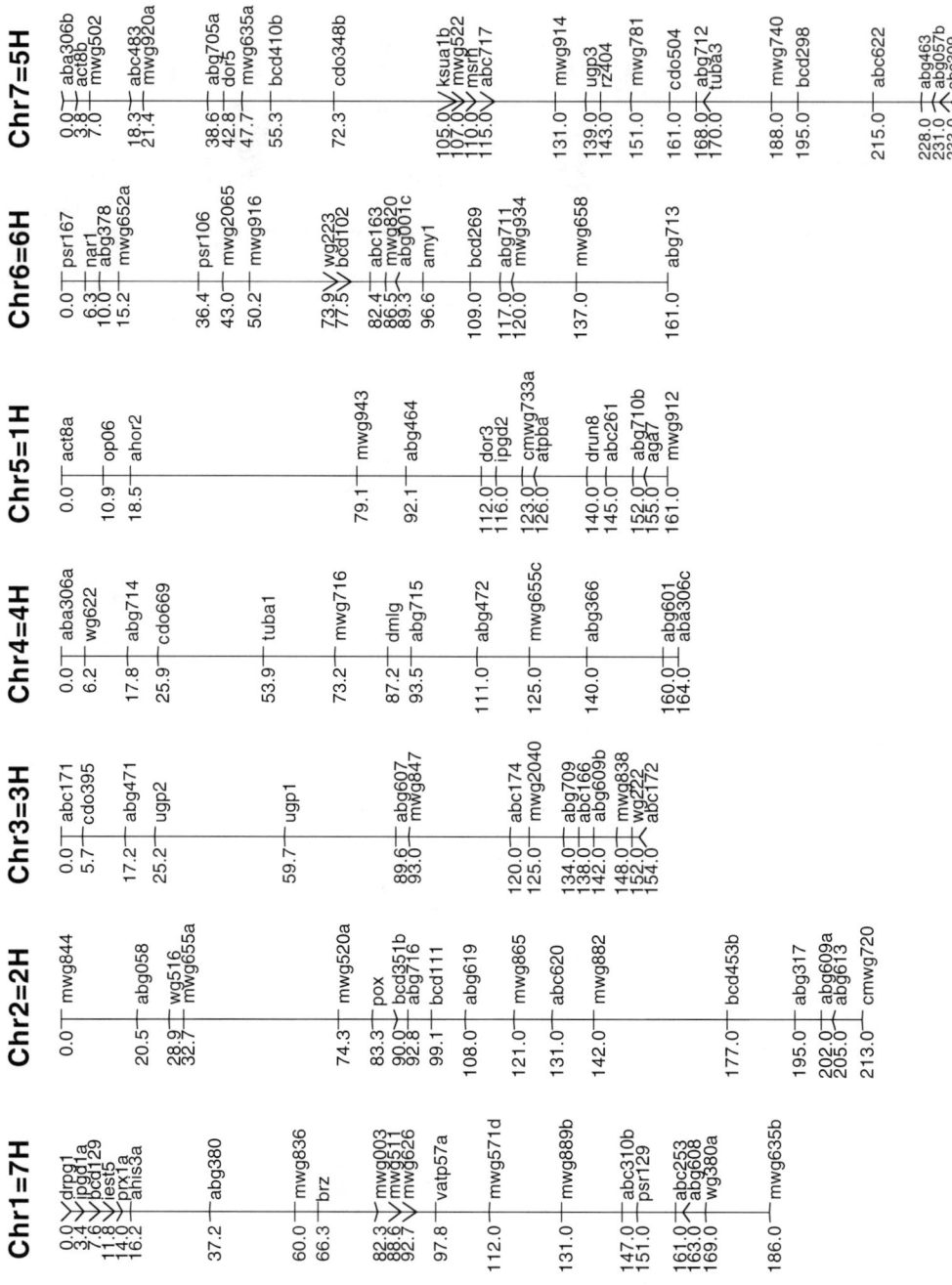

FIGURE 8.11 Barley linkage map based on the *Harington/TR306* cross. (Courtesy of Dr. N. A. Tinker, Agriculture Canada, Ottawa.)

QTL Identification Using GGEbiplot

TABLE 8.1
Groups of Selected DH Lines and their Marker Values on Chromosome 6

Group	Lines	PSR167	Nar1	ABG378	MWG652A	PSR106	MWC2065	MWG916	WG223	BCD102	ABC163	MWG820	ABG001C	Amy1	BCD269	ABG711	MWG934	MWG658	ABG713
1	HT6	1	1	1	1	1	1	1	Missing	1	1	1	1	1	1	1	1	0	0
1	HT10	1	1	Missing	1	1	Missing	1	1	1	1	1	1	Missing	1	1	1	1	1
1	HT11	1	1	1	1	1	1	1	1	1	1	1	1	1	1	1	1	0	0
1	HT29	1	1	1	1	1	1	1	1	1	1	1	1	1	1	1	1	Missing	1
1	HT34	0	1	1	1	1	1	1	1	1	1	1	1	1	1	1	1	Missing	1
1	HT43	0	1	1	1	1	1	1	1	1	1	1	1	1	1	1	1	0	0
1	HT63	1	1	1	1	1	1	1	1	1	1	1	1	1	1	1	1	Missing	0
1	HT71	0	1	1	1	1	1	1	1	1	1	1	1	Missing	1	1	1	1	0
1	HT91	0	1	1	1	1	1	1	1	1	1	1	1	1	1	1	1	0	1
1	HT130	1	1	Missing	1	1	1	1	1	1	1	1	1	Missing	1	1	1	1	0
1	HT159	1	1	1	1	1	1	1	1	Missing	1	1	1	Missing	1	Missing	Missing	Missing	0
1	HT172	1	1	1	1	1	1	0	1	1	1	1	1	Missing	1	1	1	0	1
1	HT186	1	1	Missing	1	1	Missing	1	1	1	1	1	1	1	1	1	1	1	0
2	HT44	0	0	0	0	0	0	1	1	1	1	1	1	1	1	1	1	Missing	1
2	HT68	0	0	0	0	0	0	0	1	1	1	1	1	1	1	1	1	Missing	0
2	HT80	0	0	0	0	0	1	1	1	1	1	1	1	1	1	1	1	1	1
2	HT83	0	0	0	0	0	0	0	1	1	1	1	1	1	1	1	1	1	1
2	HT100	0	0	0	0	0	0	0	Missing	1	1	1	1	1	1	1	1	1	1
2	HT301	0	0	0	0	0	0	0	Missing	0	1	1	1	1	1	1	1	Missing	0
3	HT3	0	0	0	0	0	0	0	0	0	0	0	0	0	0	0	0	0	0
3	HT9	0	0	0	0	0	0	0	0	0	0	0	0	0	0	0	0	1	0
3	HT27	0	0	0	0	0	0	0	0	0	0	0	0	0	0	0	0	0	0
3	HT31	1	0	0	0	0	0	0	Missing	0	0	0	0	0	0	0	0	Missing	0
3	HT69	0	0	0	0	0	0	0	0	0	0	0	0	0	0	0	0	0	0
3	HT94	0	0	0	0	0	0	0	0	0	0	0	0	0	0	0	0	0	0

(continued)

TABLE 8.1 (CONTINUED)
Groups of Selected DH Lines and their Marker Values on Chromosome 6

Group	Lines	PSR167	NAR1	ABG378	MWG652A	PSR106	MWG2065	MWG916	WG223	BCD102	ABC163	MWG820	ABG001C	AMY1	BCD269	ABG711	MWG934	MWG658	ABG713
3	HT103	0	0	0	0	0	0	0	0	0	0	0	0	0	0	0	0	Missing	1
3	HT114	0	0	0	0	0	0	0	1	0	0	0	0	0	0	0	0	0	1
3	HT141	0	0	0	0	0	0	0	0	0	0	0	0	0	0	0	0	0	0
3	HT144	0	0	Missing	0	0	0	0	0	0	0	0	0	Missing	0	Missing	1	0	1
3	HT175	0	0	0	0	0	0	0	0	0	0	0	0	0	0	0	0	0	1
3	HT190	1	0	0	0	1	1	1	0	0	0	0	0	0	0	0	0	0	1
3	HT102	1	1	1	1	1	1	0	Missing	0	0	0	0	0	Missing	0	0	0	0
4	HT170	1	1	1	1	1	1	1	0	0	0	0	0	0	0	0	0	0	1
4	HT183	1	1	1	1	1	1	0	0	0	0	0	0	0	0	0	0	0	0
4	HT188	1	1	1	1	0	0	1	0	0	0	0	0	1	1	0	0	1	0
4	HT52	1	0	1	1	Missing	1	0	0	0	0	1	1	0	0	0	0	0	0
5	HT62	0	0	0	1	1	Missing	1	1	1	1	1	0	Missing	1	1	0	1	0
5	HT98	0	0	Missing	0	0	0	0	0	0	1	0	0	0	0	Missing	1	0	0
5	HT155	0	0	0	0	Missing	1	1	1	1	0	0	1	1	1	0	0	0	0
5	HT316	1	1	1	0	Missing	0	0	0	0	0	0	0	0	0	1	1	0	0
5	HT815	0	0	1	1	Missing	1	1	1	1	1	0	0	0	0	0	0	0	1

8.5 QTL IDENTIFICATION VIA GGEBIPLOT

8.5.1 STRATEGY FOR QTL IDENTIFICATION

We have briefly reviewed the strategies of QTL mapping in the beginning of this chapter. The development of QTL mapping has gone through three stages: simple association between a marker and the target trait, simple map-based interval mapping, and composite map-based interval mapping. The strategy of QTL mapping in the GGEbiplot program involves two steps. It first uses simple regression to screen for markers or traits that are associated with the target trait. Instead of testing the effect of single marker locus on the target trait using t-test, the determination coefficient value is used as a measure of the association between a marker and the target trait. The threshold level of the r-squared value for the candidacy of a marker as a QTL can be calculated based on the number of entries, or it may be specified by the researcher. It is, therefore, technically flexible and practically meaningful. Both gene isolation and marker-based selection are aided by QTL mapping. For both purposes, a QTL must explain a certain proportion of the target trait variation to be practically useful. Therefore, there is advantage to using the r-squared value over the conventional t-test of the marker effect. GGEbiplot allows the researcher to gradually increase or decrease the threshold level so that a manageable number of associated markers is retained.

All markers or traits that survive the screen are then used for biplot analysis. The biplot analysis determines if the markers are linked. An isolated marker suggests the presence of a QTL around it; a group of linked markers suggests the presence of a QTL within it. Once QTL are determined for a trait, the biplot will also display the trait values of the DH lines and their allele values relative to each QTL.

8.5.2 QTL MAPPING OF SELECTED TRAITS

As mentioned earlier, the barley genomics dataset contains 145 DH lines. Based on 2000 permutations, the r-squared value between two arrays of 145 random numbers will not be greater than 2% for 95% of the permutations. Therefore, 2% could be used as a threshold level for the marker-QTL association. From the viewpoint of practical application, however, a QTL that explains less than 5% of the total variation, for example, might not be useful in practice. Therefore, we used 5% as the initial threshold level and raised it gradually as appropriate.

8.5.2.1 Yield

Eighteen of the 127 markers had associations of equal to or greater than 5%, with mean yield (YLD) across all environments (Figure 8.12). Among these, two groups of markers are visible in the biplot of PC1 vs. PC2 (Figure 8.12): one group is near the minus end of chromosome 3; the other is near the plus end of chromosome 7 (Figure 8.11). These two groups of markers are also displayed in the linear bar on the right side of the biplot. The *Harrington* allele of the QTL on chromosome 3 had negative effect on yield, as evidenced by the obtuse angles between the vectors of the markers and that of yield. The *Harrington* allele of the QTL on chromosome 7 had a positive effect on yield, as evidenced by the acute angles between the vectors of the markers and the vector of yield.

Other markers are not shown in the linear bar because their vectors are short. Short vectors do not necessarily suggest that the markers in question are less associated with the target trait. Rather, they indicate that these markers are not well represented in that biplot, and additional principal

176 GGE Biplot Analysis: A Graphical Tool for Breeders, Geneticists, and Agronomists

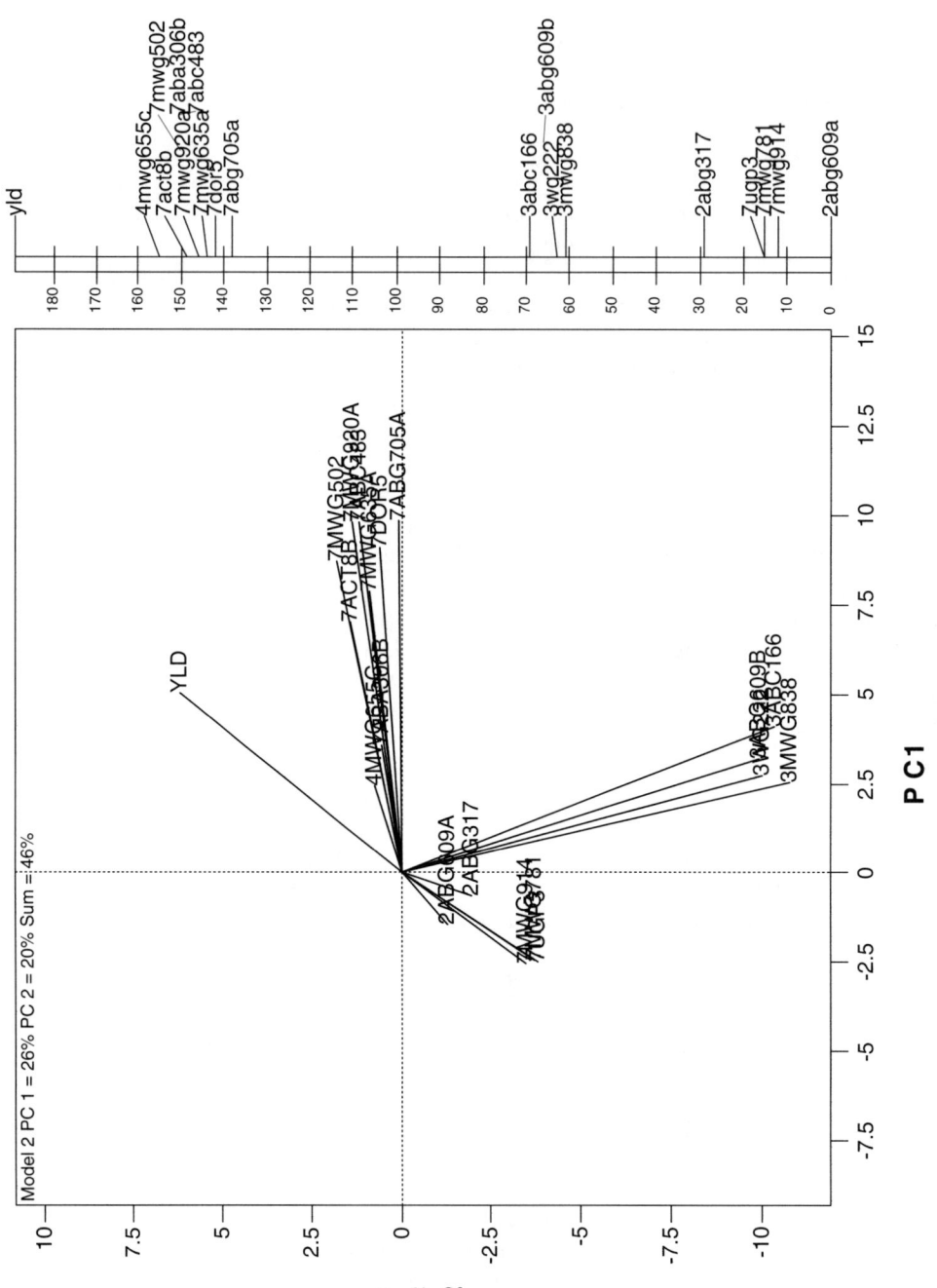

FIGURE 8.12 Markers associated with barley yield, as displayed by the biplot of PC1 vs. PC2.

components are needed to more fully explore the information. It is for this reason that GGEbiplot is designed to generate biplots from principal component axes other than PC1 vs. PC2. Two additional groups of markers become obvious in the biplot of PC3 vs. PC4 (Figure 8.13). These are two markers near the minus end of chromosome 2, and three markers in the middle part of the minus arm of chromosome 7. The alleles of *Harrington* in both regions had negative effects on yield, as their angles with yield were greater than 90°. A single marker from chromosome 4, *mwg655c*, becomes manifest only in the biplot of PC5 vs. PC6 (Figure 8.14). The *Harrington* allele of this marker had a positive effect on yield.

Thus, Figures 8.12 to 8.14 identified five chromosomal regions that possibly harbor a QTL for barley yield at the 5% association level in the *Harrington* × *TR306* mapping population: one in each of chromosomes 2, 3, and 4, and two in chromosome 7. These findings are consistent with the results from simple interval mapping (SIM), except that SIM also identifies possible QTL on chromosomes 5 and 6 (Figure 8.22).

GGEbiplot has a function called *Auto Find QTL*. When this function is invoked, markers with smaller correlation coefficients with the target trait are removed so that all remaining markers or traits are reasonably well represented by the biplot of PC1 and PC2. The method identified only two possible QTL for yield (Figure 8.15). One QTL was near the *mwg655c* marker on chromosome 4, with the *Harrington* allele having a strong positive effect on yield; the other was near the plus end of chromosome 7, around markers *aba306b* and *mwg502*. The *Harrington* alleles also had positive effects. Marker *aba306b* had a short vector because of many missing values. Markers *mwg920a* and *abg705a* are included because they are closely linked to *aba306b* and *mwg502*. These two regions were identified via interval mapping as primary QTL by Tinker et al. (1996). Note that *act8b*, which is located between *aba306b* and *mwg502*, and marker *abc483*, which is located between *mwg502* and *mwg920a* (Figure 8.11), had a smaller correlation with yield. This reveals a potential problem of using simple correlation as a criterion for including or excluding a marker as a candidate of QTL. Nevertheless, it is still a valuable tool for preliminary screening of possible QTL.

The chromosome 4 marker, *mwg655c*, had a near-90° angle with the chromosome 7 markers *mwg920a* and *abg705a*, suggesting that they are, as expected, independently inherited. The angle between *mwg655c* and chromosome 7 markers *aba306b* and *mwg502* was smaller than 90°, suggesting weak positive associations; they all had positive associations with yield.

Also shown in Figure 8.15 are the relationships between yield, KW, and lodging scores. As expected, yield is positively correlated with KW but negatively correlated with lodging. All traits and markers retained in Figure 8.15 had associations with yield equal to or greater than 10% (Table 8.2).

8.5.2.2 Heading Date

Using the *Auto Find QTL* function of GGEbiplot, three possible QTL were detected (Figure 8.16). They were located in the central part of chromosomes 1, 4, and 7. These are corresponding to the three primary QTL found through interval mapping (Figure 8.22). The *Harrington* alleles of the QTL on chromosomes 4 and 7 had positive effects on HEADINGD or later heading, as evidenced by acute angles between the vector of HEADINGD and those of the two groups of markers. The *Harrington* allele of the QTL on chromosome 1 had negative effects on HEADINGD causing earlier heading, as evidenced by the obtuse angles of the HEADINGD vector and those of the markers.

Also shown in Figure 8.16 are the positive associations between HEADINGD, maturity, and PM-group. The latter was located among the chromosome 4 markers. Later, we show that there is a major gene in this region for PM resistance.

178 GGE Biplot Analysis: A Graphical Tool for Breeders, Geneticists, and Agronomists

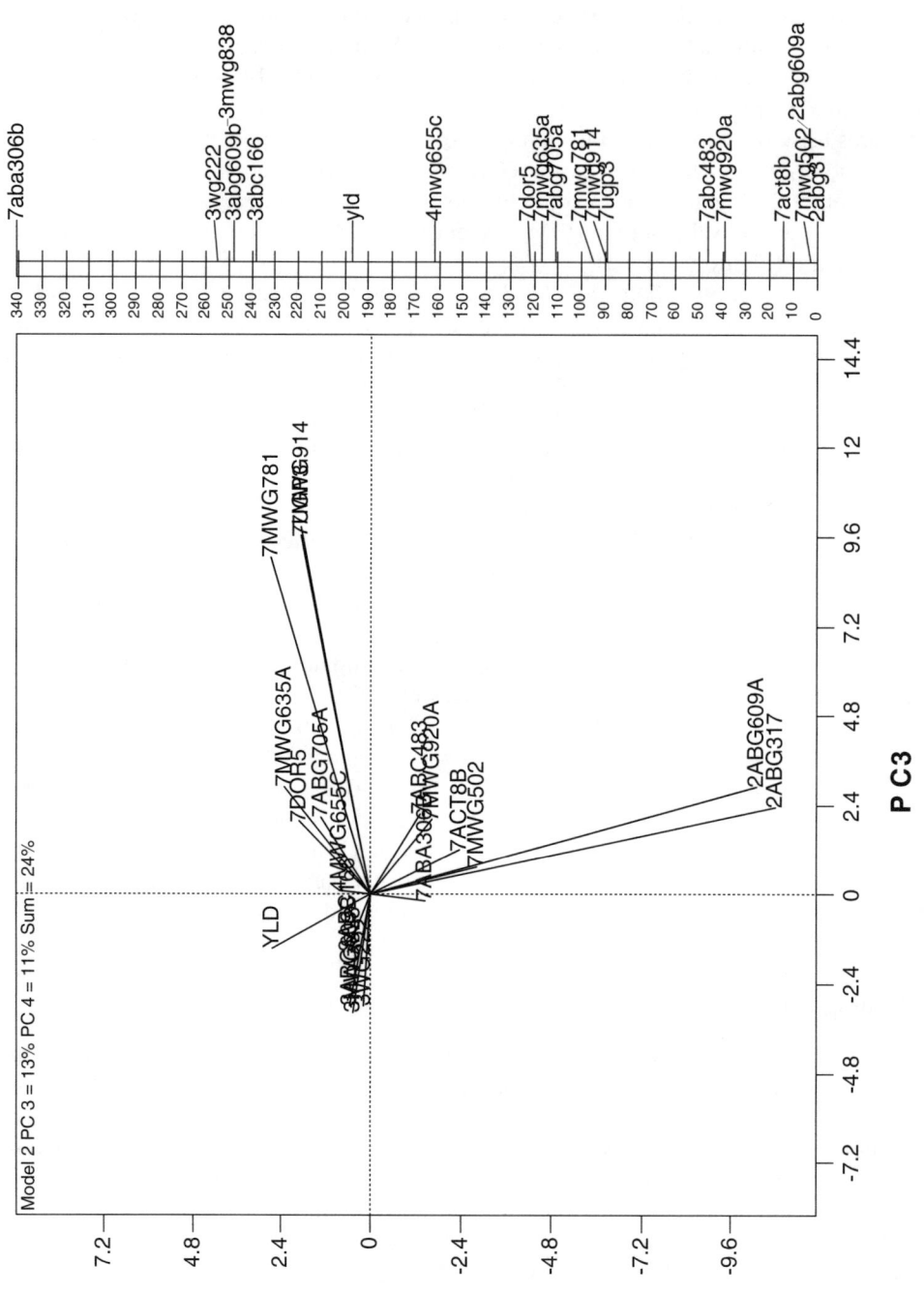

FIGURE 8.13 Markers associated with barley yield, as displayed by the biplot of PC3 vs. PC4.

QTL Identification Using GGEbiplot

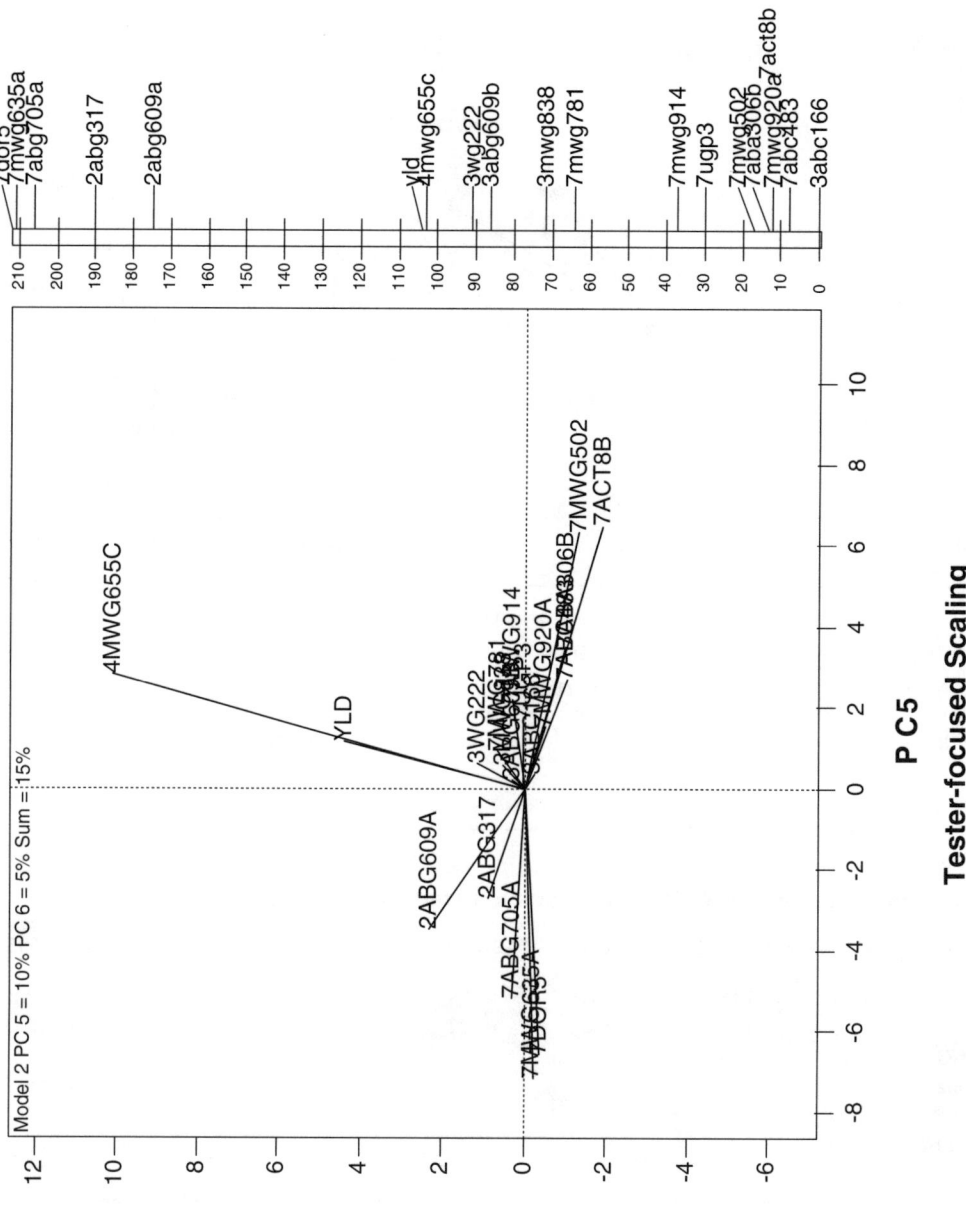

FIGURE 8.14 Markers associated with barley yield, as displayed by the biplot of PC5 vs. PC6.

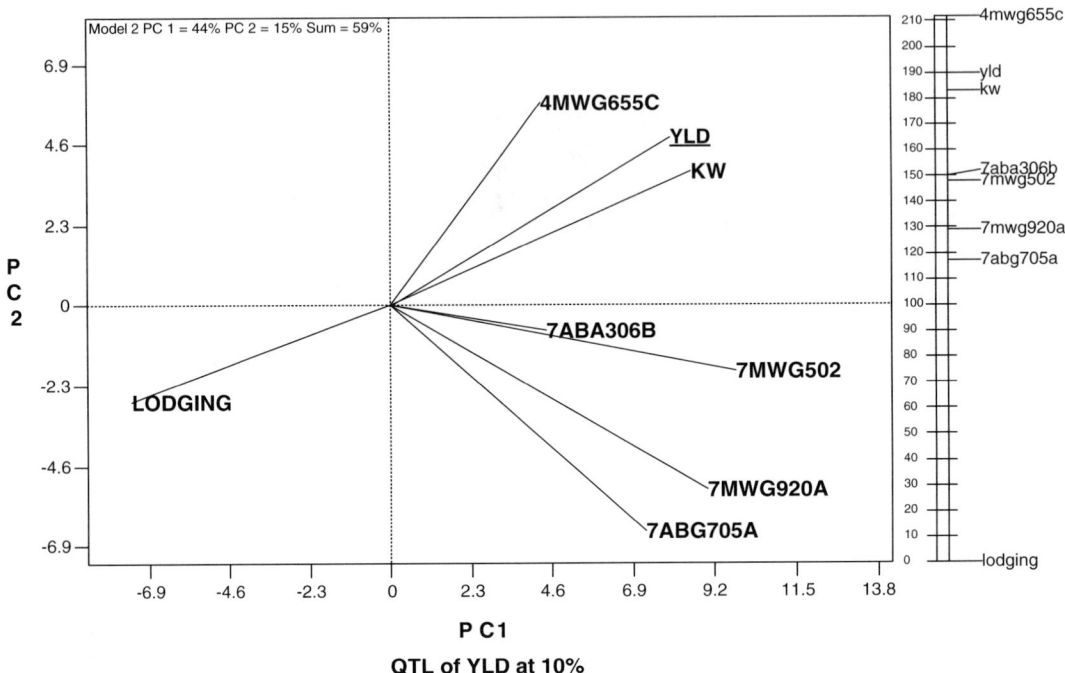

FIGURE 8.15 Markers associated with barley yield at the 10% level or higher.

TABLE 8.2
Correlation Matrix Among Yield and Associated Traits and Markers

Names	KW	LODGING	7MWG502	7ABA306B	7MWG920A	7ABG705A	4MWG655C
YLD	0.490	−0.416	0.347	0.328	0.316	0.309	0.312
KW		−0.318	0.572	0.622	0.362	0.206	0.268
LODGING			−0.436	−0.321	−0.341	−0.246	−0.213
7MWG502				0.878	0.741	0.365	0.188
7ABA306B					0.632	0.290	0.098
7MWG920A						0.667	0.210
7ABG705A							0.106

QTL Identification Using GGEbiplot

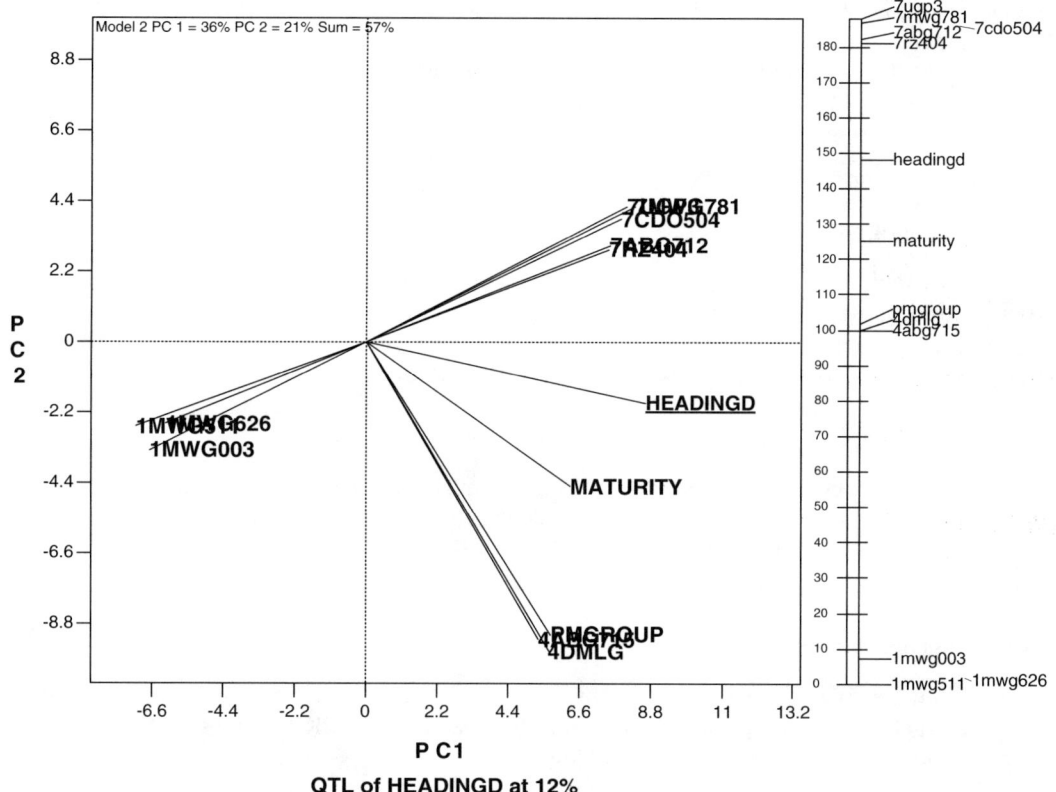

FIGURE 8.16 Three clusters of markers and one trait (maturity) associated with barley heading date at the 12% level or higher.

8.5.2.3 Maturity

Three QTL were detected for maturity: two positive QTL, in terms of the *Harrington* alleles, on chromosomes 4 and 7, and one negative QTL on chromosome 3 (Figure 8.17). Also shown in Figure 8.17 are the positive correlations between maturity and plant height, HEADINGD and PM-group, and the negative correlation between maturity and lodging. These results are fully consistent with those found through interval mapping (Figure 8.22). Comparison of Figures 8.17 and 8.16 reveals that the maturity and heading date share a common QTL on chromosome 4, which may be the reason for the positive correlation between them.

8.5.2.4 Plant Height

Two possible QTL were identified for plant height: one positive QTL, in terms of the *Harrington* alleles, on chromosome 7 and one negative QTL on chromosome 1 (Figure 8.18). These are also the primary QTL identified through interval mapping (Figure 8.22). Also shown in Figure 8.18 are positive associations between plant height, KW, and maturity. The QTL on the plus end of chromosome 7 may be responsible for these associations, since this QTL was identified for all three traits (Figures 8.17, 8.18, and 8.21).

8.5.2.5 Lodging

Two QTL were identified for lodging scores: one on chromosome 2 and the other on chromosome 7, both with negative effects causing less lodging relative to the *Harrington* alleles (Figure 8.19). Lodging score is negatively correlated with yield, KW, maturity, and the QTL on the plus end of chromosome 7 may be responsible for this interconnectedness (Figure 8.19).

8.5.2.6 Test Weight

Two QTL were detected for test weight: one on chromosome 7 and the other on chromosome 6; both had positive effects on test weight from the perspective of the *Harrington* alleles (Figure 8.20). There are positive associations between TSWT, PRO, yield, KW, and PKW, and negative associations between TSWT and lodging, XFI and AMY.

8.5.2.7 Kernel Weight

Two QTL were evident for KW; one on chromosome 7 and the other on chromosome 1 (Figure 8.21). The graph also depicts positive associations between KW and PKW, height, PLM, test weight, and yield; and a negative association with lodging.

8.6 INTERCONNECTEDNESS AMONG TRAITS AND PLEIOTROPIC EFFECTS OF A GIVEN LOCUS

We have noticed that a number of traits are associated with a QTL near the plus end of chromosome 7, represented by markers *act8b* and *mwg502* (Figures 8.15 to 8.21). On the other hand, it is common knowledge that many breeding objectives are interconnected (e.g., Yan and Wallace, 1995; Wallace and Yan, 1998; Chapter 7 of this book). The GGEbiplot function *Auto Find QTL* not only allows detection of markers associated with a target trait but also traits associated with a particular marker. Using this function, several traits are identified as associated with marker *mwg502* (Figure 8.23). The *Harrington* allele of this marker was positively associated with PRO, yield, KW, PKW, and maturity, and negatively associated with XF1 and lodging. The negative associations among yield, earliness or reverse of maturity, and XF1 (Figure 8.24) are a major challenge for brew barley breeding, and a QTL near *mwg502* may be responsible for this. This example indicates that the commonly observed negative correlations among important breeding objectives may have a genetic basis, and GGEbiplot can help reveal this.

QTL Identification Using GGEbiplot

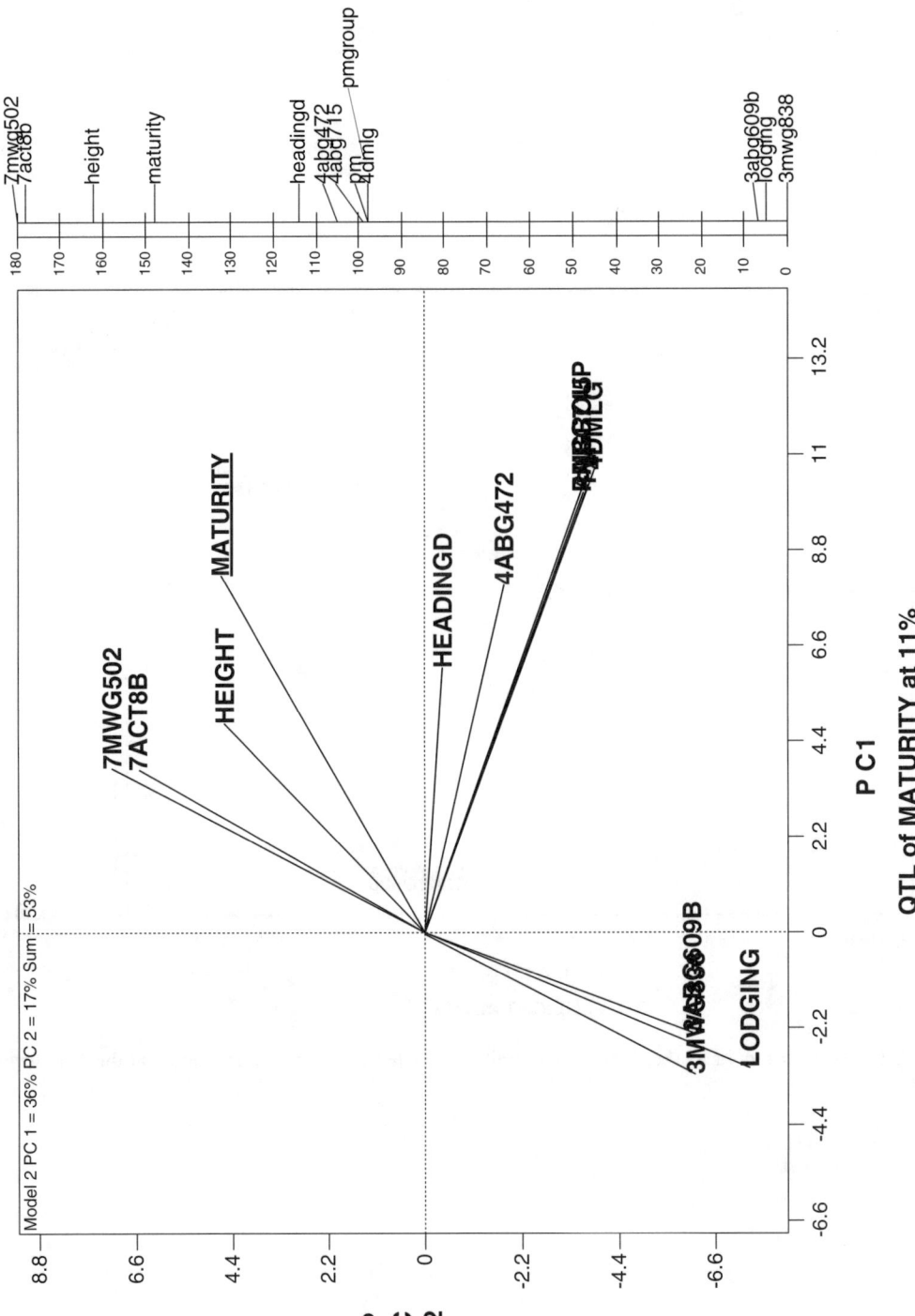

FIGURE 8.17 Three clusters of markers and certain traits associated with barley maturity at the 11% level or higher.

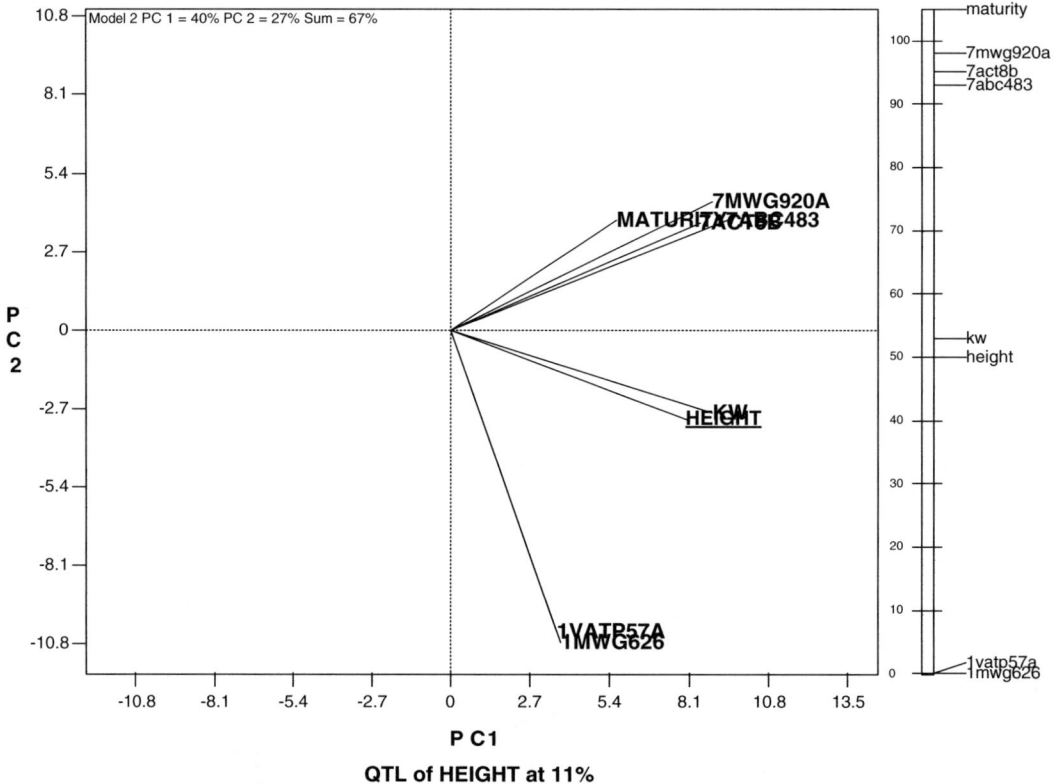

FIGURE 8.18 Two clusters of markers and certain traits associated with barley plant height at the 11% level or higher.

QTL Identification Using GGEbiplot

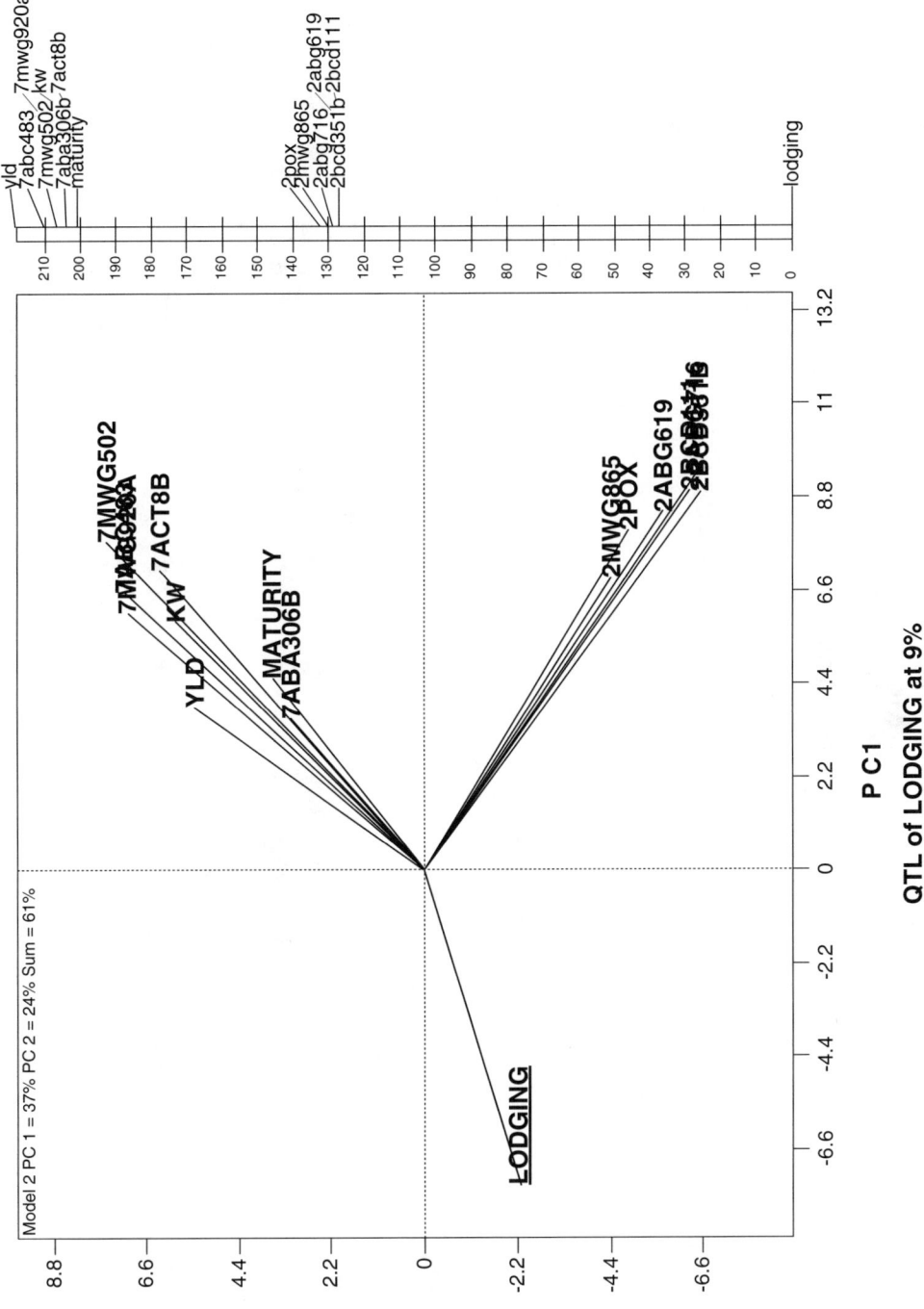

FIGURE 8.19 Three clusters of markers and certain traits associated with barley lodging scores at the 9% level or higher.

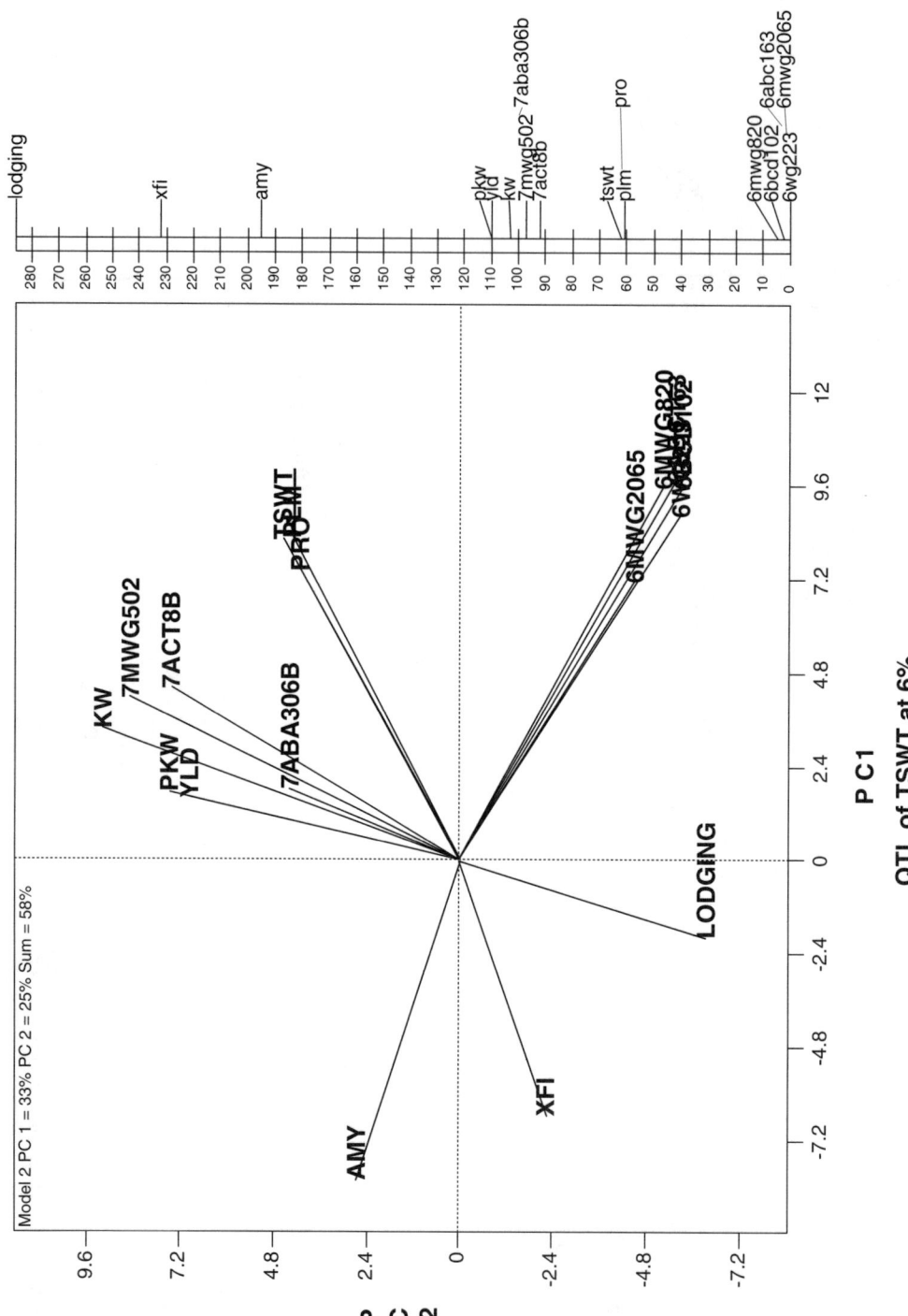

FIGURE 8.20 Two clusters of markers and certain traits associated with barley test weight at the 6% level or higher.

QTL Identification Using GGEbiplot

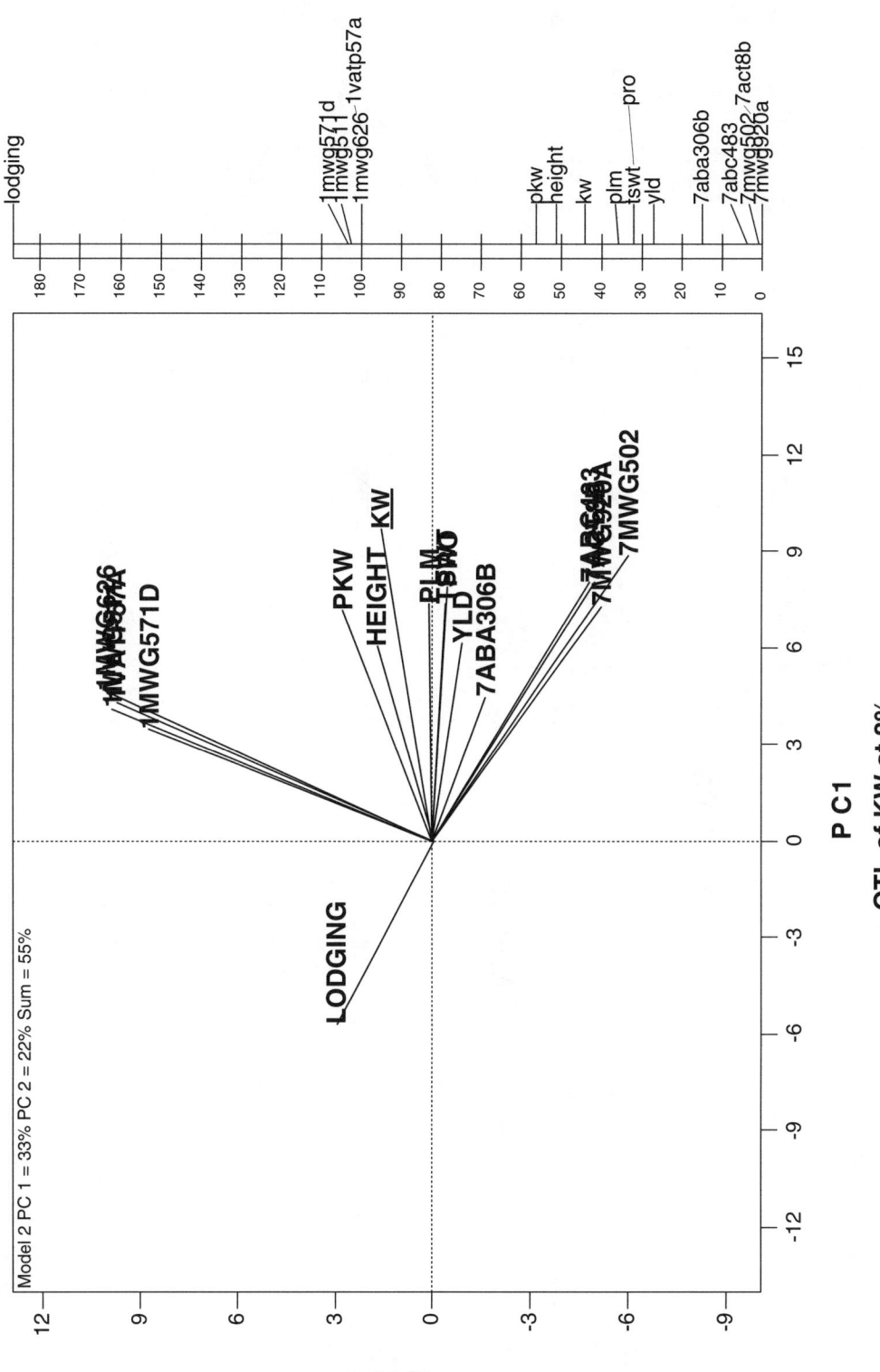

FIGURE 8.21 Two clusters of markers and traits associated with barley KW at the 8% level or higher.

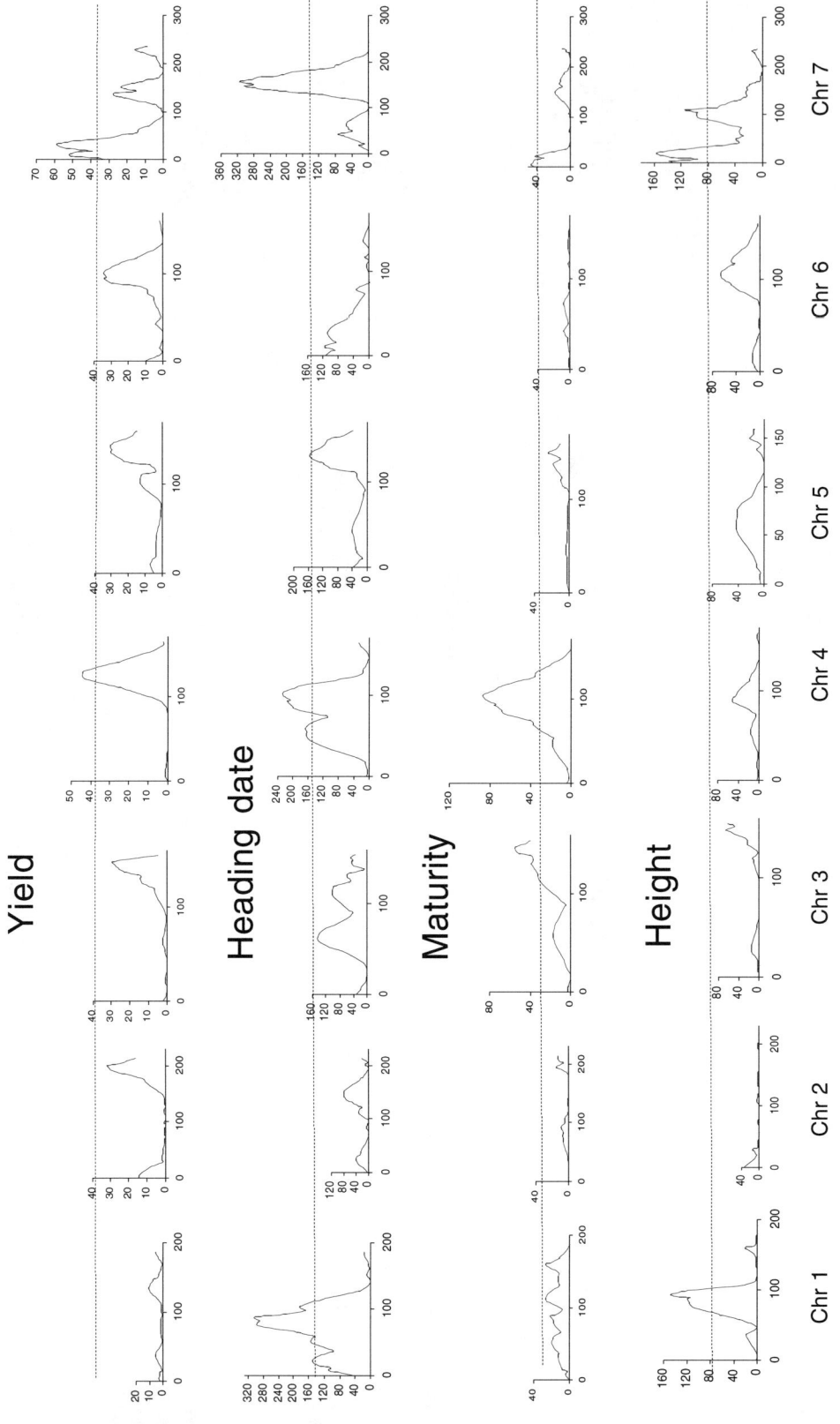

FIGURE 8.22A QTL of seven agronomic traits based on simple interval mapping. (Courtesy of Dr. N. A. Tinker, Agriculture Canada, Ottawa.)

QTL Identification Using GGEbiplot

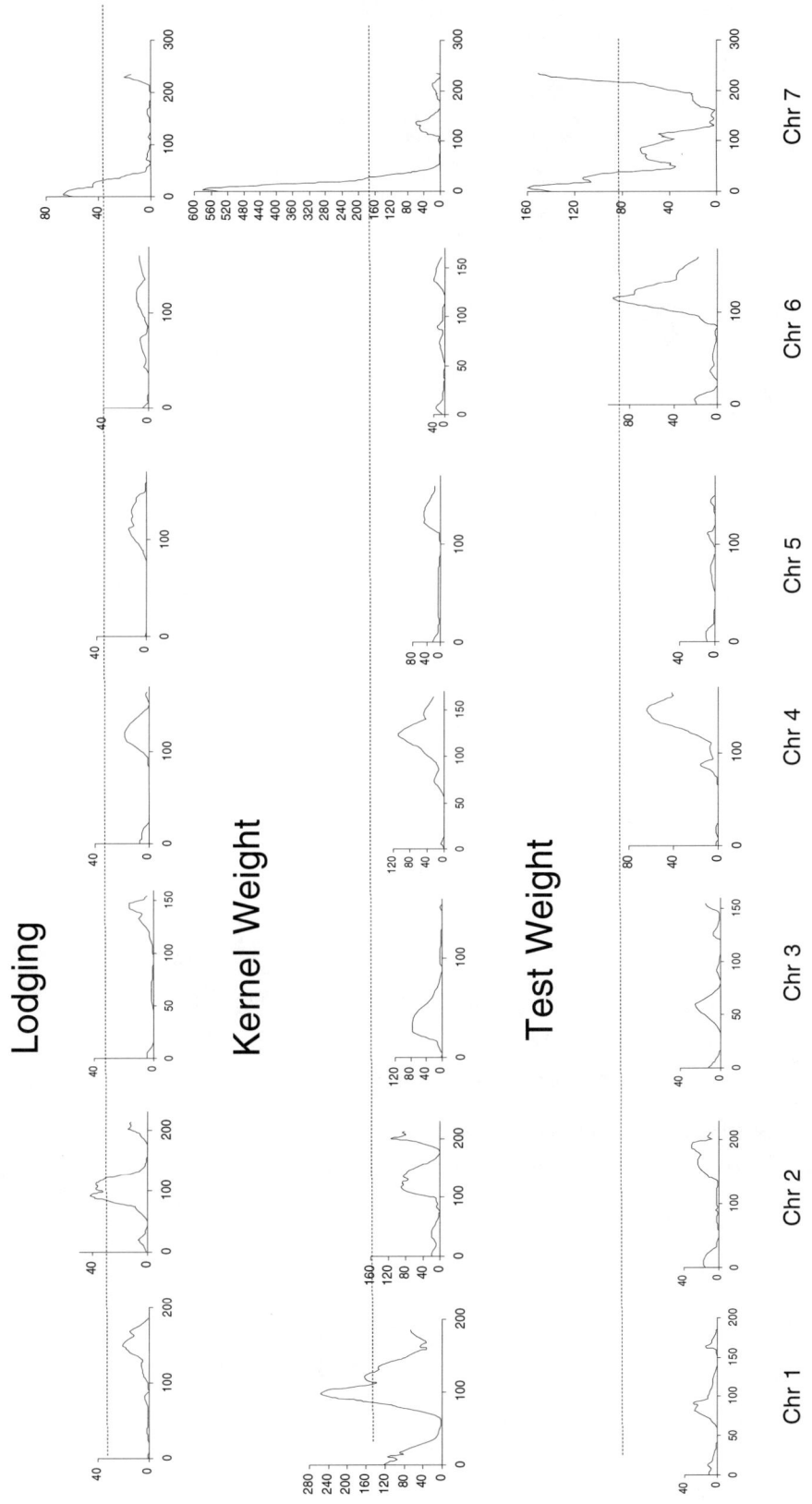

FIGURE 8.22B QTL of seven agronomic traits based on simple interval mapping. (Courtesy of Dr. N. A. Tinker, Agriculture Canada, Ottawa.)

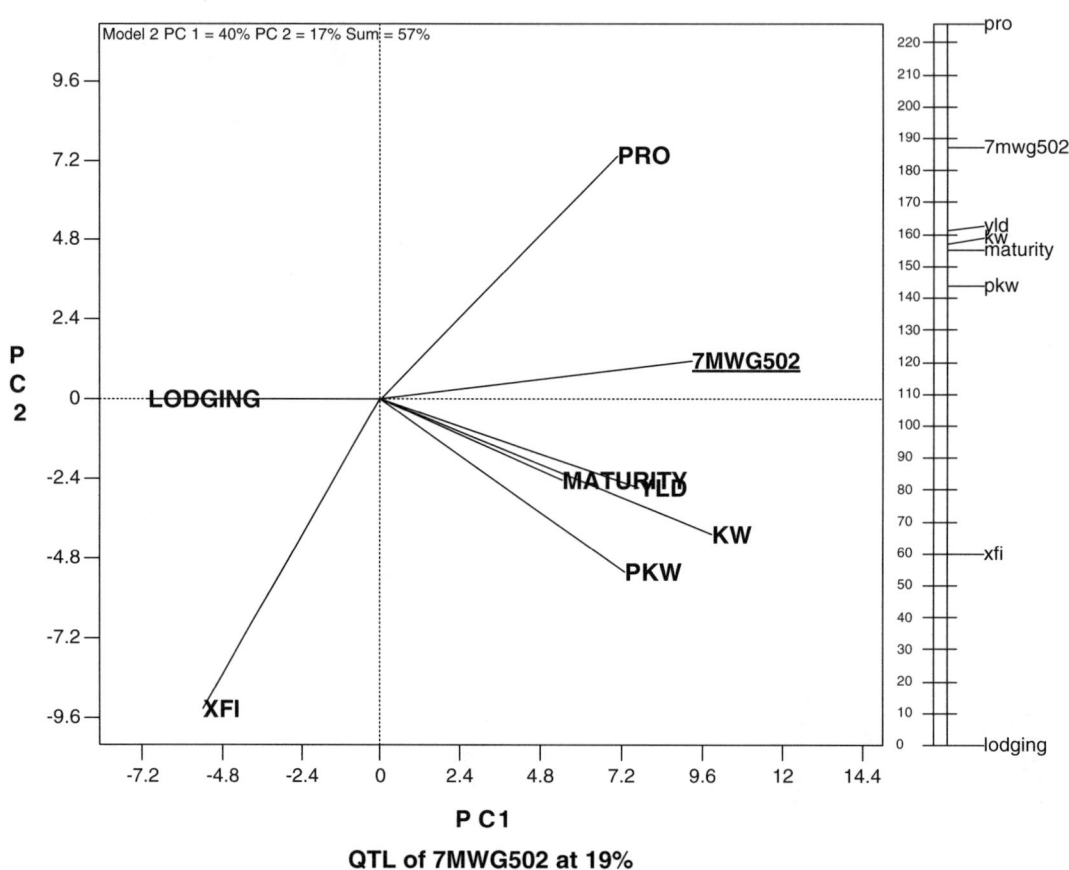

FIGURE 8.23 Traits associated with marker *mwg502* on chromosome 7.

QTL Identification Using GGEbiplot

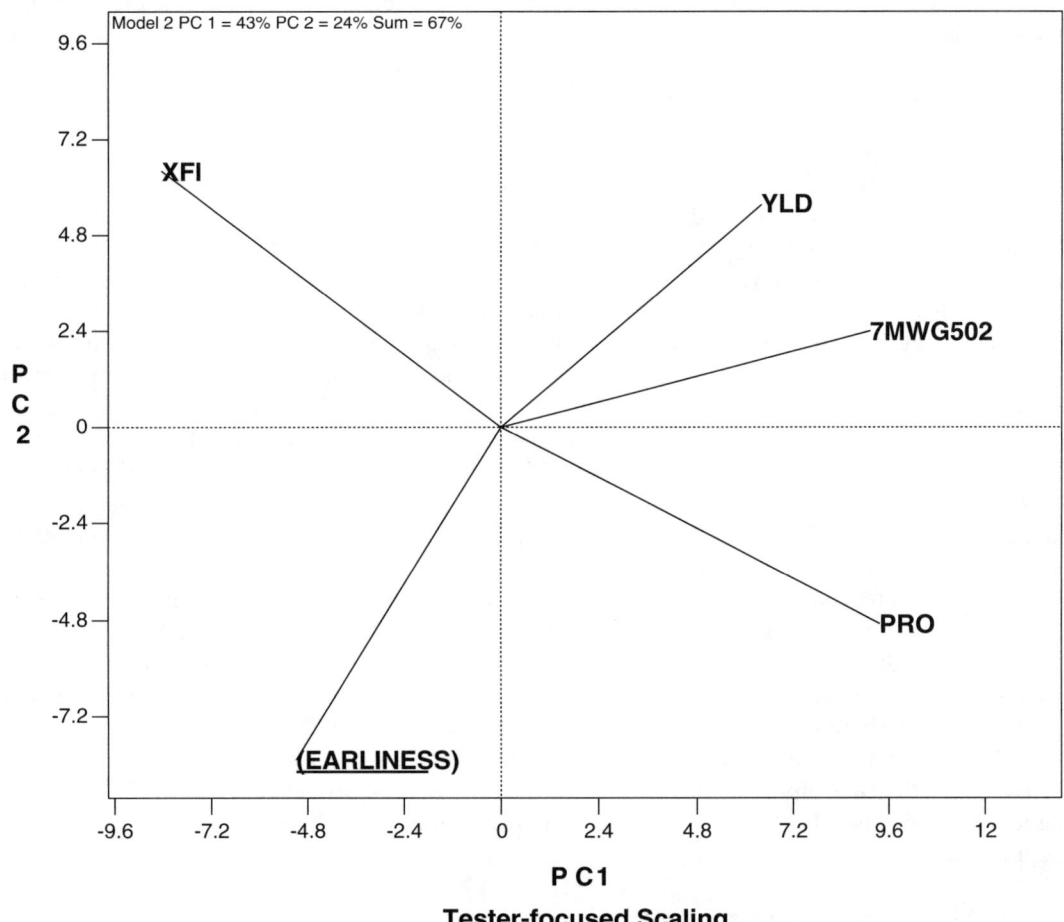

FIGURE 8.24 The association of marker *mwg502* with yield, maturity, and quality.

8.7 UNDERSTANDING DH LINES THROUGH THE BIPLOT PATTERN

In Sections 8.5 and 8.6, our focus was to show the ability of the GGEbiplot in identifying QTL for a target trait and identifying traits influenced by a single genetic marker. Also demonstrated is the superior power of GGEbiplot in studying interrelationships among traits to achieve systems understanding of the interconnectedness among breeding objectives. For that purpose, the DH lines in the biplots were deliberately hidden. As pointed out at the beginning of this chapter, it is, however, equally important to understand the genotypes and characteristics of individuals in a mapping population. In this section, we examine the usefulness of GGEbiplot in this respect using the biplot for KW and its QTL as an example.

8.7.1 Visualizing Marker and Trait Values of the DH Lines

Figure 8.25 is a biplot for KW and two groups of markers on chromosomes 1 and 7 that are associated with KW. This figure is the same as Figure 8.21 except that the traits that are associated with KW are removed and the DH lines are presented. The two groups of markers had an angle of about 90°, indicating independent inheritance; the angles between KW and the two groups of markers, however, are acute angles, indicating that KW is positively influenced by the alleles of *Harrington* in these two regions. What is of great interest is the pattern of the DH lines. They fell into four groups. Group 1 had *Harrington* alleles for both groups of markers; they, therefore, had high KW. Group 3, on the other side of the biplot origin, opposite to group 1, had *TR306* alleles for both groups of markers and had low KW. To support this interpretation, a few DH lines from each group are selected and their marker values and KW are listed in Table 8.3. All lines in Group 1 had a marker value of 1 for all markers and their KW averaged 45.1 g. On the contrary, all lines in Group 3 had a value of 0 for all markers (the *TR306* alleles) and their KW averaged 39.7 g. DH lines in Group 2 had 1s for markers (the *Harrington* alleles) on chromosome 1, but 0s for markers on chromosome 7. Their KW averaged 43.0 g. Lines in Group 4 had 1s for markers on chromosome 7 and 0s for markers on chromosome 1; their KW averaged 41.5 g. Some DH lines did not fall into any of the four groups (Figure 8.25); they are designated as Group 5 (Table 8.3). These are the lines that had crossovers within either chromosome 1 or 7 region (Table 8.3). Line 'ht56' in this latter group had the highest KW (48.2 g), even though there is a crossover between *mwg502* and *abc483*. This indicated that the QTL for KW on chromosome 7 is located closer to *mwg502* than to *abc483*. This example should convince us that the biplot method not only identifies chromosome regions that influence a target trait but also displays the marker and trait values of individual lines of a mapping population.

8.7.2 Marker Nearest to the QTL

We can take a closer look at the markers in each of the two chromosomal regions that influence KW. Figure 8.26 illustrates the interrelationships between KW and markers on chromosome 7 only. First, the order of the five markers is exactly the same as obtained via standard mapping methods (Figure 8.11). Second, KW is closest to marker *aba306b* and farthest from marker *mwg920a*, indicating that the QTL for KW is near the end of the chromosome. The vector of *aba306b* is considerably shorter than others, because of many missing values, as is reflected in Table 8.3. The vector of *act8b* is also slightly shorter than others because of some missing values (Table 8.3). Other things being equal, we would have greater confidence in taking markers that have fewer missing values as candidates for QTL. Figure 8.27 illustrates the interrelationships between KW and markers on chromosome 1. Unlike the markers on chromosome 7, the four markers on chromosome 1 are closely linked, marker *vatp57a* being closest to KW.

QTL Identification Using GGEbiplot

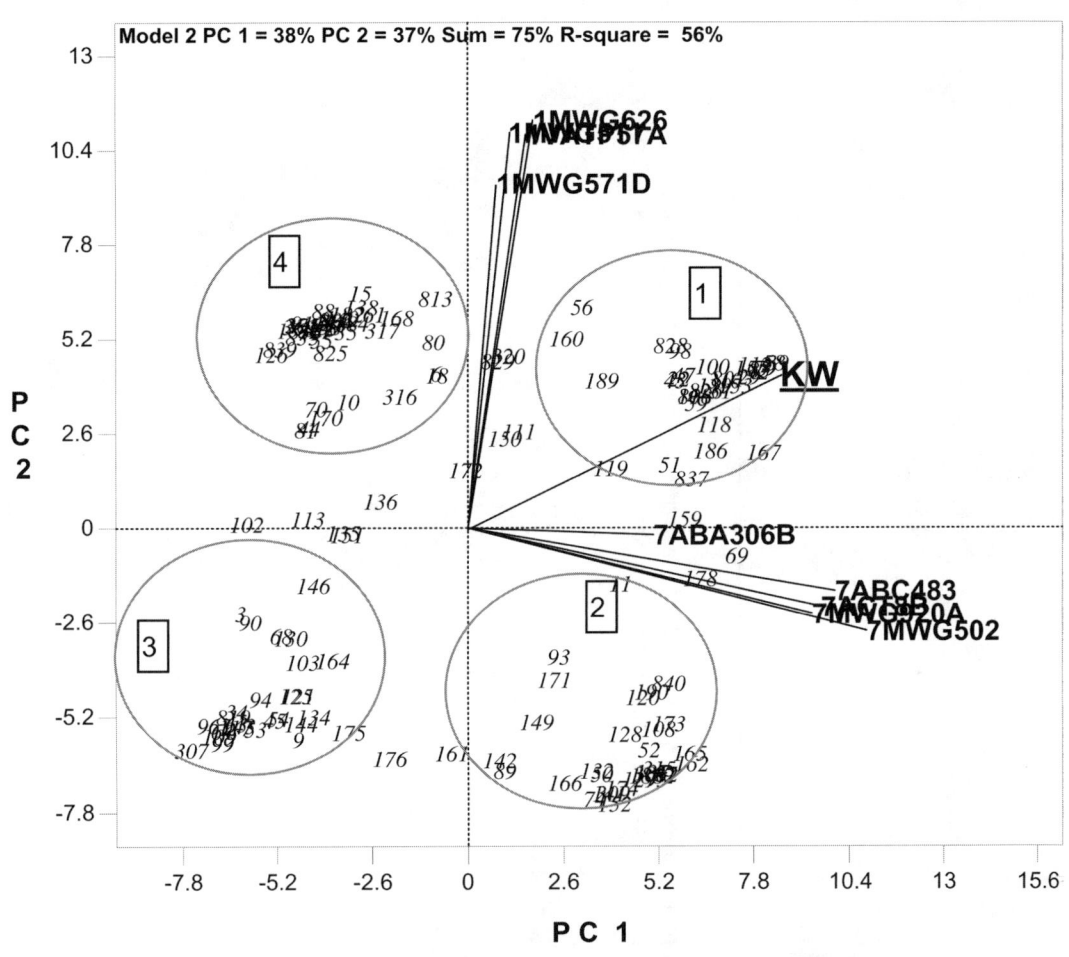

FIGURE 8.25 Two clusters of markers associated with barley KW at the 8% level or higher.

TABLE 8.3
Marker Values and KW of Five Groups of Randomly Selected DH Lines with Reference to Figure 8.25

Groups	DH lines	1MWG511	1MWG626	1VATP57A	1MWG571D	KW	7ABA306B	7ACT8B	7MWG502	7ABC483	7MWG920A
1	HT22	1	1	1	1	43.89	1	Missing	1	1	1
1	HT29	1	1	1	1	45.44	1	1	1	1	1
1	HT31	1	1	1	1	43.74	1	1	1	1	1
1	HT39	1	1	1	1	47.12	1	1	1	1	1
1	HT58	1	1	1	1	46.95	1	1	1	1	1
1	HT59	1	1	1	1	42.23	1	1	1	1	1
1	HT83	1	1	1	1	46.11	1	1	1	1	1
1	HT100	1	1	1	1	45.29	1	1	1	1	Missing
1	HT153	1	1	1	1	44.26	1	1	1	1	1
1	HT155	1	1	1	1	46.42	Missing	1	1	1	1
	Mean					45.145					
2	HT52	0	0	0	0	44.82	1	Missing	1	1	1
2	HT62	0	0	0	0	43.18	1	1	1	1	1
2	HT82	0	0	0	0	42.93	1	1	1	1	1
2	HT127	0	0	0	0	42.38	Missing	1	1	1	1
2	HT137	0	0	0	0	43.15	Missing	1	1	1	1
2	HT152	0	0	0	0	39.59	1	1	1	1	1
2	HT162	0	0	0	0	44.32	1	1	1	1	1
2	HT165	0	0	0	0	45.47	Missing	1	1	1	1
2	HT174	0	0	0	0	41.3	Missing	1	1	1	1

QTL Identification Using GGEbiplot

					Mean			
3	Mean				43.016			
3	HT34	0	0	0	40.72	0	0	0
3	HT45	0	0	0	40.78	Missing	0	0
3	HT53	0	0	0	39.6	Missing	0	0
3	HT94	0	0	0	42.19	0	0	0
3	HT96	0	0	0	38.97	0	0	0
3	HT99	0	0	0	37.25	0	0	0
3	HT109	0	0	0	37.91	0	0	0
3	HT121	0	0	0	42.78	0	0	0
3	HT307	0	0	0	36.24	0	0	0
3	HT819	0	0	0	40.14	0	0	0
4	Mean				39.658			
4	HT15	1	1	1	45.5	0	0	0
4	HT35	1	1	1	42.18	0	0	Missing
4	HT71	1	1	1	41.04	Missing	0	0
4	HT88	1	1	1	43.31	0	0	0
4	HT114	1	1	1	41.96	Missing	0	0
4	HT126	1	1	1	38.33	Missing	0	0
4	HT183	1	1	1	41.11	0	0	0
4	HT317	1	1	1	42.52	Missing	0	Missing
4	HT835	1	1	1	40.33	Missing	0	0
4	HT839	1	1	1	38.83	Missing	0	0
5	Mean				41.511			
5	HT56	1	1	1	48.20	1	1	0
5	HT136	1	0	0	43.77	Missing	0	Missing
5	HT176	0	0	0	39.53	0	1	1
5	HT178	0	1	1	44.85	Missing	1	1

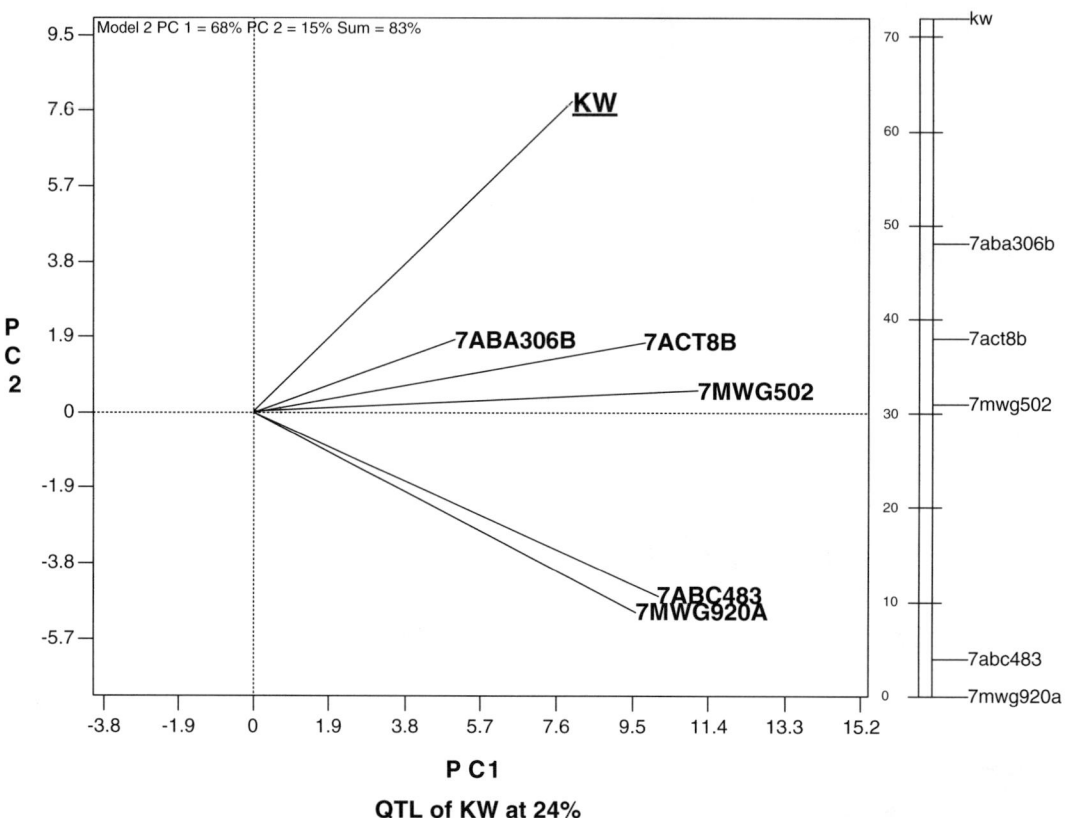

FIGURE 8.26 KW and associated markers on chromosome 7.

QTL Identification Using GGEbiplot

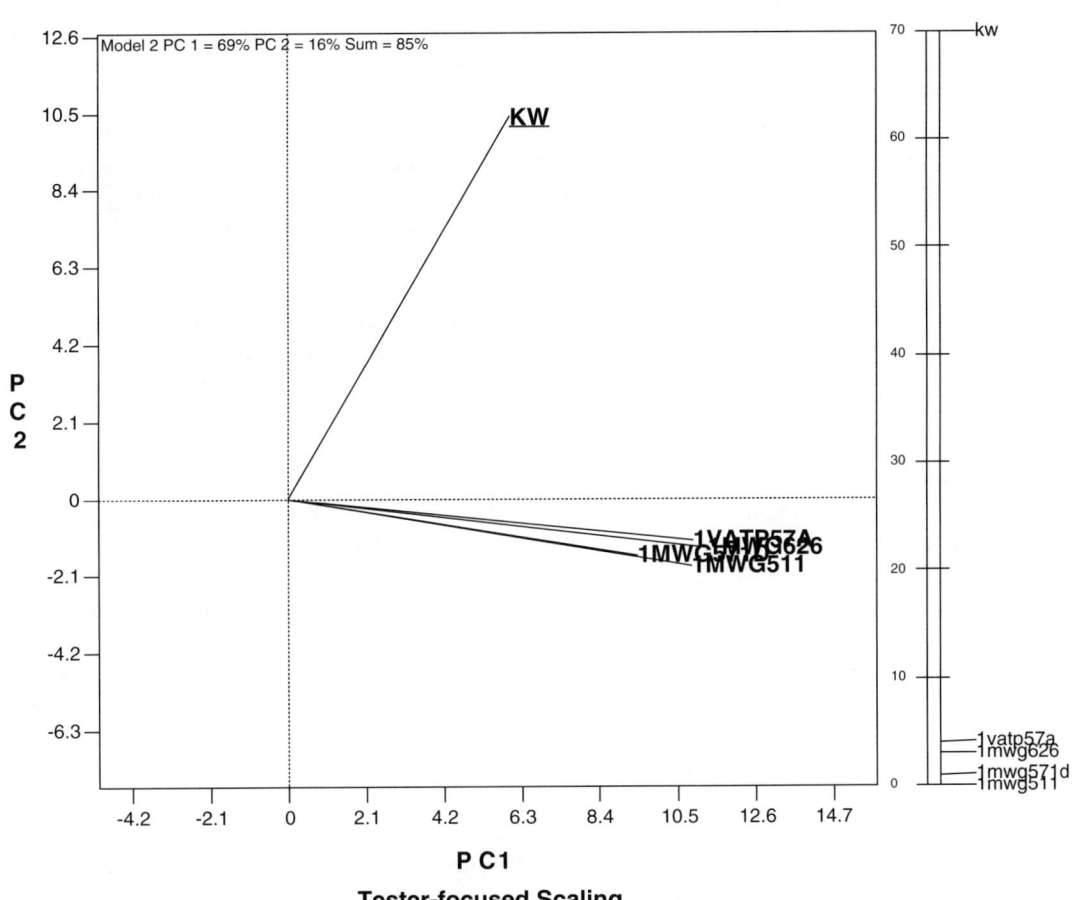

FIGURE 8.27 KW and associated markers on chromosome 1.

8.7.3 ESTIMATING MISSING VALUES BASED ON THE BIPLOT PATTERN

If we take *mwg502* on chromosome 7 and *vatp57a* on chromosome 1 as candidates of QTL for KW, the biplot will appear as represented in Figure 8.28. The DH lines fell into four distinct groups. Needless to say, lines in Group 1 have *Harrington* alleles for both markers and had highest KW. For example, line ht56 had the highest KW of 48.2 g. Lines in Group 3 had alleles of *TR306* for both markers and had the lowest KW values, with line ht307 with the lowest KW of 36.2 g (Table 8.3). Group 2 had the *Harrington* allele for *mwg502* and the *TR306* allele for *vatp57a*; and Group 4 had the *TR306* allele for *mwg502* and the *Harrington* allele for *vatp57a*. Of greatest interest are the outlier lines. All lines, except ht15 and ht813, had missing values for *mwg502* or *vatp57a* or both (Table 8.4).

Line ht118 had a missing value for marker *vatp57a* (Table 8.4). But since its KW was in the range of Group 1 (Figure 8.28), it must share the marker values of Group 1. That is, its value for *vatp57a* should be 1. Table 8.4 indeed indicates that the two flanking markers of *vatp57a* for line ht118 had a value of 1. Therefore, line ht118 could not have a *vatp57a* value of 0 unless double crossovers occurred around it, the probability of which is extremely small.

Similarly, line ht151 had a missing value for marker *mwg502*. Since it belongs to Group 4 relative to KW whose value for *mwg502* was 0, ht151 should also have a value of 0 for this marker. This speculation is confirmed in Table 8.4, where both flanking markers of *mwg502*, *act8b* and *abc483*, had the value of 0. Line ht164 had a missing value for *mwg502*, but it should be 0 since its KW is in the range of Group 3.

Lines ht108, ht128, ht159, and ht173 all had a value of 1 for *mwg502* and had a missing value for *vatp57a*. Since their KW values belong to Group 2, they, therefore, should have a value of 0 for *vatp57a*. This speculation is also supported by the values of the flanking markers (Table 8.4).

Not all missing values can be accurately determined, however. Line ht146 had a missing value for *vatp57a*. Its KW was in the range where Groups 3 and 4 overlapped, and the two groups have different values for *vatp57a*. Therefore, it is not easy to decide ht146's value for *vatp57a*.

Line ht103, located near the biplot origin, missed both markers. Since its KW is in the range where Groups 2 and 4 overlapped, the only conclusion can be that its values for the two markers are different. This is not supported by the values of the flanking markers, however. On the basis of the flanking markers for this DH line, both markers should have a value of 0 (Table 8.4). But its KW seemed to be higher than that for Group 3, which had a value of 0 for both markers. Also, lines ht15 and ht813 had the Group 4 genotype, but their KW values are higher than those for Group 4. These exceptions may be interpreted as errors in KW measurement. Alternatively, it is possible that the QTL area is actually located closer to markers other than those that were examined.

Figure 8.26 indicates that marker *act8b* is more closely linked than *mwg502* to a KW QTL, but *act8b* had more missing values than *mwg502*. After examining Figure 8.28, we are now in a better position to assess the marker values of the DH lines for *act8b* (Figure 8.29). Again, there are four distinct groups of DH lines whose KW and marker values for *vapt57a* and *act8b* can be easily determined based on the biplot pattern: Group 1: high KW, 1 (*vapt57a*), and 1 (*act8b*); Group 2: medium KW, 0, and 1; Group 3: low KW, 0, and 0; and Group 4: medium KW, 1, and 0. But there are more outliers in Figure 8.29 than in Figure 8.28 as a result of more missing values for *act8b* than for *mwg502*.

In Figure 8.29, lines ht44, ht55, and ht81 are located between Groups 3 and 4 because of missing values for *act8b*. Since their KW belonged to Group 4, they should have a value of 0 for act8b. This is supported by the values of the flanking markers of *act8b* (Table 8.4). Lines ht45, ht53, ht54, ht164, and ht166 are located between Groups 2 and 3 because of missing values for *act8b*. Since their KW values are in the range of Group 3, they should have a value of 0 for *act8b*. Although the data do not allow testing of this speculation for ht164 and ht166, the speculation for lines ht45, ht53, and ht54 is supported by values of flanking markers (Table 8.4).

QTL Identification Using GGEbiplot

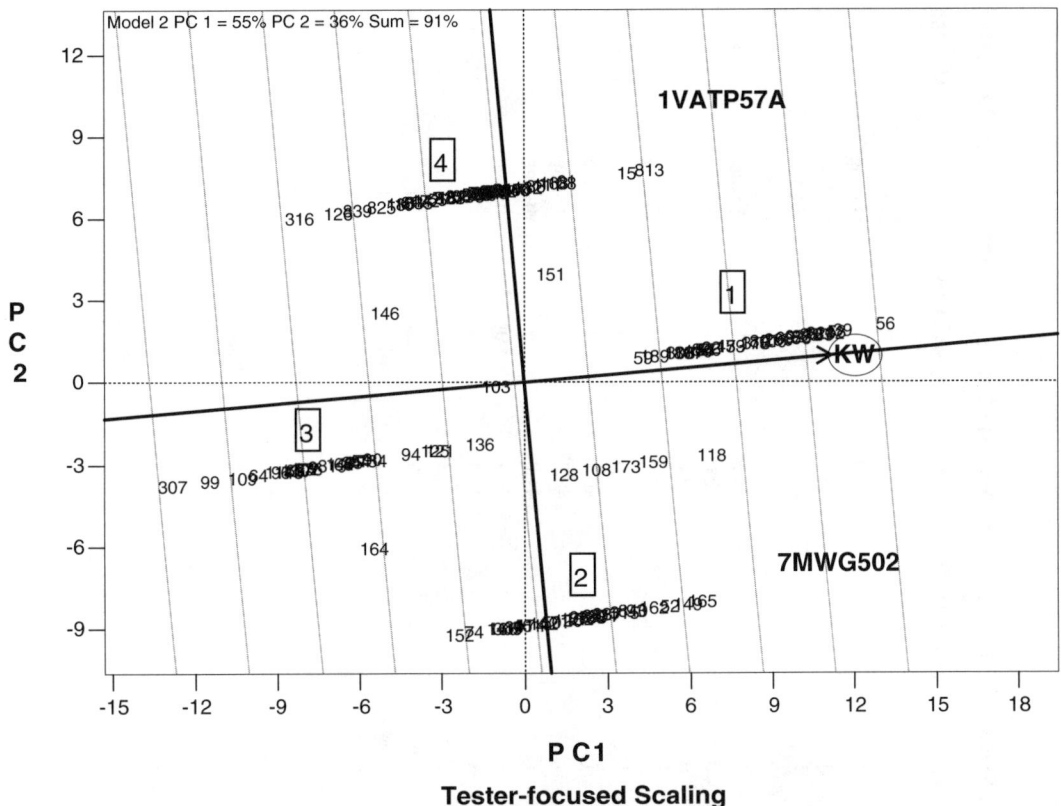

FIGURE 8.28 The variation in KW as explained by markers *vapt57a* on chromosome 1 and *mwg502* on chromosome 7.

TABLE 8.4
KW and Marker Values of Selected DH Lines with Missing Values for Marker vatp57a, act8b, or mwg502

DH LINES	1MWG626	1VATP57A	1MWG571D	KW	7ABA306B	7ACT8B	7MWG502	7ABC483
HT15	1	1	1	45.5	0	0	0	0
HT22	1	1	1	43.9	1	Missing	1	1
HT43	1	1	1	43.7	1	Missing	1	1
HT44	1	1	0	40.5	0	Missing	0	0
HT45	0	0	0	40.8	0	Missing	0	0
HT47	1	0	1	44.3	1	Missing	1	1
HT50	0	0	0	42.0	1	Missing	1	1
HT52	0	0	0	44.8	1	Missing	1	1
HT53	0	0	0	39.6	0	Missing	0	0
HT54	0	0	0	41.0	0	Missing	0	0
HT55	1	1	1	40.9	0	Missing	0	0
HT56	1	1	1	48.2	1	Missing	1	1
HT61	1	1	1	44.0	0	Missing	0	0
HT81	1	1	0	40.4	0	Missing	0	0
HT103	0	Missing	0	41.3	0	0	Missing	0
HT108	0	Missing	0	41.8	1	1	1	1
HT118	1	Missing	1	44.6	Missing	1	1	1
HT128	0	Missing	0	41.0	Missing	1	1	1
HT136	1	0	0	43.8	Missing	0	0	0
HT146	0	Missing	1	40.3	Missing	0	0	0
HT151	0	1	1	41.8	Missing	0	Missing	0
HT159	1	Missing	0	43.2	Missing	1	1	1
HT164	0	0	1	39.4	Missing	Missing	Missing	0
HT166	0	0	0	40.7	Missing	Missing	1	1
HT168	1	1	1	43.6	Missing	Missing	0	0
HT173	0	Missing	0	42.5	1	1	1	1
HT813	1	1	1	46.0	Missing	Missing	0	0

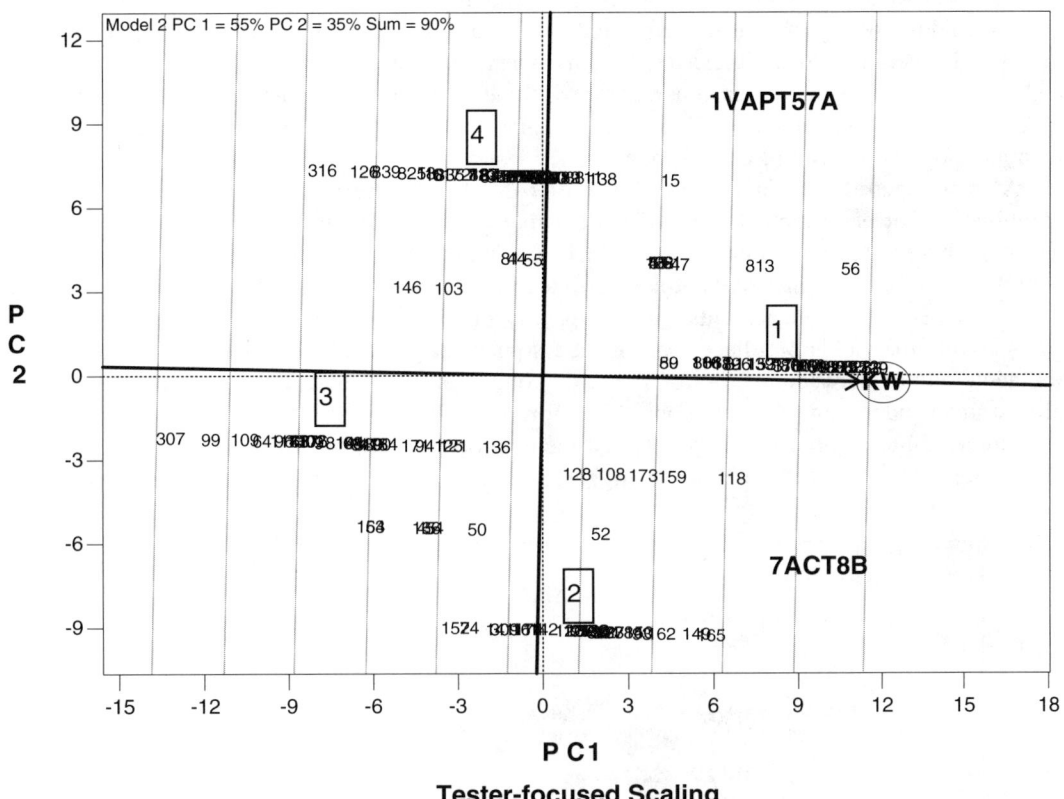

FIGURE 8.29 The variation in KW as explained by markers *vapt57a* on chromosome 1 and *act8b* on chromosome 7.

Lines ht22, ht43, ht47, ht61, and ht168 are undiscernibly clustered between Groups 1 and 4, because of a missing value for marker *act8b*. Since their KW is just between that of Groups 1 and 4, it is difficult to determine their *act8b* values. Table 8.4 data suggest the marker value of *act8b* for ht22, ht43, and ht47 to be 1, but it should be 0 for ht61 and ht168.

Although it is not possible to accurately ascertain all missing values, this section should suffice to clarify that the biplot pattern can be used to fill in most of the missing marker values with considerable confidence.

8.8 QTL AND GE INTERACTION

Using GGEbiplot, we have identified QTL for various barley agronomic traits. These QTL are specific to this mapping population and to the environments under which the traits are measured. They should not be regarded as universal and independent of mapping populations and of environments. A locus can be identified as a QTL only when polymorphism exists. The failure to validate a QTL in another population or in another environment should not be viewed as evidence that the QTL in question does not exist. It would be naïve to try to find QTL that are independent of mapping populations and of environments.

As mentioned in Section 8.2, yield data of 145 DH lines were obtained in 28 year–location combinations or environments across North America. This structure allows us to examine the genotype-by-environment interactions (GEI) for yield. Assuming that all environments are equally important, the environment-standardized GGE biplot (model 2, Chapter 4) is presented as Figure 8.30. The environments can be visually divided into four groups as outlined by the ovals. While the yields of adjacent groups are somewhat positively correlated, as evidenced by the acute angles between vectors of the two groups, Groups 1 and 3 and Groups 2 and 4 are almost independent (an angle of about 90°). Moreover, Groups 1 and 4 are negatively correlated, as evidenced by the obtuse angles between members of the two groups. As a result, although a single genotype (line ht829) was the highest yielding in both Groups 2 and 3 environments,

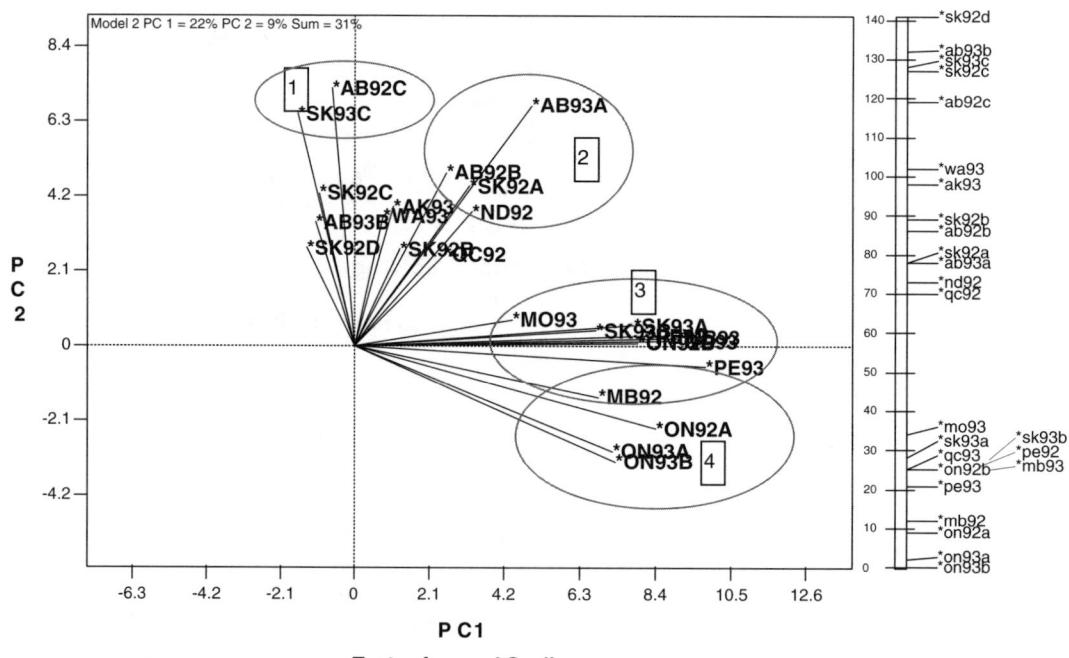

FIGURE 8.30 Vector view of the line-by-environment biplot for barley yield.

QTL Identification Using GGEbiplot

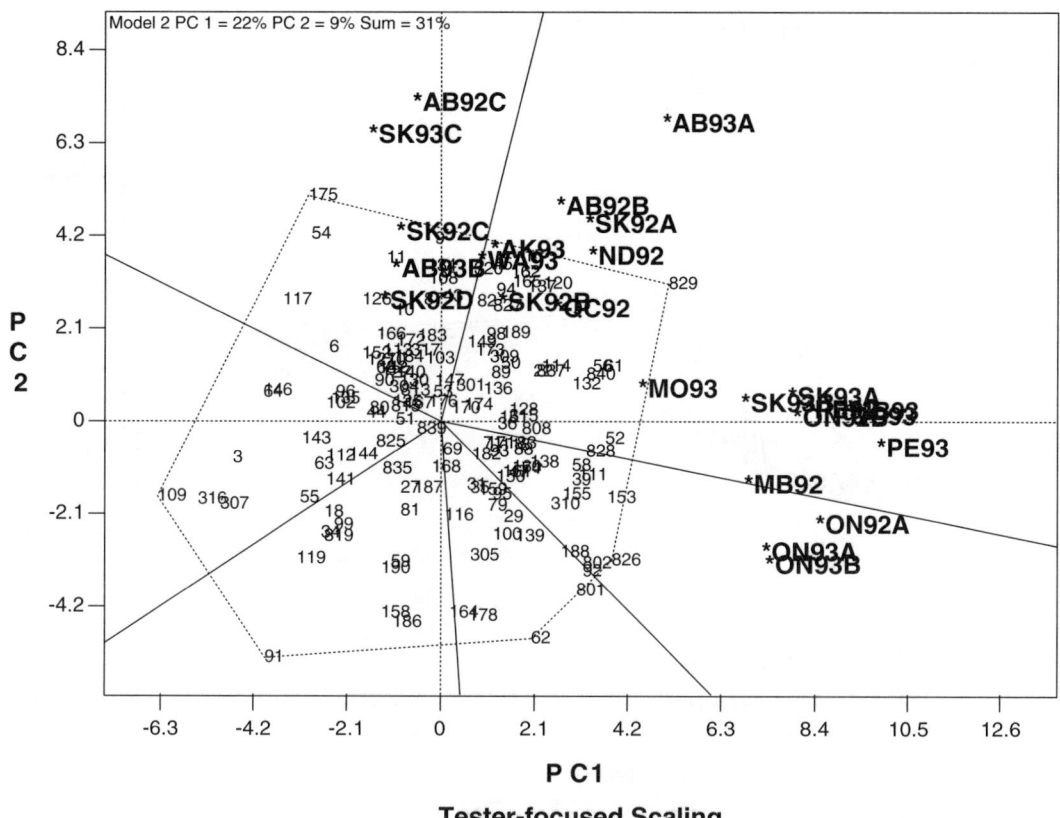

FIGURE 8.31 Polygon view of the line-by-environment biplot for barley yield.

Groups 1 and 4 had different lines with highest yields: ht175 was highest in Group 1 environments, and line ht826 was highest in Group 4 environments (Figure 8.31). Although there is no simple geographical basis for the grouping, all environments from Alberta (AB) are in Groups 1 and 2 and all environments from Ontario (ON) are in Groups 3 and 4. So Groups 1 and 2 may represent environments in the western part of North America, and Groups 3 and 4 those of the eastern part of North America.

Given the large GEI, it would not be prudent to expect QTL identified in one environment for yield to be valid for other environments, even for the same mapping population. We now examine the QTL using four representative environments, one from each of the four groups of environments. These environments are SK93B, AB93A, MB93, and ON92A.

Three QTL for yield were suggested in environment SK93B (Figure 8.32). They are located on chromosomes 1, 2, and 7. Each QTL explained about 5% of yield variation in this environment; none of them explained more than 8% of the yield variation. The *Harrington* allele for the QTL on chromosome 2 had positive effect on yield, whereas *Harrington* alleles for the other two QTL had negative effects on yield. Only one environment, AB92C, is correlated with SK93B at the 5% probability level (not shown), indicating that SK93B was poor in representing other environments.

Three QTL for yield were suggested in environment AB93A (Figure 8.33) to be on chromosomes 3, 4, and 7. Each QTL explained at least 8% of the yield variation in this environment. The *Harrington* alleles of the two QTL on chromosomes 4 and 7 had positive, whereas that for the QTL on chromosome 3 had negative contributions to yield in AB93A. Yield in AB93A was closely related with yield in the other four environments and the mean yield across all environments (Figure 8.33).

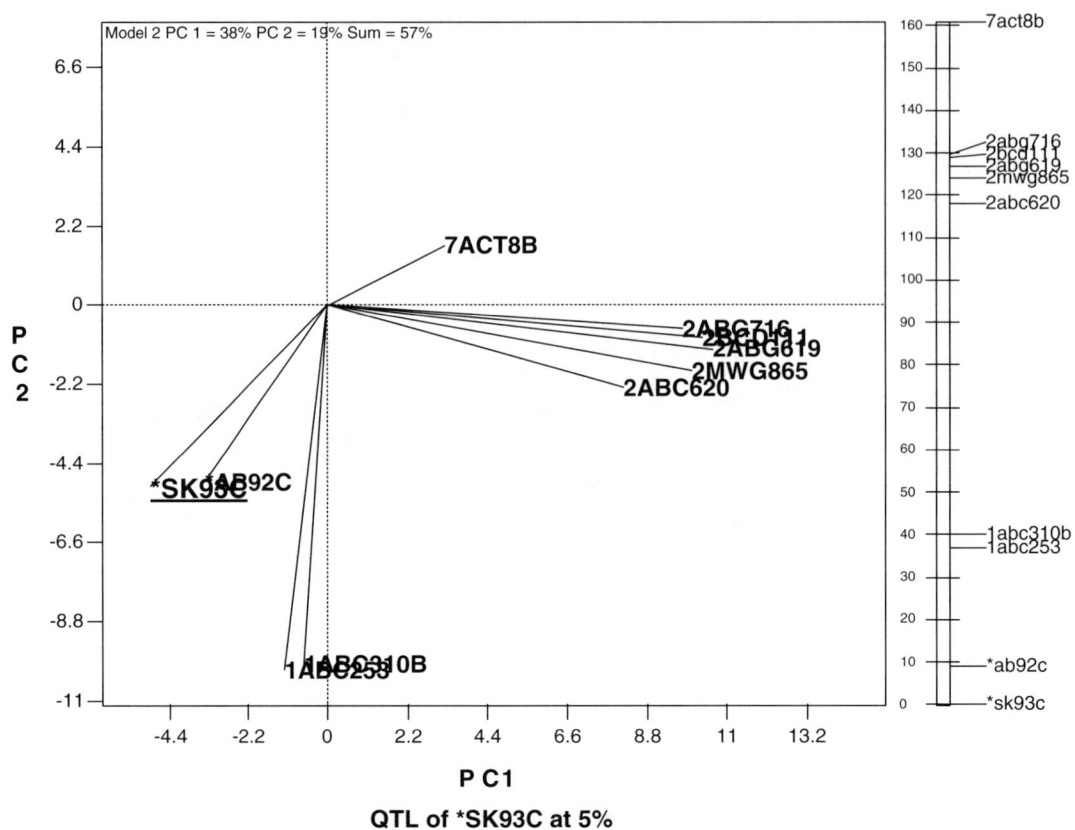

FIGURE 8.32 Markers associated with barley yield in environment SK93C.

Five QTL were identified for yield in MB93 at the 8% level (Figure 8.34): one QTL on each of chromosomes 1, 3, and 4, and two QTL on chromosome 7. The *Harrington* alleles of the QTL on chromosomes 1, 4, and one end of chromosome 7 had positive contributions to yield in MB93, as evidenced by the acute angles between the marker vectors and the vector of MB93. The *Harrington* alleles for the other two QTL had negative contributions to yield in this environment, as evidenced by the obtuse angles. Eleven environments and mean yield across all environments are positively correlated with MB93 (Figure 8.34), indicating that this environment was quite representative.

Four possible QTL were identified for environment ON92A at the 8% level (Figure 8.35): one on chromosome 4, one on chromosome 6, and two on chromosome 7. While the *Harrington* allele of the QTL near *mwg655c* on chromosome 4 and that near *abg705a* on chromosome 7 had positive effect on yield, the *Harrington* allele near *bcd269* on chromosome 6 and that near *ugp3* on chromosome 7 had negative effect. Yield in ON92A also positively correlated with yields from 11 other environments, indicating that this environment is also relatively representative.

Table 8.5 summarizes the yield QTL identified on the basis of four individual environments, and those based on YLD across all environments. The QTL identified in SK93C at the 5% association level was not seen in any other environments. Only one QTL, near *mwg655c* on chromosome 4, was identified in three of the four environments, which is also identified for YLD across environments. It can therefore be concluded that this QTL is relatively stable across environments whereas others are highly sensitive to environmental changes.

In summary, this section outlines an approach for studying QTL-by-environment interaction and its use in plant breeding. First, genotype-by-environment data are visualized using GGE biplot, which reveals, among other things, the magnitude of GEI and interrelationship among environments.

QTL Identification Using GGEbiplot

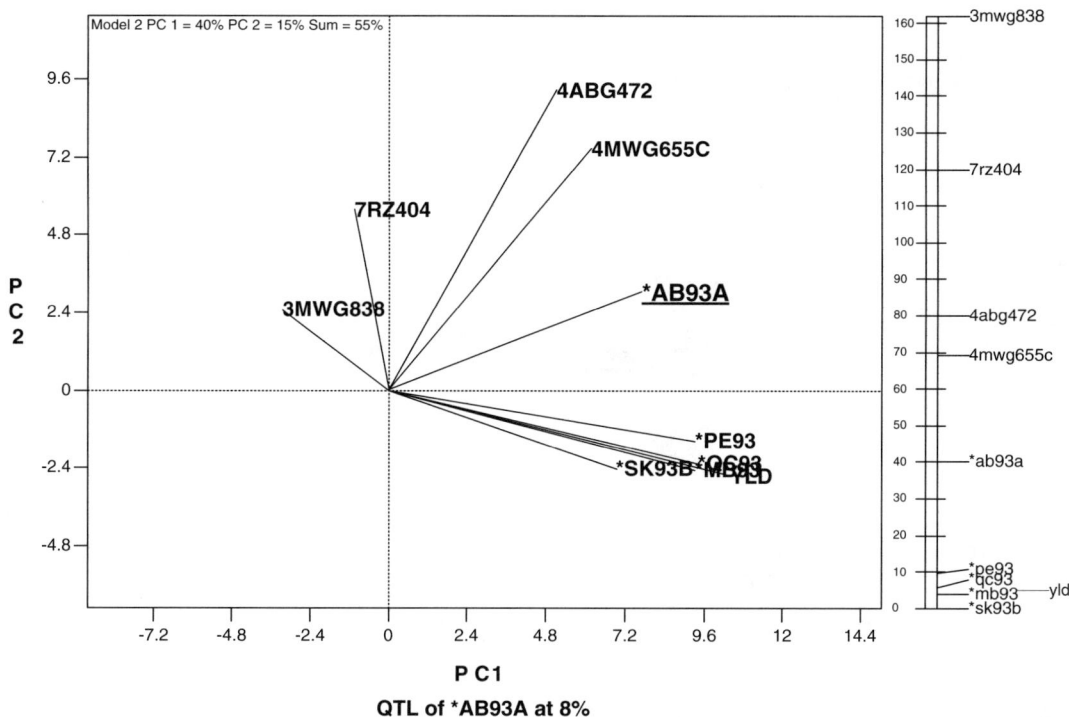

FIGURE 8.33 Markers associated with barley yield in environment AB93A.

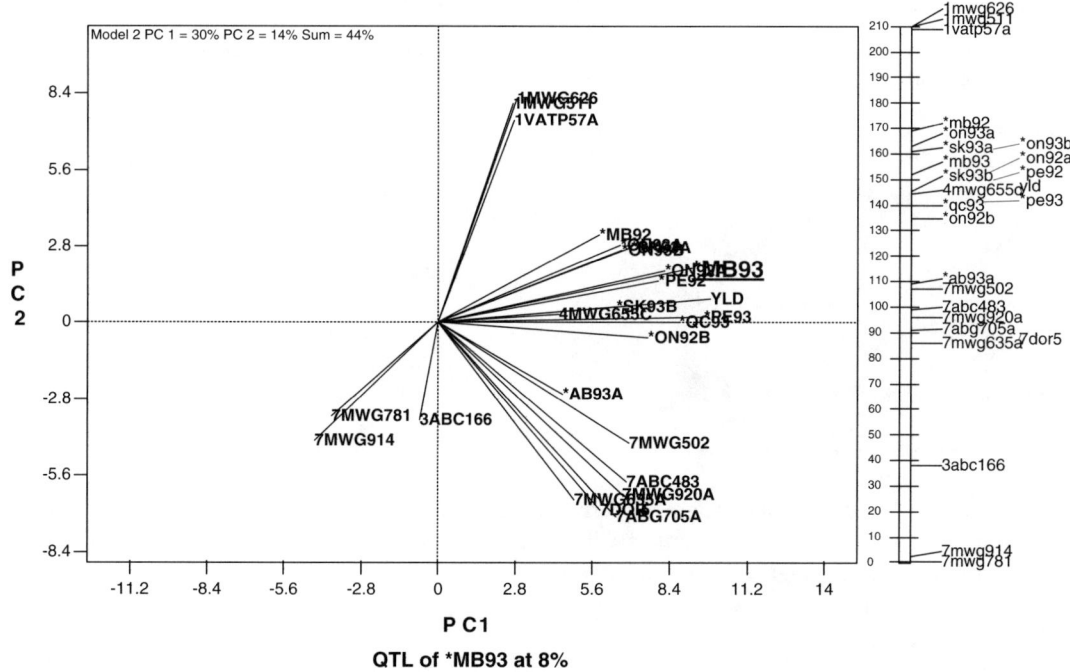

FIGURE 8.34 Markers associated with barley yield in environment MB93.

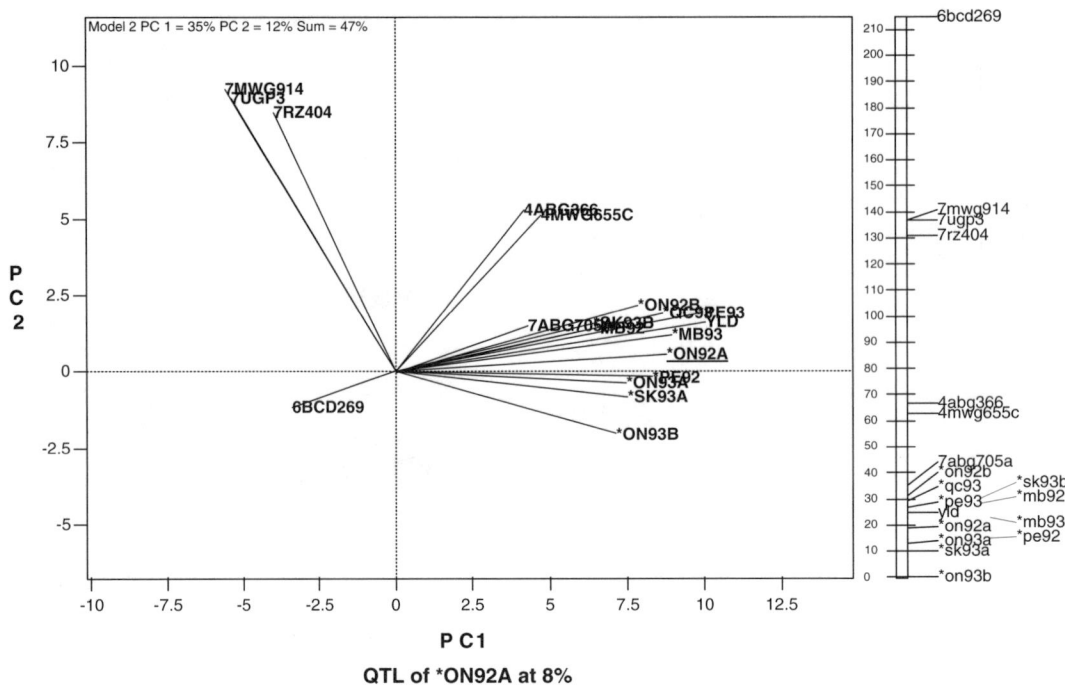

FIGURE 8.35 Markers associated with barley yield in environment ON92A.

TABLE 8.5
QTL of Yield-Based Evaluation in Different Environments

Environment	QTL of Yield with Positive Effects by the Harrington Alleles	QTL of Yield with Negative Effects by the Harrington Alleles
SK93C	1abc253–1abc310b	7act8b; 2abg716–2bcg111–2abg619–2mwg865
AB93A	4mwg655c–4abg472; 7rz404	3mwg838
MB93	4mwg655c; 7mwg502–7abc483; 1mwg511–1vapt57a	3abc166; 7mwg781–914
ON92A	4mwg655c; 7abg705a	7mwg914–7rz404; 6bcd269
Mean across all 28 environments	4mwg655c; 7aba306b–7mwg502	None

Second, a few but distinct environments are selected and QTL analyses are conducted for each individual environment. Finally, results are compared and common QTL are identified, which may be good candidates for marker-based selection.

9 Biplot Analysis of Diallel Data

SUMMARY

Diallel crosses have been used in genetic research to determine the inheritance of a trait among a set of genotypes and to identify superior parents for hybrid or cultivar development. Conventional analysis of diallel data is limited to partitioning the total variation attributable to crosses into general combining ability (GCA) of each parent and specific combining ability (SCA) of each cross. The SCA effects are just residuals not explained by the GCA effects; they are cross specific and do not provide much information on parents. The biplot approach of diallel data analysis introduced in this chapter allows a much better understanding of parents. For a given set of data, the following information can be easily visualized: 1) the GCA effect of each parent; 2) the SCA effect of each parent (not cross); 3) the best crosses; 4) the best testers; 5) the heterotic groups; and 6) genetic constitutions of parents with regard to the trait under investigation.

Diallel crosses represent matings made in all possible combinations among a set of genotypes. They have been widely used in genetic research for investigating inheritance of quantitative traits among a set of genotypes. There are four types of diallel mating schemes (Griffing, 1956): 1) method 1 — diallel cross with parents and reciprocals; 2) method 2 — diallel cross with parents but without reciprocals; 3) method 3 — diallel cross with reciprocals but without parents; and 4) method 4 — diallel cross without parents and reciprocals. Reciprocals are made for the purpose of detecting any maternal effect. We confine our discussion to method 2 diallel cross, although other types of diallel crosses can be easily accommodated. Conventionally, analysis of diallel cross data is conducted to partition total genetic variation into GCA of the parents and SCA of the crosses.

In this chapter, we will use two sets of diallel data (Tables 9.1 and 9.2) to demonstrate the biplot approach to diallel analysis. The first dataset is to demonstrate the general steps and utilities of biplot analysis for diallel data. The second dataset is used to exemplify analysis of a large dataset for which a biplot of the first principal component (PC1) vs. the second principal component (PC2) may not be adequate. A discussion of both datasets is necessary because they contain contrasting entry × tester interaction or gene action patterns.

9.1 MODEL FOR BIPLOT ANALYSIS OF DIALLEL DATA

Buerstmayr et al. (1999) reported a diallel study of winter wheat resistance to *Fusarium* head blight (FHB). They used seven winter wheat genotypes of diverse origin and with large differences in resistance to FHB. They presented data on area under the disease progress curve and percentage of infected kernels for all seven parents and their F_1 hybrids; the two measures were highly correlated. Since no reciprocal effect was observed for this trait, means of the reciprocals for individual crosses were reported. Percentages of uninfected kernels of the seven parents and their F_1s as a measure of resistance to FHB, are given in Table 9.1. The GCA effects of the parents are presented in the last column of the table, which is simply the row means minus the grand mean. Thus, we see parents F and G have high GCA, A and B have low GCA, and C, D, and E have intermediate GCA. This is the only piece of information on the parents that can be obtained using the conventional analysis. All other information that can be derived by conventional methods is pertinent to crosses rather than parents. We will show that the biplot approach allows a much deeper understanding of the parents and of the trait under investigation.

TABLE 9.1
Resistance to FHB of Seven Winter Wheat Genotypes and their F_1 Hybrids, as Measured by Percentage of Uninfected Kernels Upon Inoculation

	Testers								
Entries	A	B	C	D	E	F	G	MEAN	GCA
A	27.5	35.7	46.4	53.7	33.3	64.9	43.3	43.5	−7.3
B	35.7	37.5	46.2	40.8	51.9	45.6	57.5	45.0	−5.8
C	46.4	46.2	38.7	49.1	50.4	55.6	69.4	50.8	0.0
D	53.7	40.8	49.1	51.2	49.4	48.1	57.5	50.0	−0.8
E	33.3	51.9	50.4	49.4	42.5	63.1	68.9	51.4	0.6
F	64.9	45.6	55.6	48.1	63.1	60.0	63.1	57.2	6.4
G	43.3	57.5	69.4	57.5	68.9	63.1	43.7	57.6	6.8
Mean	43.5	45.0	50.8	50.0	51.4	57.2	57.6	50.8	0.0

Note: The genotype codes are A = Alidos, B = 81-F3–79, C = Arina, D = SVP–72017–17–5–10, E = SVP-C8715–5, F = UNG–136.1, and G = Ung–226.1.

Source: Data based on Buerstmayr, H. et al., *Euphytica*, 110:199–206, 1999.

In a diallel cross data, a parent is both an entry and a tester. The model used for biplot analysis of diallel data is the tester-centered principal component analysis. It was labeled as Equation 4.5 and is presented again below:

$$\hat{Y}_{ij} - \mu - \beta_j = g_{i1}e_{1j} + g_{i2}e_{2j} + \varepsilon_{ij}$$

where $\hat{Y}_{ij}$ is the expected value of the cross between entry i and tester j; μ is the grand mean; β_j is the main effect of tester j; g_{i1} and e_{1j} are called the primary effects for entry i and tester j, respectively; g_{i2} and e_{2j} the secondary effects for entry i and tester j, respectively; and ε_{ij} is the residue not explained by the primary and secondary effects. A biplot is constructed by plotting g_{i1} against g_{i2} and e_{1j} against e_{2j} in a single scatter plot. Using GGEbiplot software, a biplot will be generated in about one second after the data are read (Figure 9.1). The parents are presented in lowercase italics when they are viewed as entries; they are in regular uppercase when viewed as testers. The biplot explained 77% (46 and 31% by PC1 and PC2, respectively) of the total variation, which, in conventional analyses, would be partitioned into GCA effects of the parents and SCA effects of the crosses. The guidelines cross at the biplot origin (0,0). What we can learn from this biplot follows.

9.2 GENERAL COMBINING ABILITY OF PARENTS

Figure 9.2 is the average tester coordination (ATC) view of the biplot brought up by the *Average Tester Coordination* function of GGEbiplot. The small circle represents an average tester, which is defined by the average PC1 and PC2 values of all testers. The line passing through the biplot origin and the average tester, with an arrow pointing to the average tester, is called the average tester axis or ATC abscissa and the line that passes through the origin and is perpendicular to the average tester axis is called average tester ordinate or ATC ordinate.

GCA and SCA are defined here as properties of the entries. The GCA effects of the entries are approximated by their projections onto the ATC abscissa. Lines parallel to the ATC ordinate help rank entries relative to GCA effects. Thus, entries *f* and *g* (UNG–136.1 and UNG–226.1), both being derivatives of Sumai 3, a very resistant wheat cultivar of Chinese origin, had the highest

Biplot Analysis of Diallel Data

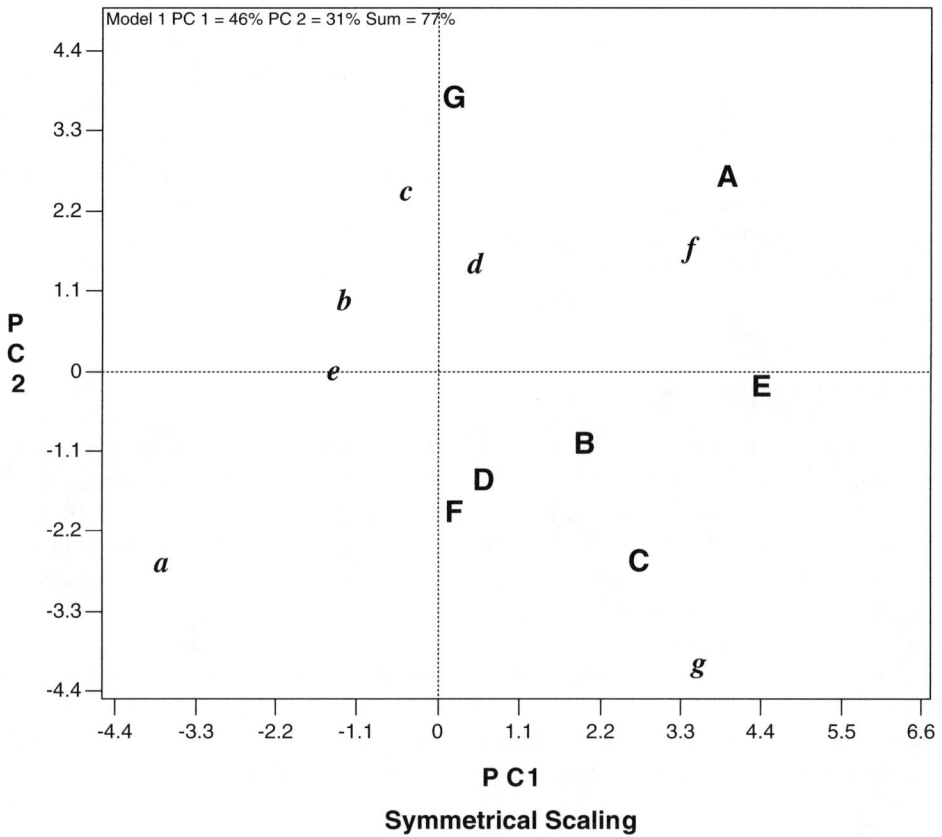

FIGURE 9.1 Biplot based on the wheat FHB research data. The seven parents are in lowercase when viewed as entries and in uppercase when viewed as testers.

GCA effects, whereas entry *a* (Alidos, the most susceptible parent) had the lowest GCA effect. The GCA effects of the entries are in the order of: $g \approx f > d > c > b \approx e > a$. Note that this order is highly consistent with that of the last column in Table 9.1 except that E was misplaced. We see later that E is something special; it is the best tester among the tested parents (see Section 9.5). The correlation between the GCA effects and the projections onto the ATC axis is 0.926.

9.3 SPECIFIC COMBINING ABILITY OF PARENTS

As was mentioned earlier in this chapter, in conventional diallel cross analysis, the term specific combining ability or SCA is associated with crosses. It is merely a residual that is not explained by the GCA effects and does not shed any light on the parents. The biplot, however, allows viewing of the SCA effects of the *parents*. If, as just demonstrated, the ATC abscissa approximates the GCA effects of the entries, then the ATC ordinate, which is orthogonal to the GCA effects, must approximate the SCA of the entries, which represents the tendency of an entry to produce superior hybrids with some, but not all, testers. Since the testers are located both above — A and G — and below — B, C, D, and F — the ATC abscissa (Figure 9.2), a greater projection toward either direction means greater SCA effects. Thus, entries *g* and *a*, below the ATC abscissa, had the greatest SCA effects or largest projections onto the ATC ordinate. Above the ATC abscissa, entries *c* and *f* had large SCA effects relative to entries *b* and *d*. Entry *e* had the smallest SCA effect since it had smallest projection onto the ATC ordinate.

9.4 HETEROTIC GROUPS

The tester vector view, brought up by the *Show/Hide Tester Vectors* function, helps group testers into heterotic groups (Figure 9.3). Since all testers have positive scores for PC1, the testers are grouped primarily by PC2 scores. Three groups of testers are obvious: A and G above the PC2 guideline, and B, C, D, and F below the PC2 guideline. Moreover, above the guideline, testers A and G interacted positively with entries *b*, *c*, *d*, and *f*; but they interacted negatively among themselves. Similarly, below the guideline, testers B, C, D, and F interacted positively with entries *a* and *g*; the four testers also interacted negatively among themselves. The PC2 guideline serves as a mirror —the same thing is displayed both above and below it. This interaction pattern clearly indicates heterosis in crosses (A and G) × (B, C, D, and F). Thus, A and G and B, C, D, and F are two different heterotic groups. Tester E does not seem to belong to either heterotic group (Figure 9.3). Rather, it is a tester that can effectively reveal the GCA effects of the parents, which is discussed next.

9.5 THE BEST TESTERS FOR ASSESSING GENERAL COMBINING ABILITY OF PARENTS

An ideal tester for revealing the GCA effects of entries should fulfill two criteria: it should be representative of all testers and, at the same time, be most discriminating of the entries. Based on this definition, an ideal tester must be located on the ATC axis to be representative of all testers; its vector should be the longest of all testers to be most discriminating. Such a tester is indicated by the center of the concentric circles in Figure 9.4. Although such an ideal tester may not exist in reality, it can be used as a reference to compare the real testers. The concentric circles are drawn for this purpose with the hypothesized ideal tester at the center (Figure 9.4). The closer a tester is to this ideal center, the more desirable it is. Clearly, tester E was the best tester in this dataset, as it was very close to the ideal tester. On the contrary, G was the poorest tester as it is the least representative of all testers. Figure 9.4 can be brought about by a single click on the GGEbiplot function *Compare With...the Ideal Tester* (Chapter 6).

Biplot Analysis of Diallel Data

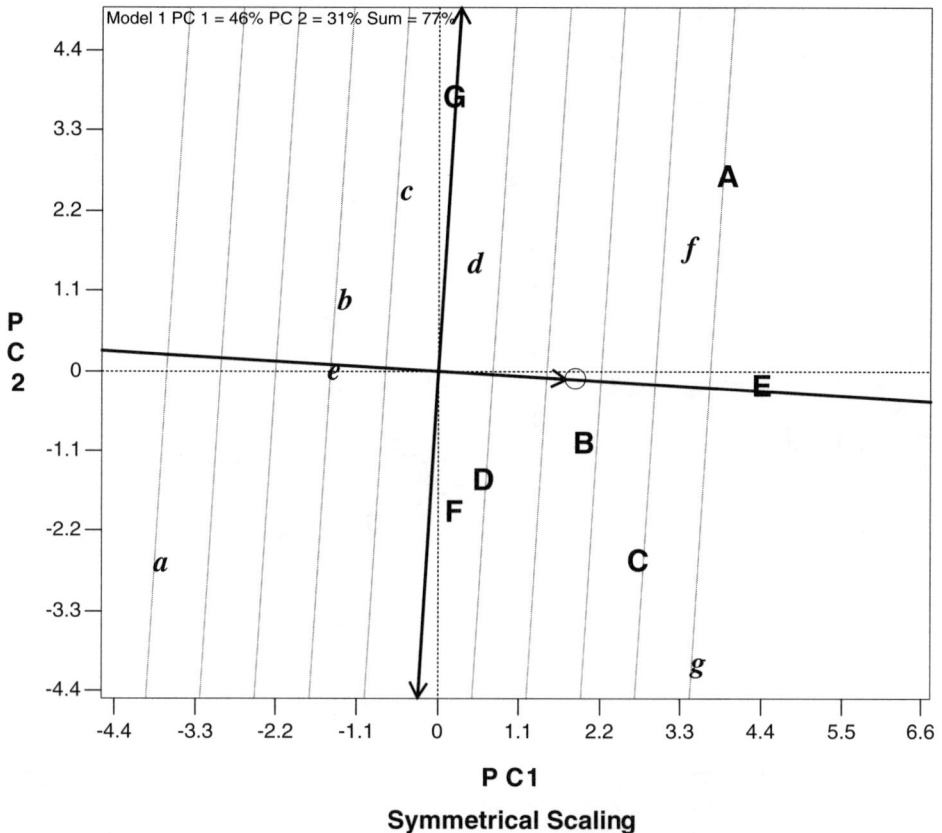

FIGURE 9.2 Average tester coordination (ATC) view of the biplot based on the wheat FHB research data. The small circle represents the average tester. Entries are in lowercase italics and testers are in uppercase.

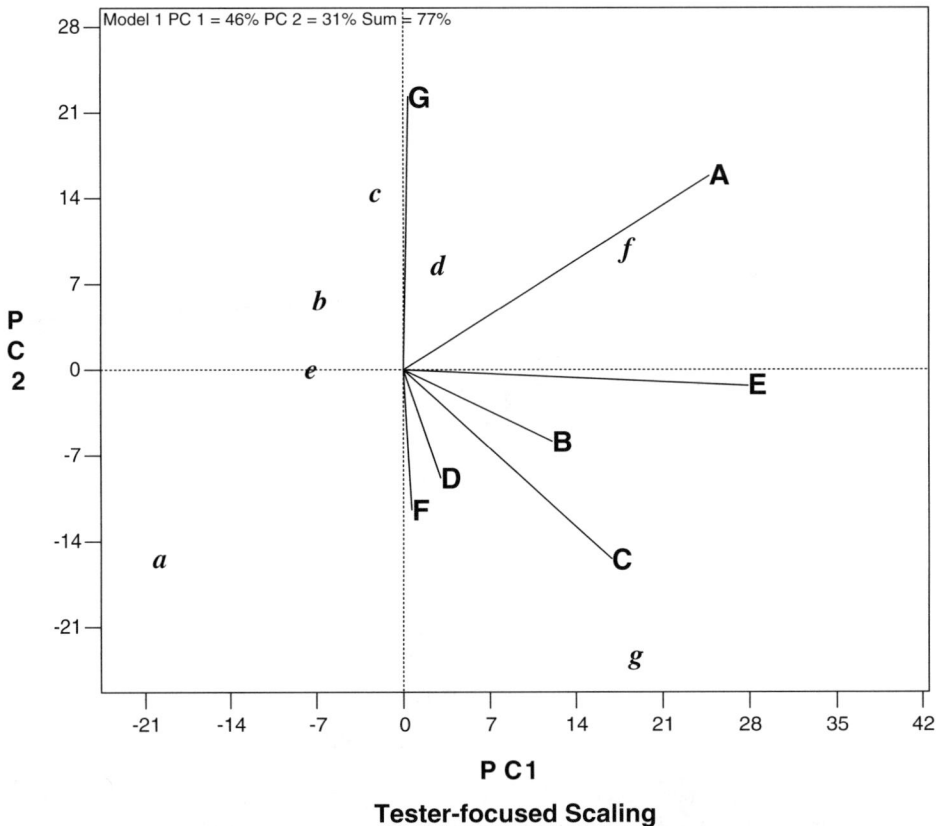

FIGURE 9.3 The tester vector view of the biplot based on the wheat FHB research data. The seven parents are in lowercase when viewed as entries and in uppercase when viewed as testers.

Biplot Analysis of Diallel Data

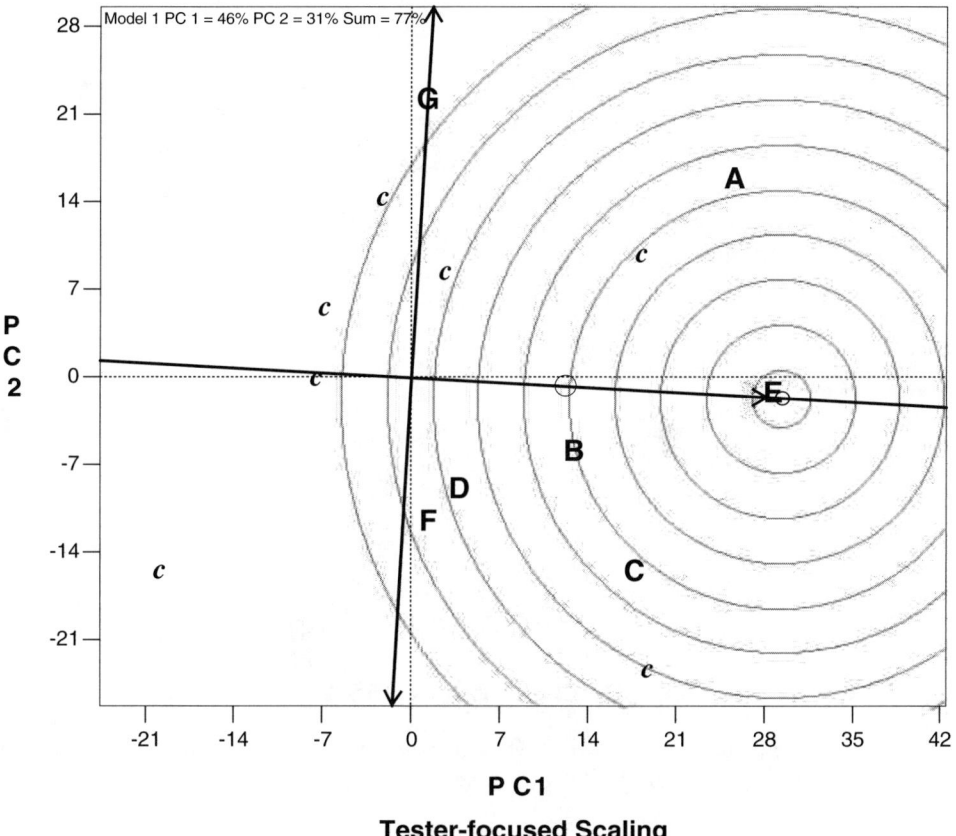

FIGURE 9.4 Visual evaluation of the parents as testers. The concentric center represents the ideal tester, which is the most discriminating of entries and has no preference for mating partners, i.e., with zero specific combining ability. The seven parents are in lowercase italics when viewed as entries and in uppercase when viewed as testers.

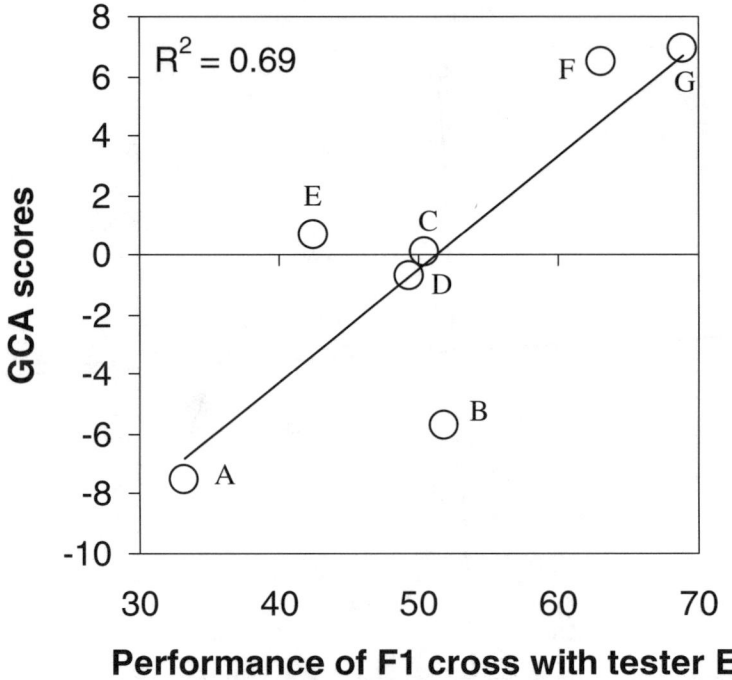

FIGURE 9.5 Measured F_1 values of each parent when crossed with tester E, plotted against the measured GCA of each parent. This figure reveals that the F_1 with tester E is a good indicator of the GCA of an entry.

Biplot Analysis of Diallel Data

That tester E is the best tester implies that the GCA effect of an entry can be reasonably assessed by the value of its hybrid with tester E. To verify, F_1 hybrid between each entry and tester E is plotted against the GCA effect of the entry (Figure 9.5). The two are highly correlated except entries *b* and *e* are off the regression line. The performance of B×E was above expectation based on the overall pattern of the data.

9.6 THE BEST CROSSES

The polygon view of the biplot (Figure 9.6), brought up by the GGEbiplot function *Show/Hide the Polygon*, helps reveal the best crosses between the entries and the testers. The polygon consists of straight lines connecting the entries positioned furthest from the biplot origin such that all other entry markers are contained within the polygon. Lines perpendicular to each side or its extension of the polygon are drawn from the biplot origin to divide the biplot into sectors. In Figure 9.6, the biplot was divided into four sectors, with entries *a*, *c*, *f*, and *g* as the vertex entries, and are referred to as sector *a*, sector *c*, sector *f*, and sector *g*, respectively. No tester fell in the *a* sector, suggesting that entry *a* was not the best mating partner with any of the genotypes. Moreover, this suggests that entry *a* produced the poorest hybrids with some or all of the testers. The original data (Table 9.1) did show that parent *a* per se was poor and it produced the poorest hybrids with B, E, and G, which is consistent with the prediction from Figure 9.6. The original data also indicated that entry *a* was the best mating partners for F (Table 9.1), which is not reflected in Figure 9.6. This illustration suggests entry *a* to be the second best mating partner of F, following entry *g*. Drawing a vector for tester F will facilitate visualization of this aspect.

A single tester, G, fell in the *c* sector, indicating that entry *c* was the best mating partner with G. That is, the cross C×G was predicted to be the best of all crosses involving G, which is consistent with the data (Table 9.1). Moreover, since parent C as a tester was not in sector *c*, the cross C×G must be heterotic, i.e., better than both parents (C×C and G×G). Had tester C fallen in sector *c*, the combination C×C, i.e., pureline C would be the best among all crosses involving C, and consequently, heterosis between C and any other parents would not be possible.

A single tester, A, fell into sector *f*, indicating that entry *f* was the best mating partner for A, which is consistent with the data. Since tester F was not in the *f* sector, the cross F×A was heterotic. Testers B, C, D, E, and F fell in the same sector, i.e., sector *g*, suggesting that entry *g* was the best mating partner for these testers. This is almost exactly the case (Table 9.1). Since G was not in the *g* sector, all crosses between genotype G and B, C, D, E, and F were heterotic.

To summarize, the following seven F_1 hybrids were among the best in this FHB study (Figure 9.6): C×G in sector *c*; F×A in sector *f*; and, G×B, G×C, G×D, G×E, and G×F in sector *g*. Note that the cross C×G appeared twice in the above list: in sector *g*, G was predicted to be the best mating partner with C, among others; and in sector *c*, C was predicted to be the best partner with G. Thus, C and G are mutually identified as the best partners; therefore, the cross C×G should be the best of all possible combinations, which is consistent with the results (Table 9.1). In addition, tester E was almost on the line that separates sectors *g* and *f*. Therefore, F should also be a good parent for crossing with E, as can be verified from the data (Table 9.1).

9.7 HYPOTHESIS ON THE GENETIC CONSTITUTION OF PARENTS

In conventional analyses of diallel data, interpretation of the genetic constitutions of parents with regard to the trait under investigation is not attempted until the F_2 generation. The biplot provides a unique means to visualize the interrelationships among parents, and thus allows hypotheses to be formulated on the genetic constitution of parents.

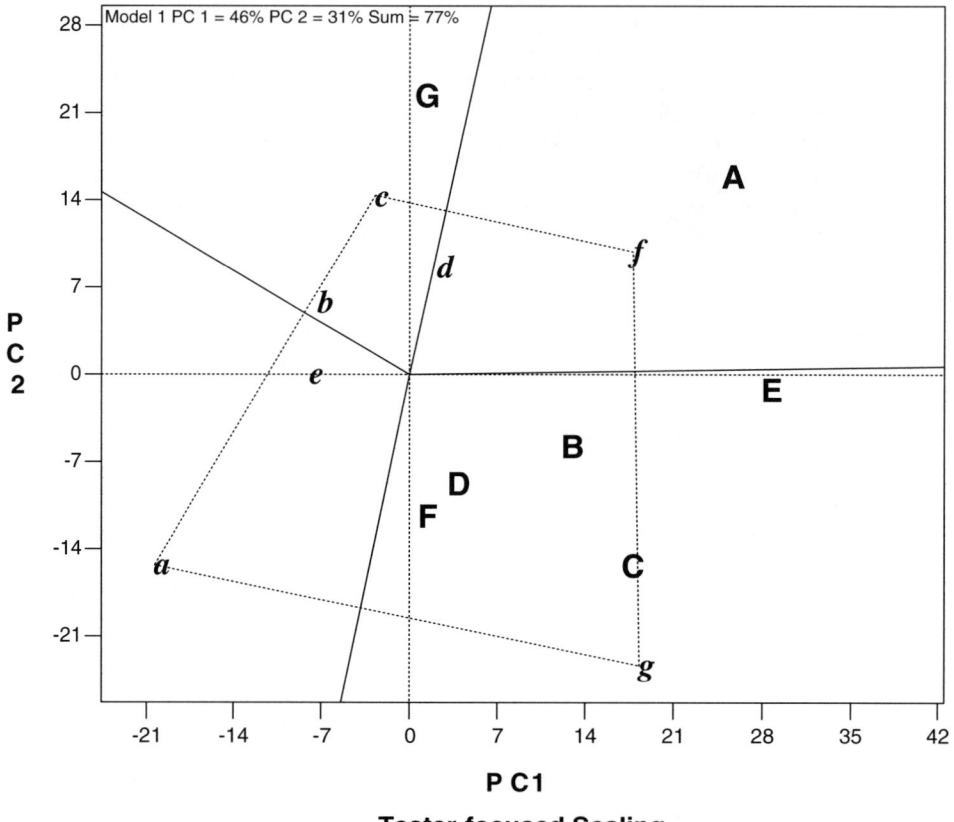

FIGURE 9.6 Polygon view of the biplot, showing the best crosses among all possible combinations. The best crosses are *g* by C, B, D, E, and F in the *f* sector and *f* by A, and E in the *f* sector. The parents are in lowercase italics when viewed as entries and in uppercase when viewed as testers.

All testers had positive PC1 scores (Figure 9.7), implying that the interaction between the entries and the testers displayed by PC1 is of non-crossover type or proportional interaction (Crossa and Cornelius, 1997; Yan et al., 2000). Thus, the entry PC1 scores should approximate the GCA effects, as discussed in Section 9.2. Figure 9.7 indicates three groups relative to GCA. Group 1 contains entry *a* only, with the smallest GCA; Group 2 includes entries *b*, *c*, *d*, and *e*, with intermediate GCA effects and minor differences among them; and Group 3 includes entries *f* and *g*, with the largest GCA. To explain the differences in GCA, we hypothesize that Group 2 had an additive gene (*A1*) relative to Group 1 i.e., entry *a*, and Group 3 had an additional gene (*A2*) relative to Group 2.

PC2 displays the nonproportional interactions between entries and testers, as the testers assumed different signs (Crossa and Cornelius, 1997; Yan et al., 2000). Specifically, PC2 displays positive interactions between two heterotic groups: A and G as one group and B, C, D, and F as the other. If we assume that heterosis arises from the accumulation of different dominant genes, then the two groups must have different dominant FHB resistance genes that are designated as *D1* and *D2* (Figure 9.8). Entry *e* is located right on the PC2 guideline, implying that there was no nonproportionate interaction between *e* and either of the two heterotic groups. This can be explained by one of the three hypotheses: 1) entry *e* carries neither *D1* nor *D2*; 2) entry *e* carries both *D1* and *D2*; and 3) entry *e* carries a resistance gene that is different from both *D1* and *D2*. The first hypothesis is the least tenable because it cannot explain the heterosis observed between entry *e* and other parents. The second and the third hypotheses are equally satisfactory in explaining the heterosis between *e* and other parents, but the third hypothesis must invoke an additional gene.

Figure 9.9 integrates Figures 9.6 to 9.8 to formulate the genotypes of the seven parents and to explain their performance as purelines and the performances of their hybrids. The hybrids between G and B, C, D, and F were among the best hybrids relative to FHB resistance because they each integrated the four resistance genes (*A1, A2, D1,* and *D2*). The same can be said for the cross A×F.

The formulations in Figure 9.9 allow some general discussion on the nature of a superior parent and a superior tester. Parents F and G are regarded as superior, because they had both high GCA and SCA. They had high GCA by having more resistance genes, and they had high SCA by having resistance genes different from those in the other heterotic groups. Parent E was regarded as a superior tester for identifying parents with high GCA, because it has a gene that is different from all the existing resistance genes in the other parents. Superior hybrids combine all or most of the resistance genes through one of the two pathways: 1) both parents exhibit high GCA but belong to different heterotic groups, and 2) one high-GCA parent and one superior tester.

Caution must be used when reading this section, however. The hypotheses on the genetic constitutions of the parents with regard to resistance to FHB are highly speculative. These hypotheses have yet to be subjected to critical testing. Nevertheless, we are excited about the capability of the GGEbiplot to allow such hypotheses to be formulated, given the importance of hypotheses in scientific discovery.

9.8 TARGETING A LARGE DATASET

Butron et al. (1998) reported on percentage of yield loss in a 10-parent maize (*Zea mays* L.) diallel cross following artificial infestation with corn pink stem borer (PSB), *Sesamia nonagrioides*. The yield loss data were converted into tolerance to PSB, measured as the yield of infested plants as percentage of noninfested plants (Table 9.2). We use this dataset to illustrate biplot analysis of a large diallel dataset.

218 GGE Biplot Analysis: A Graphical Tool for Breeders, Geneticists, and Agronomists

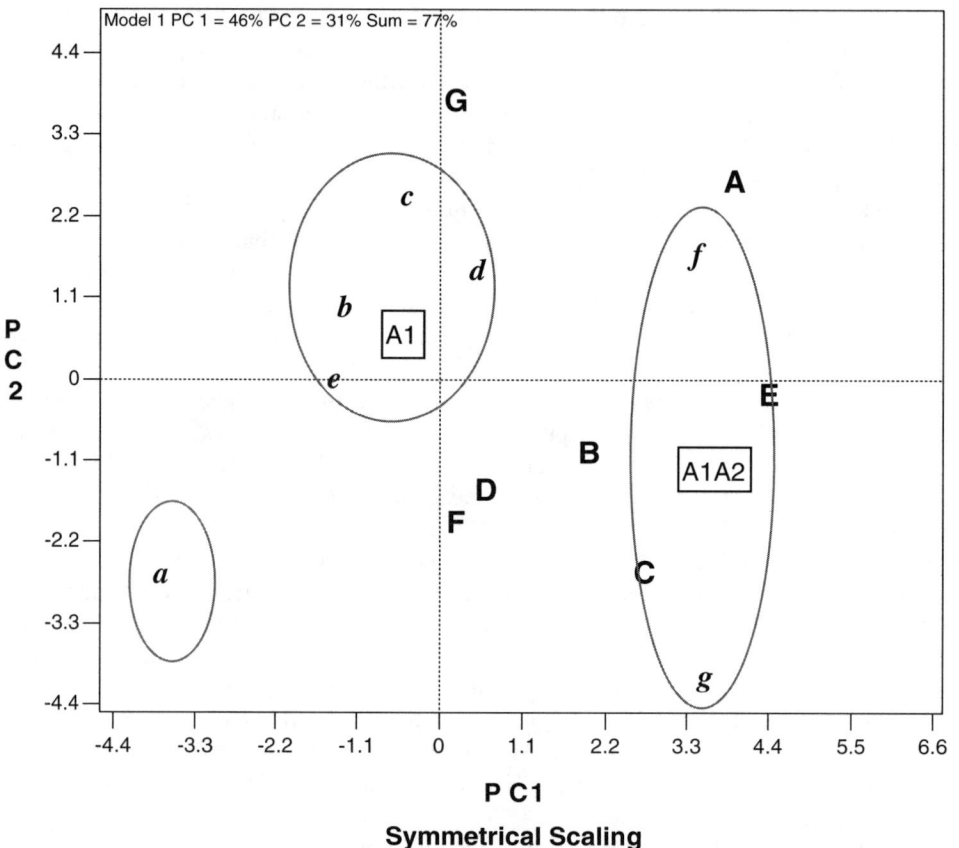

FIGURE 9.7 The proposed genotypes of the parents based on PC1. Parents when viewed as entries are in lowercase italics and in uppercase when viewed as testers.

Biplot Analysis of Diallel Data

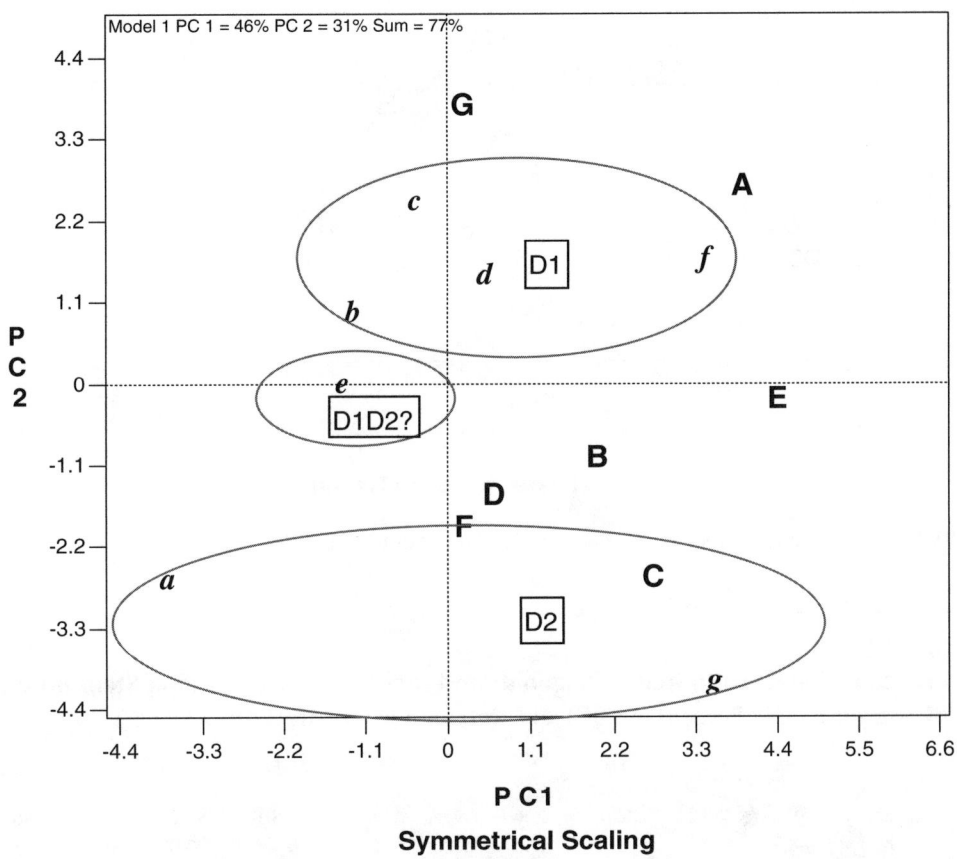

FIGURE 9.8 The proposed genotypes of the parents based on PC2. Parents when viewed as entries are in lowercase italics and in uppercase when viewed as testers.

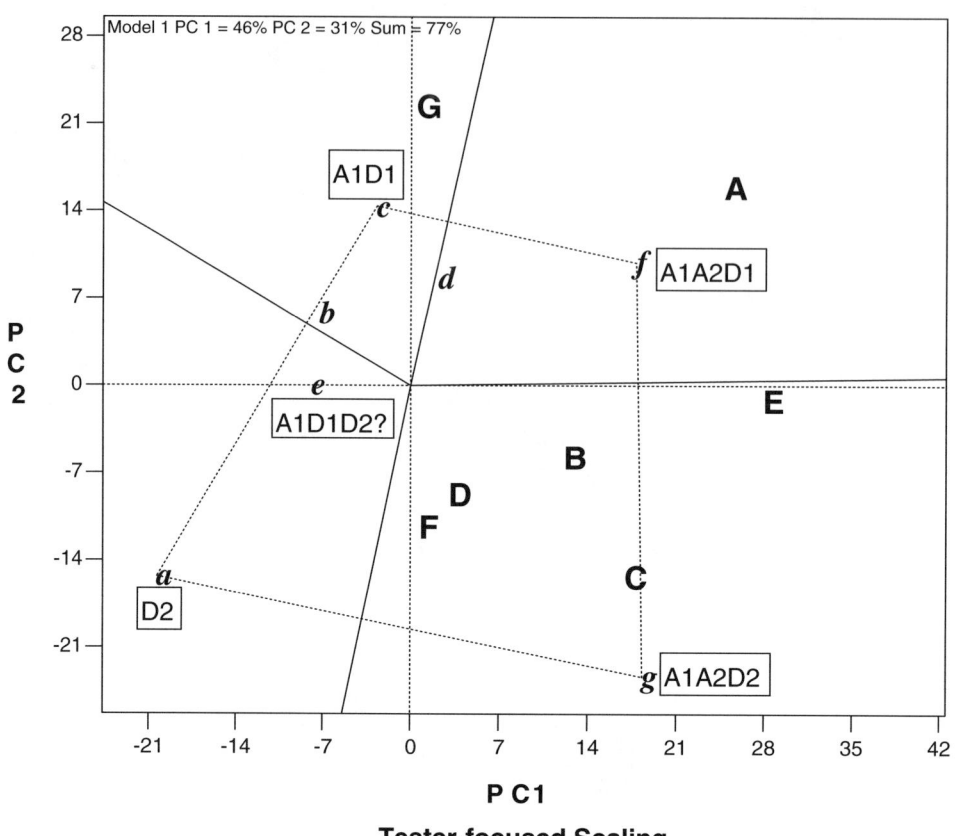

FIGURE 9.9 Possible genotypes of the parents based on both PC1 and PC2.

TABLE 9.2
Tolerance of Ten Corn Inbreds (Diagonal) and their F_1 Hybrids to Pink Stem Borer, as Measured by the Percentage of Yield Retained after Infestation

	A	B	C	D	E	F	G	H	I	J	Mean
A	85.8	89.1	86.3	82.0	86.6	92.4	82.9	88.1	84.2	86.0	86.3
B	89.1	88.3	79.7	72.4	89.6	78.5	97.6	84.8	83.0	89.8	85.3
C	86.3	79.7	89.9	88.3	97.2	86.0	72.2	92.5	74.9	82.6	85.0
D	82.0	72.4	88.3	-[a]	91.1	91.4	76.5	90.4	76.8	81.5	83.4
E	86.6	89.6	97.2	91.1	81.7	83.3	86.3	94.9	83.9	86.5	88.1
F	92.4	78.5	86.0	91.4	83.3	88.4	88.9	87.0	76.8	83.8	85.6
G	82.9	97.6	72.2	76.5	86.3	88.9	70.7	97.7	75.8	83.9	83.2
H	88.1	84.8	92.5	90.4	94.9	87.0	97.7	87.6	99.9	83.9	90.7
I	84.2	83.0	74.9	76.8	83.9	76.8	75.8	99.9	92.9	82.1	83.0
J	86.0	89.8	82.6	81.5	86.5	83.8	83.9	83.9	82.1	78.0	83.8

Note: The codes of the inbreds are: A = A509; B = A637; C = A661; D = CM105; E = EP28; F = EP31; G = EP42; H = F7; I = PB60; and J = Z77016.

[a] Missing cell replaced by its column average for completing the calculation.

Source: Data based on Butron, A. et al., *Crop Sci.*, 138:1159–1163, 1998.

Biplot Analysis of Diallel Data

9.8.1 Shrinking the Dataset by Removing Similar Parents

The polygon view of the biplot based on the 10 × 10 diallel data (Table 9.2) explained 63%, 37% by PC1 and 26% by PC2, of the total variation (Figure 9.10). This is considerably smaller than that explained by the biplot for the FHB dataset (Figure 9.1). Less variation explained by the biplot implies that some predictions based on the biplot will be less accurate. Therefore, it would be a good strategy to try to reduce the data size by removing redundant parents. Figure 9.10 suggests that A and J, C and D, and E and F are pairs of parents that are similar both as entries and testers. Therefore, parents J, D, and F (alternatively, A, C, and E) can be removed from the data without losing critical information.

9.8.2 The Best Crosses

Figure 9.11 is the polygon view of the biplot based on the subset of data with parents D, F, and J removed. It explains 73%, (53% + 20%), of the total variation. It indicates that entry *h* was the best mating partner of testers A, G, and I; entry *g* was the best mating partner of H and B; entry *c* was the best mating partner of E; and entries *c* and *h* were equally good partners of tester C. Therefore, the best hybrids for PSB resistance were: four hybrids involving H: H×A, H×G, H×I, H×C, G×B; and C×E. Sure enough, these are the best hybrids based on the data per se (Table 9.2).

9.8.3 GCA and SCA

The ATC view brought up by the GGEbiplot function *Average Tester Coordination* helps in visualizing the GCA and SCA effects of the parents (Figure 9.12). The ATC axis happens to coincide with the PC1 axis. Thus, the PC1 scores of the entries approximate their GCA effects, with $r = 0.953$. The highest PC1 entry is *h*, followed by *b*, *e*, *a*, *i*, *c*, and *g*. Entry *c* had the highest SCA; it interacted positively with itself and E but negatively with others.

9.8.4 Best Tester

The best tester in this subset was parent G, as it is closest to the ideal tester represented by the center of the concentric circles (Figure 9.13). All other parents are poor as testers, either due to lack of discriminating ability (A and E), or lack of representativeness (C and I). Parents B and H are particularly poor as testers as they have negative PC1 scores, which means that parents that produce good hybrids with B and H tend to have low, rather than high, GCA. The existence of different signs for PC1 scores is an indication of large SCA or interaction effects relative to GCA or additive effects (Yan et al., 2001).

9.8.5 Heterotic Groups

The tester vector view, brought about by the GGEbiplot function *Show Tester Vectors*, helps visualize the heterotic groups (Figure 9.14). The testers seem to fall into three groups: H and B in group 1; G and I in group 2; and C and E in group 3. Tester A falls between groups 2 and 3 and has a short vector. Groups 1 and 2 interacted positively to produce heterosis in terms of PSB resistance. They are two heterotic groups with different dominant resistance genes. C and E in group 3 interacted negatively with groups 1 and 2, suggesting recessive resistance genes. However, within group 3, C and E interacted positively.

9.8.6 Genetic Constitutions of Parents with Regard to PSB Resistance

From the perspective of the entries, PC1 is the contrast of *c*, *g*, and *i* vs. *b*, *e*, and *h*, although there are differences within each group. For simplicity, we assign dominant genes *R*1 and *R*2, respectively (Figure 9.15). This assignment implies that there could be heterosis in all crosses between the two groups.

222   GGE Biplot Analysis: A Graphical Tool for Breeders, Geneticists, and Agronomists

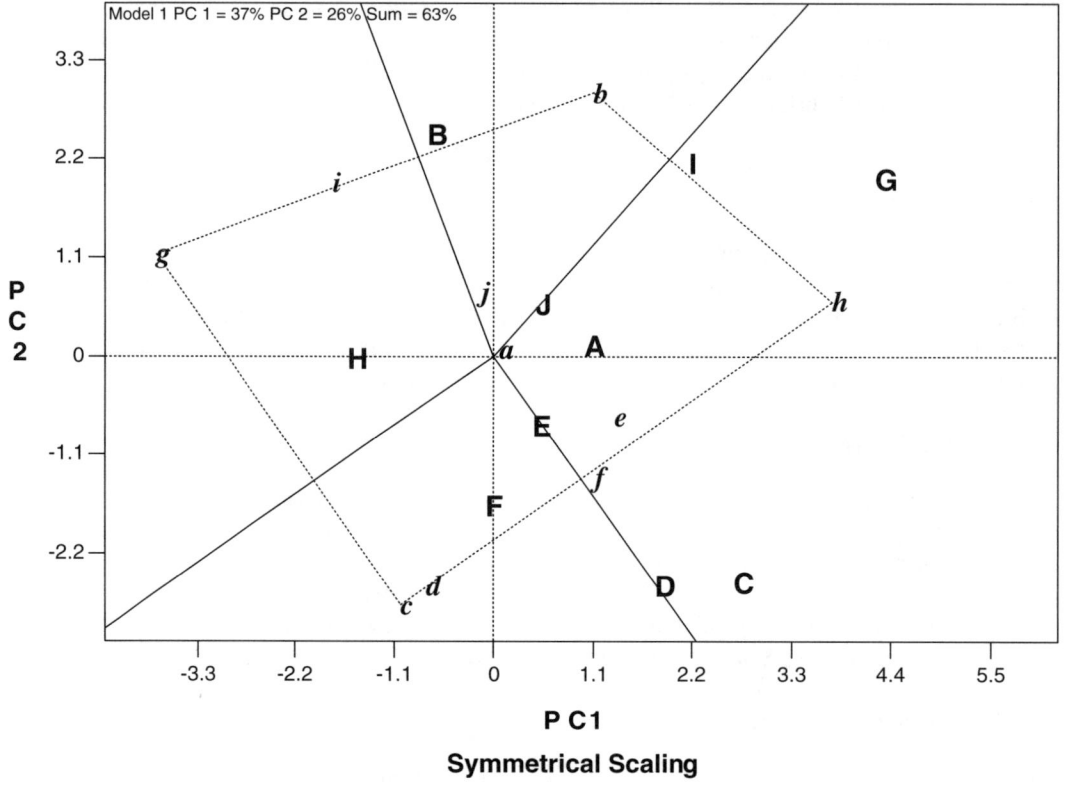

FIGURE 9.10 Polygon view of the biplot of the 10 × 10 maize diallel crosses for resistance to corn pink stem borer. The ten inbreds are in lowercase when viewed as entries and in uppercase when viewed as testers.

Biplot Analysis of Diallel Data

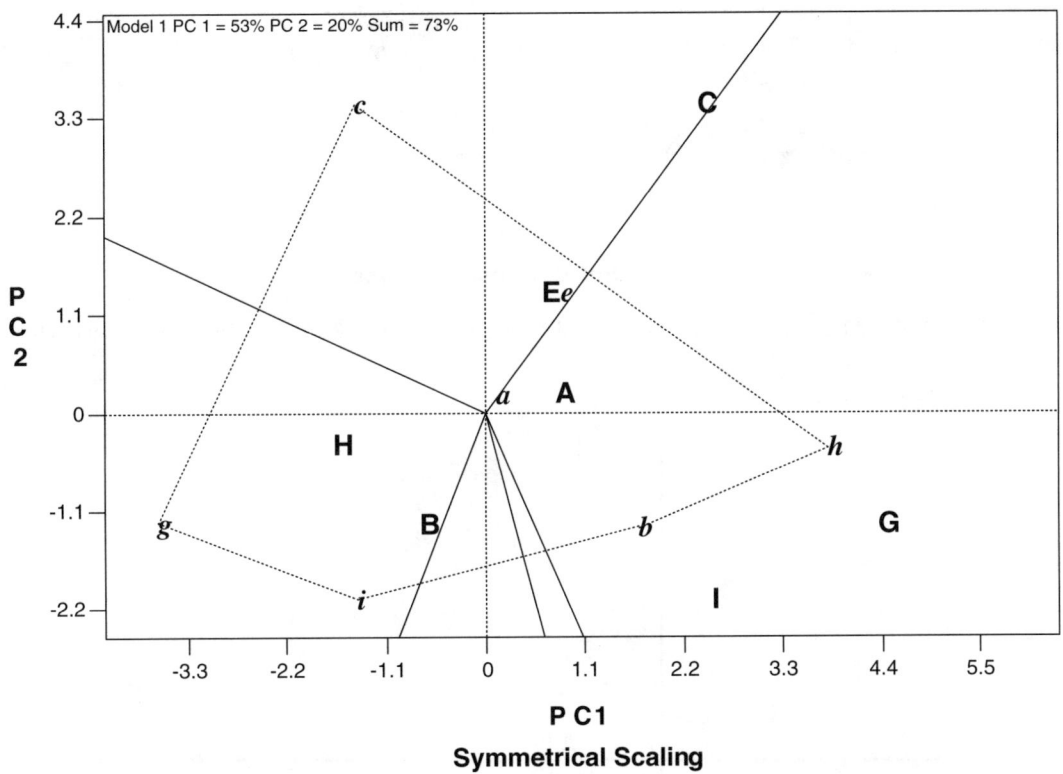

FIGURE 9.11 Polygon view of the biplot based on a subset of the maize diallel cross data; three parents (D, F, and J) were deleted from the data. Parents when viewed as entries are in lowercase italics and in uppercase when viewed as testers.

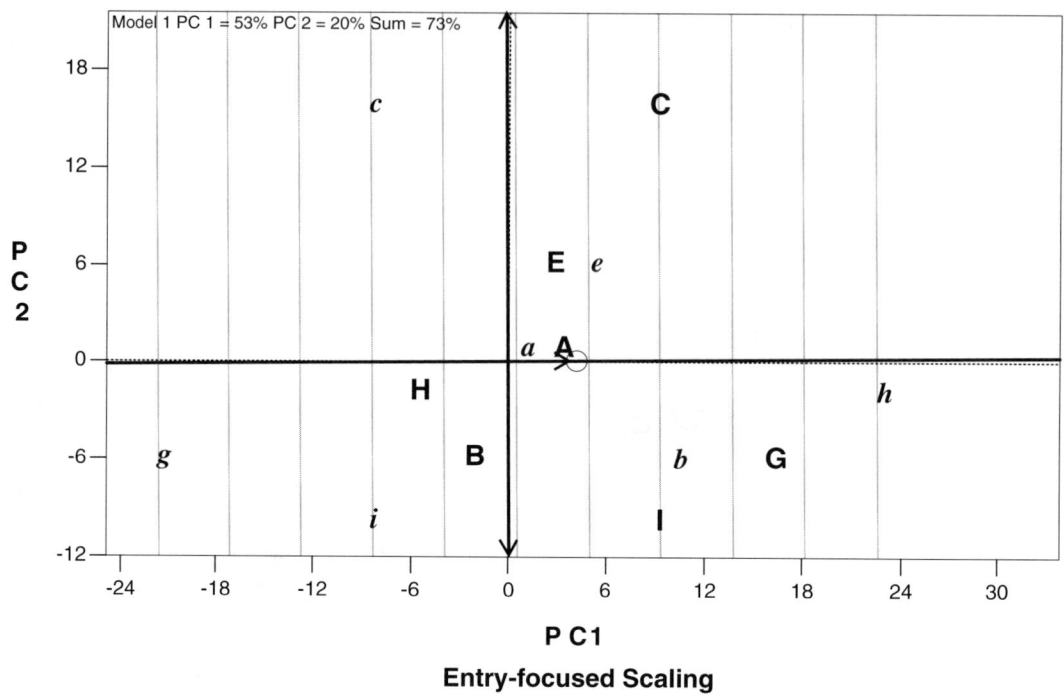

FIGURE 9.12 The ATC view of the biplot based on the maize diallel cross subset, showing the GCA of the entries.

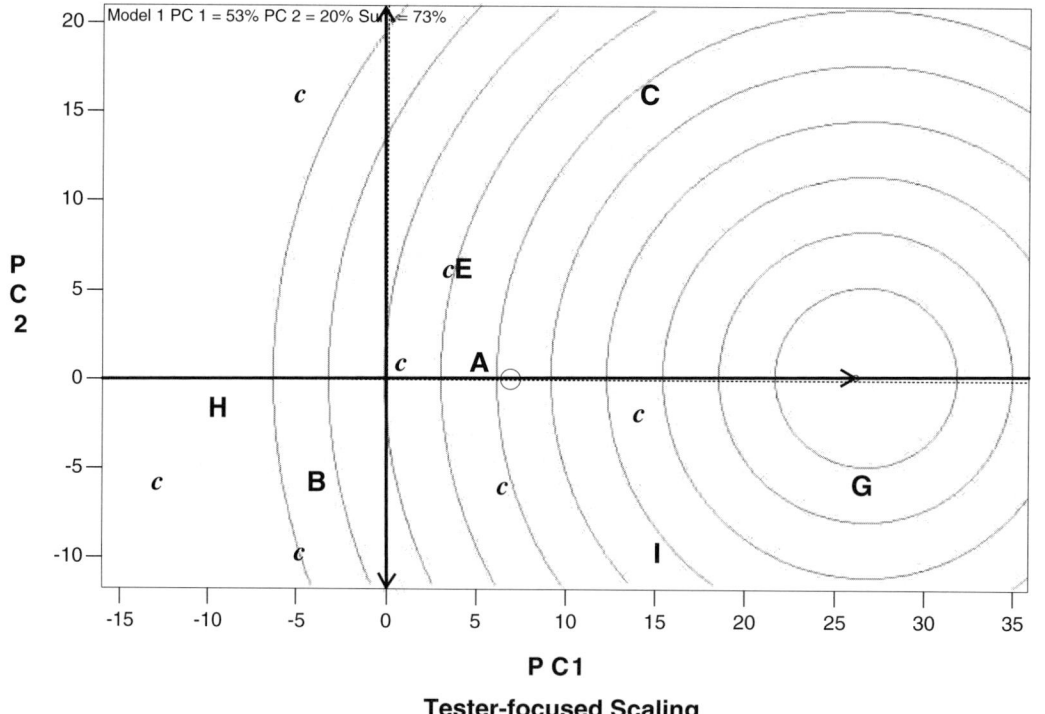

FIGURE 9.13 The ATC view of the biplot based on the maize diallel cross subset, comparing testers with the ideal tester, which is presented by the concentric center. The ideal tester is the most discriminating of entries and has no preference for mating partners.

Biplot Analysis of Diallel Data

PC2 is the contrast between c and e vs. g, i, b, and h (Figure 9.16). Interestingly, these two groups interacted negatively between groups and positively within groups, suggesting that recessive resistance genes are involved. Therefore, $r3$ and $r4$ are assigned to the two groups, respectively. Parent A is near the biplot origin, indicating lack of interaction with any of the two groups. It is, therefore, assigned the double recessive genotype $r3r4$.

Integrating Figures 9.11, 9,15, and 9.16 gives a complete picture of the PSB resistance phenotypes of the parents and their hybrids, and possible interpretations from the perspective of genetic constitutions (Figure 9.17). Again, it should be pointed out that these interpretations are only hypotheses, which have to be critically tested.

9.9 ADVANTAGES AND DISADVANTAGES OF THE BIPLOT APPROACH

The main purpose of conducting diallel analyses is to obtain information on the parents: their genetics and their potential or probability of generating superior hybrids or pureline cultivars if used as parents. In conventional diallel-cross analyses, understanding of the parents relies solely on GCA. The GCA effects are reliable for parent evaluation only if SCA effects of the crosses are negligibly small, however. If there are considerable SCA effects, GCA alone will be of limited use. Thus, we are taught in statistics and quantitative genetics courses that if there are significant interaction effects, the analysis should be focused on individual crosses rather than main effects of the parents, i.e., GCA. On the contrary, the biplot approach allows understanding of the parents through two dimensions: GCA and SCA. Since principal components are estimated through least-squares methods, PC1 always explains at least as much variation as the GCA effects, and any variation explained by PC2 is additional to GCA. Therefore, theoretically there is no doubt that a biplot of PC1 vs. PC2 is at least as effective as the conventional approach in achieving an understanding of the parents. This is the first advantage of the biplot approach over the conventional method.

A second obvious advantage of the biplot approach is its graphical presentation of the diallel data: "A picture is worth a thousand words." From a biplot, the GCA and SCA of the parents, the best crosses, the best testers, and the heterotic groups are immediately displayed. With the help of GGEbiplot, visualization of these aspects of a dataset occurs within seconds. Indeed, the biplot approach provides a nonsubstitutable means for revealing patterns of a diallel cross dataset. The GGEbiplot software also has a function *Diallel Without Parents*, which generates a biplot after the parent per se values are removed. Although results with or without parents are generally close, this provides another perspective for examining the data.

A third advantage of the biplot approach follows. Since the biplot displays a complete picture of the interrelationships among parents, it provides a unique opportunity or possibility to peek into the genetic constitutions of the parents using the F_1 rather than the F_2 generation. Although the hypotheses on the genetics of parents are preliminary at this stage, since no experimental verifications have been attempted so far, we are quite confident about the relative genetic differences among the vertex parents. Research on testing hypotheses based on the biplot pattern will have great merit. Once verified, this would be a great asset for genetics research.

The only disadvantage of the biplot approach we can think of is its lack of a measure of uncertainty. If two parents are located close to one another in the biplot, we are sure that they have similar genetic behaviors. We cannot quantify the difference, however. Also, we cannot be sure if the two parents are significantly different if they look different in the biplot. Therefore, conventional statistics, such as variance analysis on GCA and SCA, are still necessary. If both GCA effects and SCA effects were nonsignificant, there would be no need to conduct biplot analysis. In reality, however, this would be very rare, since diallel crosses are expensive, and the parents are usually carefully selected. Once either GCA or SCA is known to be significant, the lack of a measure of uncertainty will no longer be a problem, because the difference between any two parents can be

226 GGE Biplot Analysis: A Graphical Tool for Breeders, Geneticists, and Agronomists

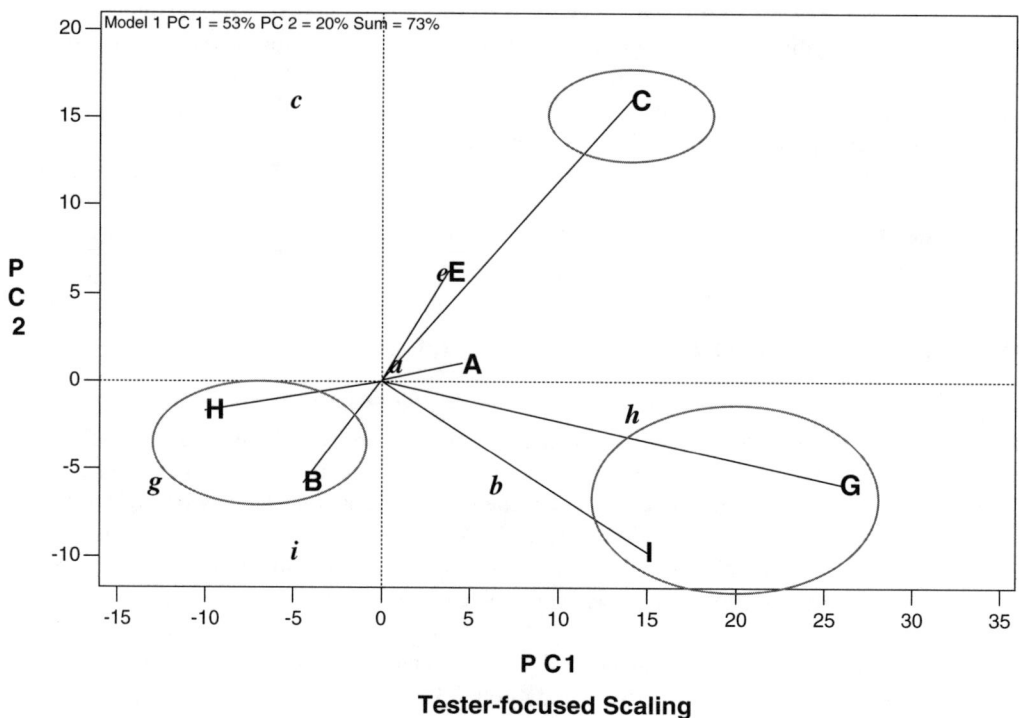

FIGURE 9.14 The tester vector view of the biplot based on the maize diallel cross subset, showing groups of testers.

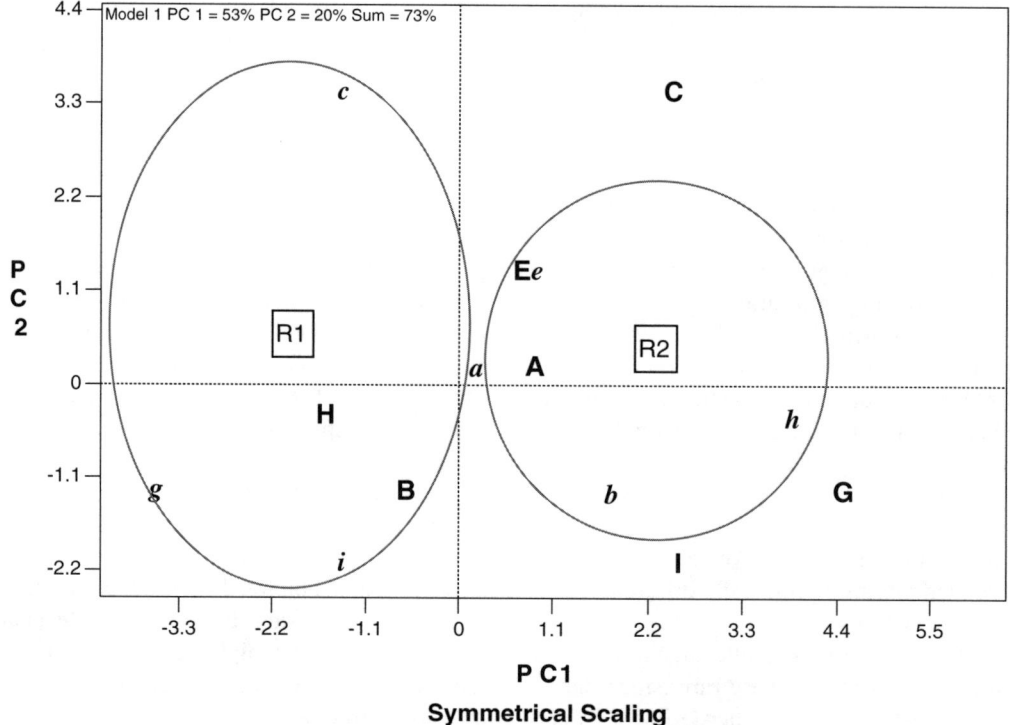

FIGURE 9.15 The proposed genotypes of the entries based on PC1.

Biplot Analysis of Diallel Data

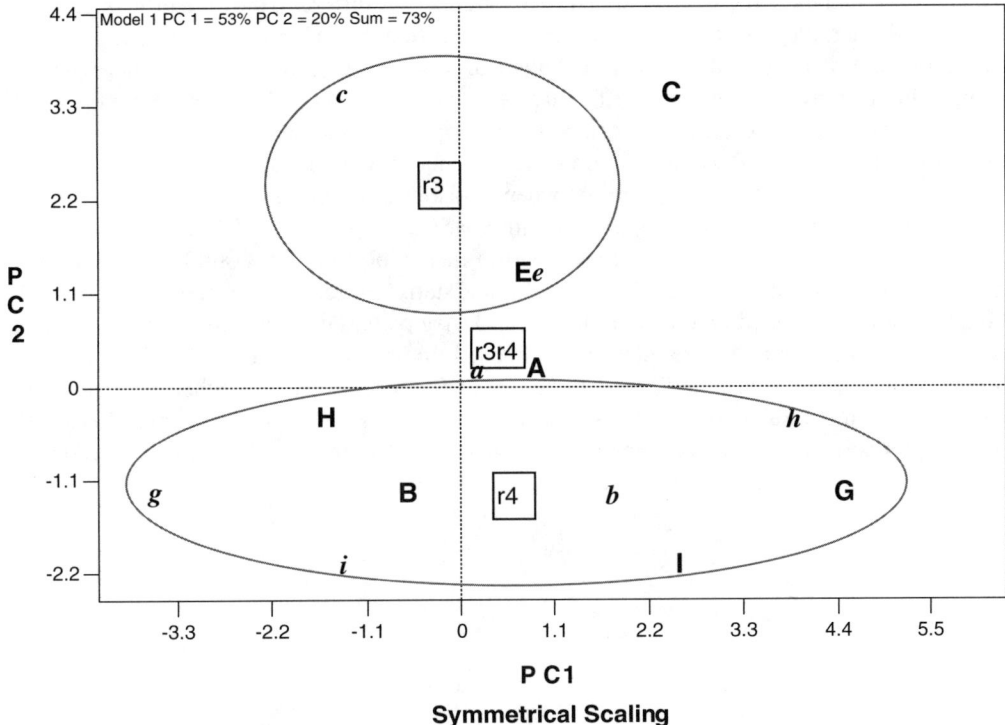

FIGURE 9.16 The proposed genotypes of the entries based on PC2.

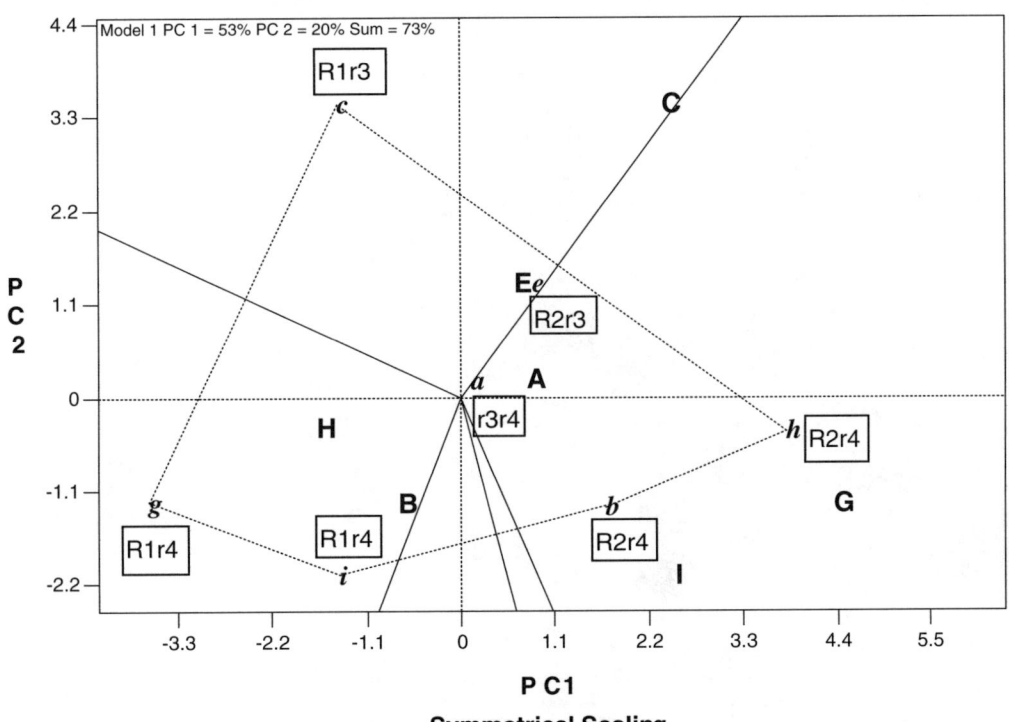

FIGURE 9.17 Proposed genotypes of the entries based on both PC1 and PC2. Minor genes may exist to differentiate between entries g and i and between entries b and h.

visualized with reference to the biplot size. If two parents look different, they are probably different. If they look similar, they are probably not very much different. Unfortunately, decision-making in scientific research is heavily dependent on statistical tests. If two things look similar, what is the point in trying to prove that they are different? Yes, statistical tests make us more confident about our conclusions. But the bottom line is that all decisions are subjective: it is subjective to choose among many testing methods, and it is subjective to choose a significance level.

Someone has said, "the relationship between statistics and agriculture is like that between a lamp post and a drunk — it is for support, not illumination." It is our belief that researchers heavily rely on statistical tests partly because they have no means of getting a complete picture of their problem, research, or dataset. In the light of a biplot, statistical tests may become less crucial for making decisions. Seasoned breeders do not make decisions based on statistical tests, not because tests are not available, but because they are not needed. In most cases, the need for a statistical test for a particular hypothesis is an indication of lack of confidence by the researcher in the hypothesis; but when this is the case, statistical tests will not dramatically increase confidence. Nonetheless, the GGEbiplot software is now equipped to do conventional statistics. This will allow researchers to examine their data both via the biplot and conventional way.

10 Biplot Analysis of Host Genotype-by-Pathogen Strain Interactions

SUMMARY

In the previous chapters, we have demonstrated biplot analyses of genotype-by-environment (GE) data (Chapter 5), genotype-by-trait data (Chapter 7), quantitative trait loci (QTL)-mapping data (Chapter 8), and diallel-cross data (Chapter 9). Another important and fascinating area of research related to plant breeding is genotype-by-pathogen interaction. Breeding for resistance to various diseases and pests has been a most important objective in most breeding programs, and an understanding of host genotype-by-pathogen strain or pest interaction for a given disease or pest is of vital importance for successful breeding.

Breeding for disease resistance is an important part of any plant-breeding project. Effective breeding for disease resistance relies on a good understanding of the host–pathogen relationships. This chapter demonstrates application of biplots in the analysis of host genotype-by-pathogen strain interactions using two contrasting datasets. The first dataset is on barley resistance to net blotch, which shows strong host genotype-by-pathogen strain specificities. In this scenario, the main effects of genotypes and strains are of little importance. It is the specific genotype-by-strain interaction that is important. The second dataset is on wheat resistance to *Fusarium* head blight (FHB), which shows no clear host–pathogen specificity. The analysis should, therefore, be concentrated on identifying the best resistant cultivars and the most virulent strains that differentiate the cultivars.

10.1 VERTICAL VS. HORIZONTAL RESISTANCE

Conceptually, there are two distinct types of host genotype-by-pathogen strain relationships: race-specific and nonspecific. In specific genotype-by-strain interaction systems, a gene-for-gene hypothesis describes that it is the interaction or compatibility between host genotype and pathogen genotype that determines whether or not a disease symptom will develop (Flor, 1946). There has been debate on the gene-for-gene interaction mechanism (Vanderplank, 1991; Johnson and Knott, 1992), but we will not delve into it. Race-specific resistance is also called vertical resistance because the response to pathogens is more or less qualitative, and there is a clear-cut differentiation between resistant and susceptible responses. A genotype is either resistant or susceptible to a strain, depending on the genetics of both the cultivar and the pathogen. Diseases caused by obligate parasites, e.g., cereal rusts (Knott, 1990, 2000) and powdery mildew (PM), are good examples of vertical resistance genotype-by-strain system. In contrast, in nonspecific genotype-by-strain interaction systems, response of host genotypes to pathogen strains is quantitative; resistant genotypes tend to be resistant to, and susceptible genotypes tend to be susceptible to, all strains of the pathogen in question, hence the term horizontal resistance. Many diseases caused by hemisaprophytes, which can live on both living and dead tissues, belong to this category.

Understanding whether resistance to a given pathogen is race-specific (vertical) or nonspecific (horizontal) is a prerequisite to selection of breeding strategies. For nonspecific resistance, since there is no clear race differentiation, universal resistance can be sought and selected for. All virulent strains of the pathogen can be used in selection, but some strains may be more effective in discriminating host genotypes. For race-specific resistance, no universal resistance can be sought since each resistance gene is effective against a specific race. Resistance to multiple strains can be achieved, however, via pyramiding different resistance genes.

Technically, whether a host–pathogen relation is race-specific or nonspecific is detected by the presence or absence of crossover interactions between host genotypes and pathogen strains. As already demonstrated in previous chapters, genotype main effect and genotype-by-environment (GGE) biplot is thus far the best method for visualizing important crossover interactions. In this chapter, we will demonstrate the use of GGEbiplot to study genotype-by-strain interactions using published data of barley net blotch and wheat FHB.

10.2 GENOTYPE-BY-STRAIN INTERACTION FOR BARLEY NET BLOTCH

Barley net blotch, caused by *Pyrenophora teres* Drechs, is an important disease throughout the world. Gupta and Loughman (2001) provided a report on current virulence of this disease on barley in Western Australia. They collected 79 net blotch isolates from infected leaf samples across all barley-growing regions of Western Australia, nine historical isolates collected and lyophilized during 1975 to 1985, and one isolate from Eastern Australia. These isolates that were were tested on 47 barley lines. On the basis of seedling responses, the pathogen isolates were clustered into 8 groups and the barley lines were clustered into 13 groups (Table 10.1). Gupta and Loughman (2001) discussed the relationships among the barley line groups as well as among the net blotch isolate groups. Let's see what biplots can tell us about these relationships.

10.2.1 Model for Studying Genotype-by-Strain Interaction

The model for constructing a genotype-by-strain biplot is the basic model for GGE biplot described in Equation 4.5, which is given below. All definitions are the same except that the terms genotype and environment used in Chapter 4 are replaced with entry and tester, respectively.

TABLE 10.1
Mean Responses of 13 Barley Line Groups (LG) to 8 *Pyrenophora teres* Isolate Groups (IG) in Scales from 0 to 9, with Larger Values Indicating Greater Susceptibility

Barley Groups	Net Blotch Groups of Isolates							
	IG86	IG75	IG87	QNB85I	IG88	IG84	IG48	IG78
CAMEO	6.0	4.3	5.8	5.0	4.9	6.1	4.0	4.7
CLIPPER	5.2	5.0	5.5	5.0	6.4	6.9	4.0	3.5
BETZES	7.7	6.3	7.3	5.0	6.9	6.4	5.0	5.2
LG30	7.6	7.1	7.9	8.0	7.1	7.4	7.2	6.5
LG10	2.1	2.0	1.9	2.0	1.9	2.0	4.9	4.0
LG16	1.8	1.8	1.9	1.9	2.1	2.0	2.3	1.9
LG32	2.3	2.1	2.3	2.8	2.9	2.8	2.9	2.6
LG8	2.3	2.0	2.0	2.0	2.3	2.2	7.8	6.8
LG31	3.6	2.7	2.7	2.0	3.6	3.6	7.4	6.4
LG17	4.7	4.6	4.8	4.0	4.9	5.0	3.9	4.2
LG22	5.1	4.1	5.0	8.0	4.7	5.0	4.6	4.7
LG33	4.5	3.6	4.5	4.7	4.2	4.1	4.6	4.4
LG34	2.8	2.7	3.5	4.0	3.8	3.8	3.0	2.6

Source: Data from Gupta, S. and Loughman, R., *Plant Dis.*, 82:960–966, 2001.

$$\hat{Y}_{ij} - \mu - \beta_j = g_{i1}e_{1j} + g_{i2}e_{2j} + \varepsilon_{ij}$$

where:

$\hat{Y}_{ij}$ = the expected value for entry i in combination with tester j,
μ = the grand mean of all entry–tester combinations,
β_j = the main effect of tester j,
g_{i1} and e_{1j} = the first principal component (PC1) scores for entry i and tester j, respectively,
g_{i2} and e_{2j} = the second principal component (PC2) scores for entry i and tester j, respectively, and
ε_{ij} = the residue for each entry–tester combination not explained by PC1 and PC2.

A GGE biplot is constructed by plotting g_{i1} against g_{i2}, and e_{1j} against e_{2j} in a single scatter plot.

In MET data (Chapter 5) and genotype-by-trait or marker data (Chapters 7 and 8), the genotypes are naturally the center of research, and hence are regarded as the entries in the entry-by-tester two-way data. In a genotype-by-strain two-way data, however, the question of which factor is the entry is determined by the researcher's interest. If the primary focus is in the resistance or susceptibility of host genotypes, genotypes should be treated as entries; in the contrary, if the focus is on the virulence or avirulence of pathogen strains, strains should be treated as entries. Since we are interested in both genotypes and strains, the genotype-by-strain data should be examined in both ways. This is a challenging task using conventional methods; but using the GGEbiplot software, switching between the two modes is achieved quickly by clicking on the function *Tester/Entry Switch Roles*.

10.2.2 BIPLOTS WITH BARLEY LINES AS ENTRIES

To study the resistance or susceptibility of the barley line groups to various isolates, we use the barley line groups (LG) as entries (Figure 10.1). The biplot generated using GGEbiplot (Figure 10.1) contains the following background information: 1) barley LG are treated as entries and isolate groups (IG) as testers, as evidenced by the setting that LG are in italics and testers are in uppercase; 2) the tester-centered data model, i.e., Equation 4.5, is used as indicated by model 1; 3) the singular value is partitioned symmetrically into entry and tester eigenvectors as indicated by the biplot title, symmetric scaling, below the biplot; and 4) the biplot explained 94% of the total variation of the tester-centered data. Thus, the biplot explained most of the total variation, which is an indication that some of the testers are closely related.

Recall from the previous chapters, the polygon is drawn on barley LG that are located away from the biplot origin such that all other barley groups are contained in the polygon. Lines perpendicular to the sides of the polygon are drawn, starting from the biplot origin, to divide the biplot into sectors. There is a vertex LG for each sector, so the sectors can be named after the vertex LG. For example, there are four sectors in Figure 10.1: LG30, LG8, LG16, and Clipper. These four LG are the extremes of all LG; they are either the most resistant or most susceptible LG to one or more of the IG. Figure 10.1 also indicates the following — LG31 is similar to LG8; LG32 is similar to LG16; LG22 and Cameo are similar to Clipper; LG10 is intermediate between LG8 and LG16; LG34 is intermediate between LG16 and Clipper; and Betzes (hidden) is intermediate between Clipper and LG30. The biplot pattern among the LG is based on their responses to the IG that fall into two clusters relative to their discrimination of the LG: IG48 and IG78 as one cluster and the other six IG as the other.

An interesting property of the polygon view is that in a sector, the vertex LG has the highest values or susceptibility for all IG within that sector. Thus, LG30 was most susceptible to both clusters of IG; and LG8 was equally susceptible to the isolate cluster consisting of IG48 and IG78. LG8 differs from LG30 in that LG8 is less susceptible, or more resistant, to the isolate cluster consisting of six IG.

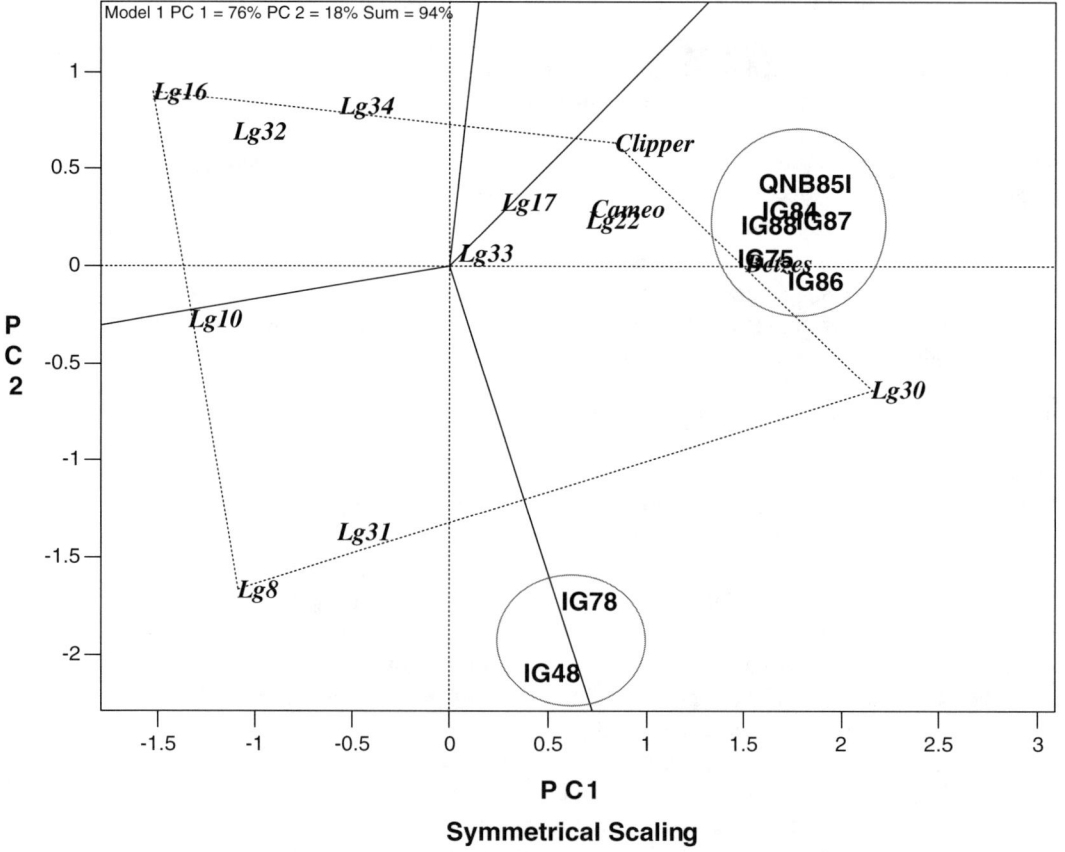

FIGURE 10.1 GGE biplot based on disease scores indicating susceptibility of 13 barley cultivars or LG inoculated with 8 net blotch IG. The barley lines are used as entries — in italics — and the IG are used as testers — uppercase.

Since the disease responses were scored such that a larger value means greater susceptibility (Table 10.1), Figure 10.1 is most appropriate for comparing the susceptibility among barley lines. Using the GGEbiplot function *Reverse Sign of All Testers*, Figure 10.1 will be transformed into Figure 10.2 instantly, which is more appropriate for comparing the resistance among barley lines. The IG are underlined in Figure 10.2 to indicate that their signs were reversed. Figure 10.2 reveals exactly the same line-by-isolate interaction pattern as Figure 10.1. It explicitly indicates, however, that the barley LG16, clustered with LG32 and LG34, was most resistant to both clusters of IG. Barley LG8, clustered with LG31, was as resistant as LG16 to the large cluster of isolates but was less resistant than LG16 to IG48 and IG78. Since barley LG10 — partially hidden — is located between LG16 and LG8, it is also equally resistant to the large cluster of isolates. Both Figures 10.1 and 10.2 indicate that LG30 was most susceptible, and LG16 most resistant, to all groups of isolates, which can be verified from Table 10.1.

There is no doubt that the six IG in the large cluster of isolates were similar in differentiating the barley lines. The question, "Are there sub-clusters among them?" can be examined by deleting IG48 and IG78. The GGEbiplot function *Run Subset by Removing Testers* provides a convenient way to study a subset of the original data. With the removal of IG48 and IG78, Figure 10.1 is transformed into Figure 10.3. The difference between isolate QNB85i (from Eastern Australia) and other isolate groups (IG75, IG84, IG86, IG87, and IG88; all from Western Australia) becomes obvious. Figure 10.3 reveals that QNB85i differs from the Western Australian isolates in that it separated barley LG22 from Cameo and Clipper; LG22 is more susceptible to QNB85i than Cameo and Clipper. LG22 is as susceptible as LG30 to QNB85i but is not as susceptible to the other IG.

Again, the barley lines can be visually compared for their resistance with the signs of the testers or IG, as in our case, reversed. Using the function *Reverse Sign of All Testers*, Figure 10.3 is transformed into Figure 10.4. It reveals that a cluster of barley lines represented by LG10 and LG16 was most resistant to all IG. Barley LG31 was as resistant as these barley lines to isolate QNB85i but was not as resistant to other IG.

Are there any major differences among IG75, IG84, IG86, IG87, and IG88? This again can be investigated by constructing a biplot by removing isolate group QNB85i (Figure 10.5). The remaining five IG separated into two apparent clusters — IG84 and IG88 as one cluster and IG75, IG86, and IG87 as the other. To both clusters of IG, however, LG30 was the most susceptible. Therefore, there were no major crossover interactions between the five IG and the barley LG. Only minor crossovers exist between the two IG clusters in virulence to barley lines Clipper and Betzes (Figure 10.5).

By reversing the signs of the testers, IG, Figure 10.5 is transformed into Figure 10.6, which does, in fact, reveal crossovers between the aforementioned two clusters of IG: barley LG10 was most resistant to IG84 and IG88, whereas LG16 was most resistant to the other three IG. Since LG10 and LG16 are closely located, the crossover is very minor. Therefore, the five Western Australian IG can be regarded as homogeneous.

In conclusion, biplots with barley lines or LG as entries revealed clear genotype-by-strain interactions. Specifically, the biplots revealed that barley lines or line groups LG30, LG8, LG16, Clipper, LG31, LG22, and Betzes interacted differentially with three clusters of net blotch IG. The first isolate cluster consists of IG48 and IG78, the second consists of a single Eastern Australian isolate QNB85i, and the third consists of five Western Australian IG: IG75, IG84, IG86, IG87, and IG88.

10.2.3 BIPLOTS WITH NET BLOTCH ISOLATES AS ENTRIES

If the main purpose is to study the virulence or avirulence of the net blotch isolates, the IG should be treated as entries (Figure 10.7). In Figure 10.7, three clusters of IG revealed in the previous discussion are immediately obvious — cluster I: IG48 and IG78 are most virulent to genotype groups LG8 clustered with LG31 and LG10; cluster II: QNB85i is most virulent to LG22; and cluster III: IG75, IG84, IG86, IG87, and IG88 are most virulent to barley cultivar Betzes.

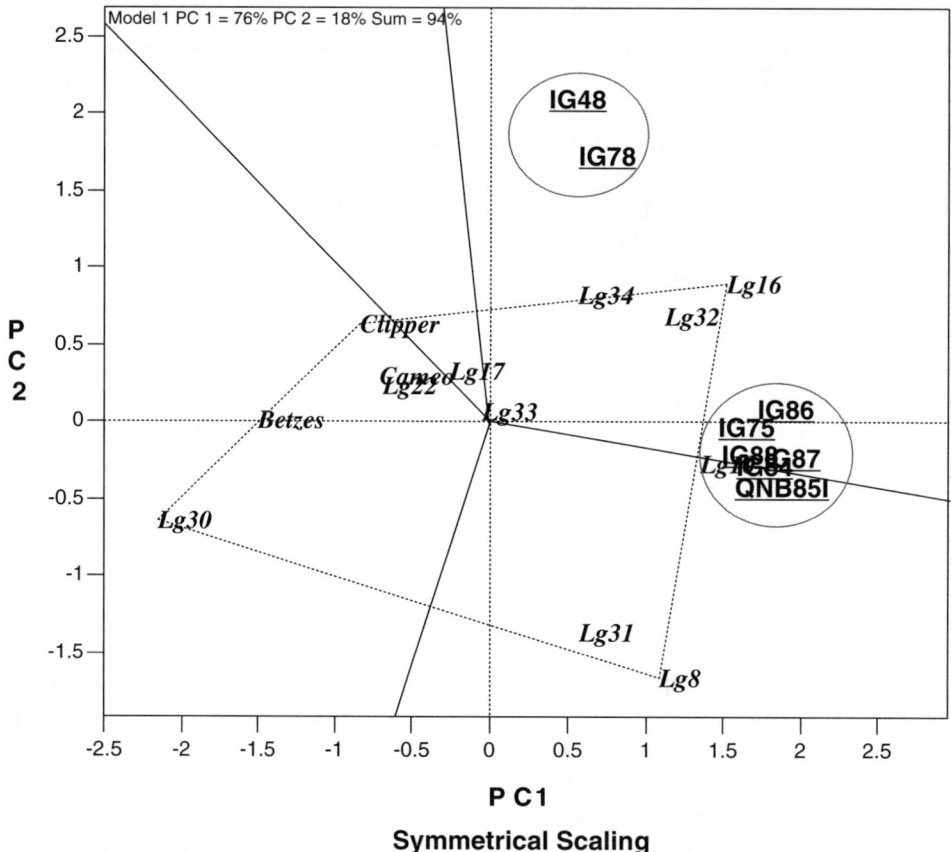

FIGURE 10.2 GGE biplot based on reversed disease scores indicating resistance of 13 barley cultivars or LG inoculated with 8 net blotch IG. The barley lines are used as entries — in italics — and the IG are used as testers — uppercase.

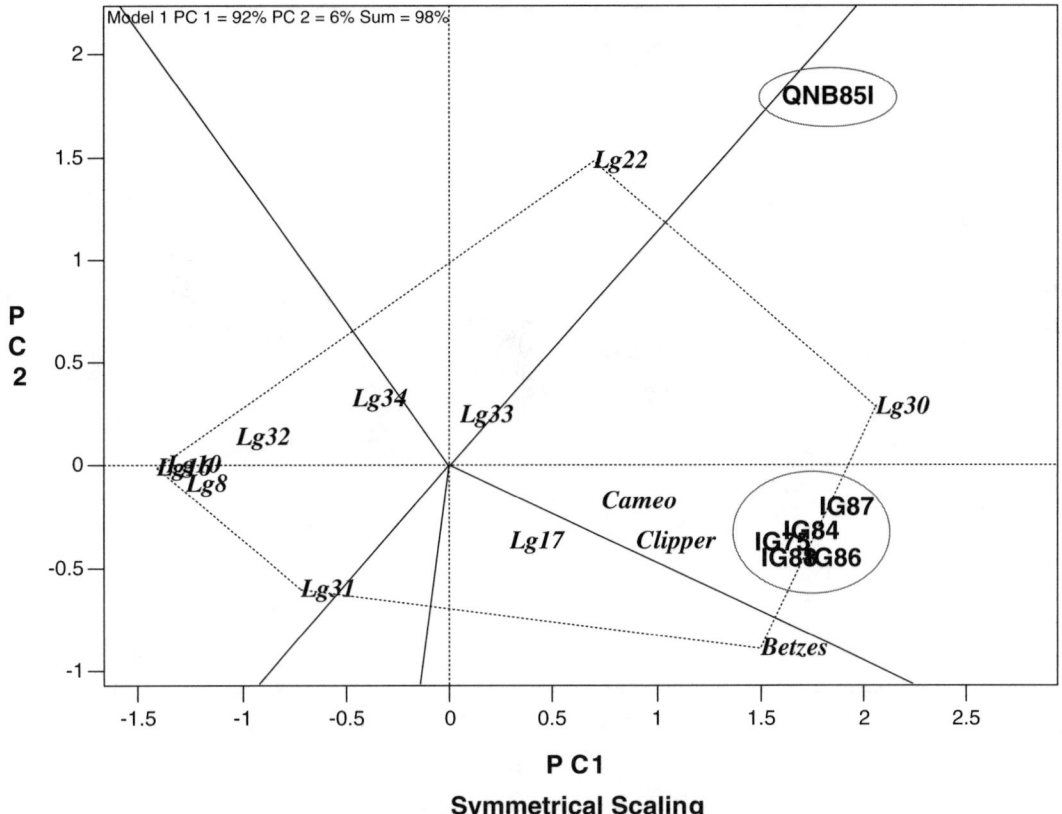

FIGURE 10.3 GGE biplot based on disease scores indicating susceptibility of 13 barley cultivars or LG inoculated with 6 net blotch IG. The barley lines are used as entries — in italics — and the IG are used as testers — uppercase.

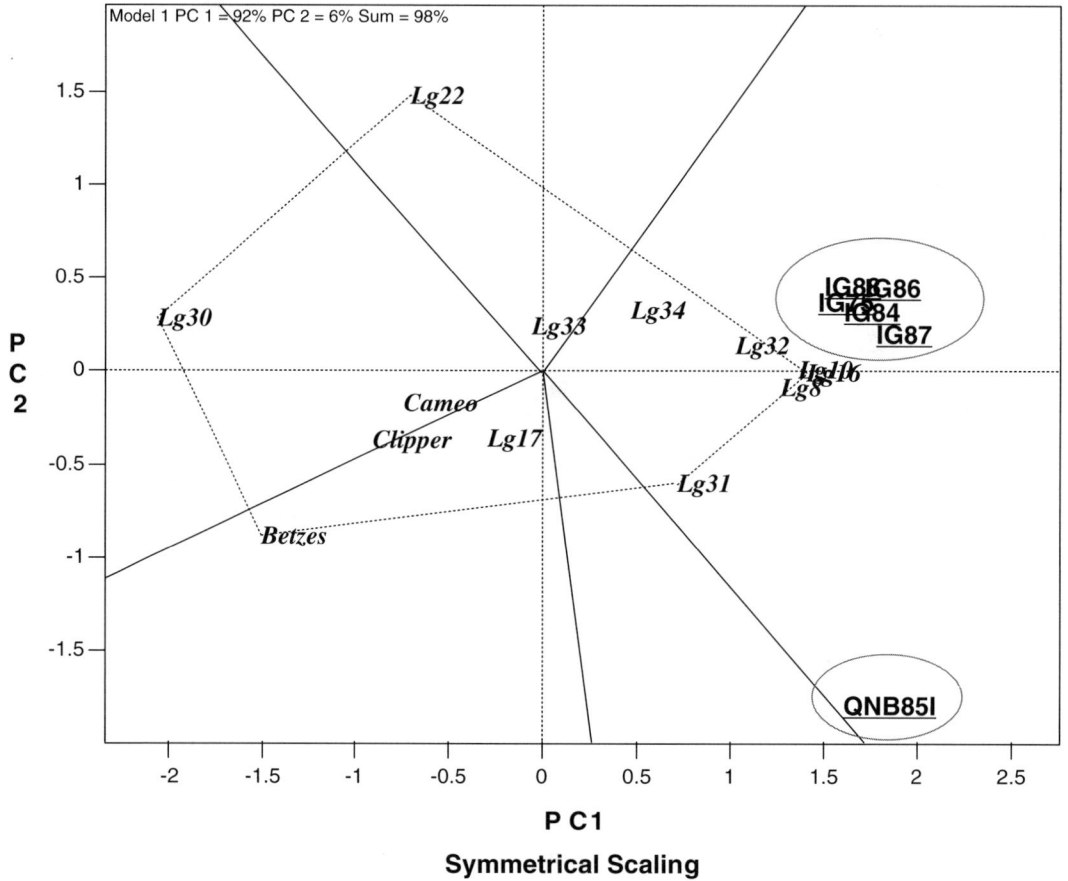

FIGURE 10.4 GGE biplot based on reversed disease scores indicating resistance of 13 barley cultivars or LG inoculated with 6 net blotch IG. The barley lines are used as entries — in italics — and the isolate groups are used as testers — uppercase.

Biplot Analysis of Host Genotype-by-Pathogen Strain Interactions   237

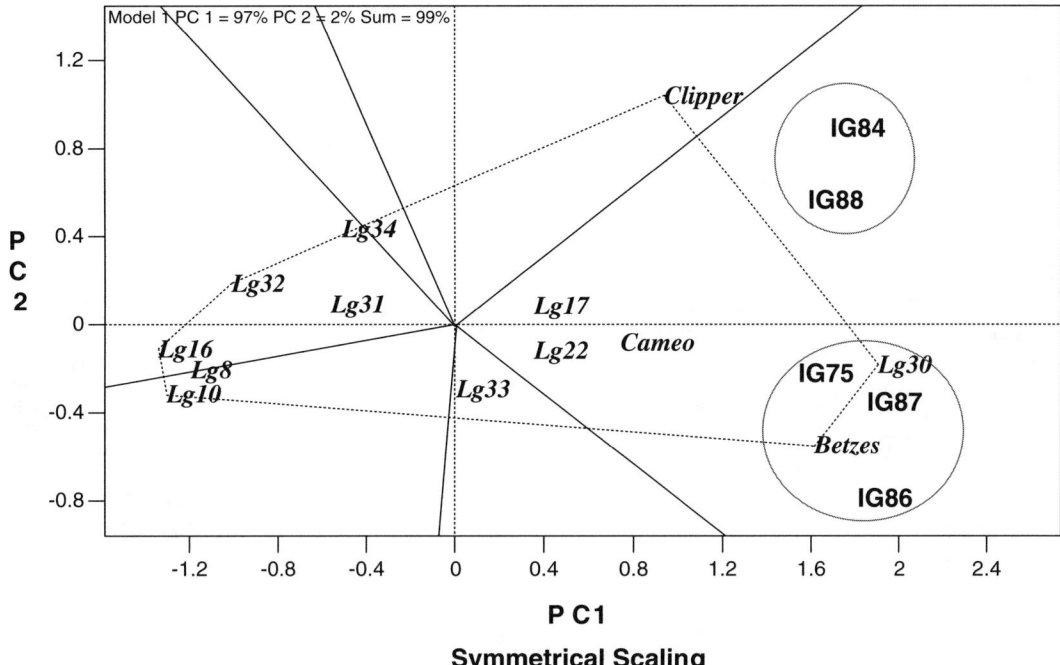

FIGURE 10.5 GGE biplot based on disease scores indicating susceptibility of 13 barley cultivars or LG inoculated with 5 net blotch IG from Western Australia. The barley lines are used as entries — in italics — and the isolate groups are used as testers — uppercase.

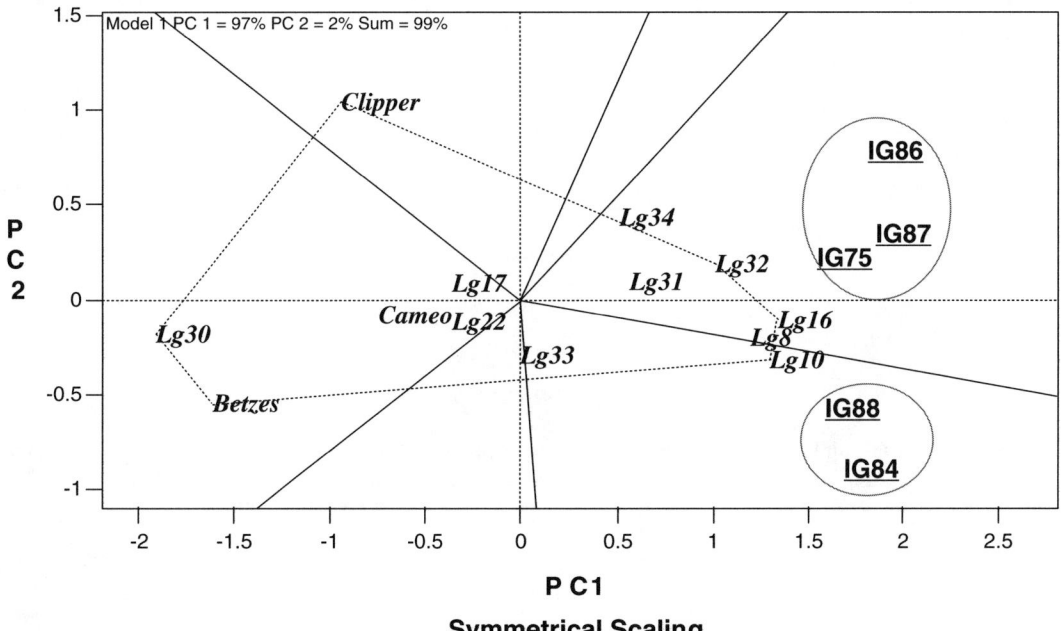

FIGURE 10.6 GGE biplot based on reversed disease scores indicating resistance of 13 barley cultivars or LG inoculated with 5 net blotch IG from Western Australia. The barley lines are used as entries — in italics — and the isolate groups are used as testers — uppercase.

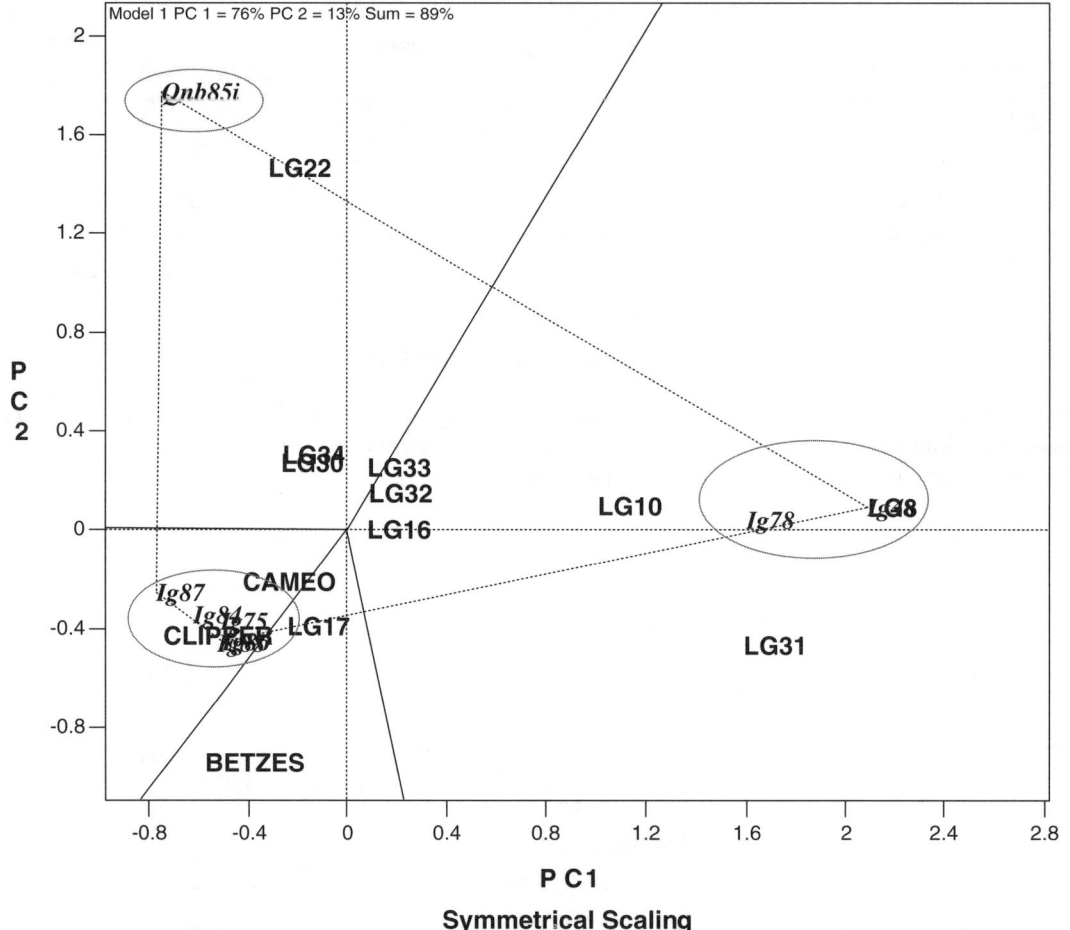

FIGURE 10.7 GGE biplot based on disease scores of 13 barley cultivars or LG inoculated with 8 net blotch IG. The IG are used as entries — in italics — and the barley LG are used as testers — uppercase.

Within cluster III, the five IG fell into three sub-clusters — cluster 1: IG86 and IG87 is most virulent to cultivar Betzes; cluster 2: IG84 and IG88 is most virulent to cultivar Clipper; and cluster 3: IG75 is an isolated cluster and is least virulent to all barley lines, particularly Cameo (Figure 10.8). IG84 and IG88 are similar, IG84 being more virulent to most, if not all, barley lines; IG86 and IG87 are similar, but it appears IG86 is more virulent to Betzes, whereas IG87 is more virulent to Clipper (Figure 10.8).

To summarize, both biplots using barley lines as entries and biplots using net blotch isolates as entries reveal the same relationship among the eight IG and their interactions with the barley genotypes, as summarized in Figure 10.9. The eight IG were first differentiated into two clusters: a small cluster consisting of IG48 and IG78, and a large cluster consisting of the other six IG. The major difference between the two clusters lies in their virulence to barley LG8, and to a lesser extent, LG31. The Eastern Australian isolate QNB85i was separated from the five Western Australian IG mainly due to their differential virulence to barley LG22 and LG31. The five western IG were further divided into three small clusters due to their differential virulence to cultivar Cameo, Clipper, and Betzes. A linear dendrogram such as Figure 10.9 is a common way to depict interrelationships among genotypes or isolates. Through this example, we hope readers will agree with us that a two-way biplot is much more informative.

As an aside, Figure 10.9 represents slightly different conclusions from those of the original authors (Gupta and Loughman, 2001). For example, they concluded IG86 was more similar to IG75 than to IG87, which is not seen from the biplots. The discrepancy may be due to the fact that the dendrogram presented in Gupta and Loughman (2001) was based on cluster analysis of the full dataset of 79 isolates by 47 barley lines, whereas the biplots were based on the summary data of the cluster analysis.

10.3 GENOTYPE-BY-STRAIN INTERACTION FOR WHEAT *FUSARIUM* HEAD BLIGHT

FHB has been a devastating disease of wheat throughout the world. It not only reduces grain yield and quality but also produces poisonous substances (deoxynivalenol, DON or vomitoxin) in wheat grains. Therefore, breeding for wheat FHB has become a major objective in many wheat-breeding programs. The pathogens that cause wheat FHB are *Fusarium graminearum* and *Fusarium culmorum,* which are non-obligate parasites affecting many different plant species. It is generally believed that there are no race specificities within the pathogen, though little evidence is available. Snijders and van Eeuwijk (1991) reported on this problem. They inoculated 17 wheat genotypes with 4 *F. culmorum* strains (39, 329, 348, and 438 collected from wheat seeds, culm, head, and sheath, respectively) during 1986 to 1988. Percentage of spikelets infected in inoculated population was reported (Table 10.2). Their analysis using additive main effect and multiplicative interaction (AMMI) indicated significant genotype-by-strain interaction but suggested that the interaction was mainly year-dependent.

The biplot based on data in Table 10.2, using wheat genotypes as entries, is presented in Figure 10.10. Crossover genotype-by-strain-by-year interactions are apparent, since the year–strain combinations fall into three sectors. Notably, genotype SVP13 is most susceptible to five year–strain combinations, particularly to strain 39 in 1987 (87–39); genotype Nautica was most susceptible to strain 39 in 1986 (86–39); genotype SVP14 was most susceptible to a few year–strain combinations, e.g., 86–438 and 87–329, although these year–strain combinations were not highly discriminating; they are located near the biplot origin.

When the GGEbiplot function *Reverse the Sign of All Testers* is used, the signs of the disease scores are reversed. Figure 10.10 is transformed into Figure 10.11, which allows visualization of resistance, as opposed to susceptibility of the genotypes. Compared with Figure 10.10, the crossover genotype-by-strain-by-year interactions are less apparent in Figure 10.11, as almost all year–strain combinations have fallen in the sector of SVP1. This indicates that genotype SVP1, along with a group of genotypes that are closely located, was most resistant to almost all year–strain combinations. Genotype SVP3 appears to be slightly more resistant than SVP1 to three year–strain combinations: 87–39, 87–436, and 88–348.

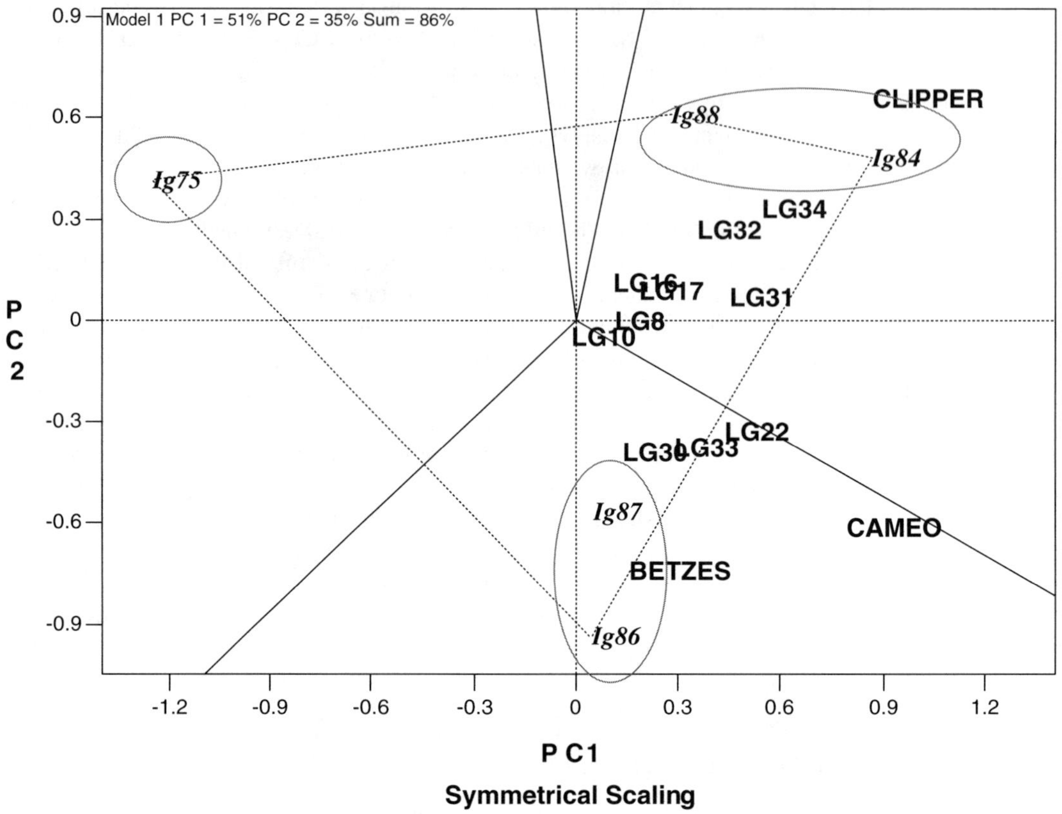

240 GGE Biplot Analysis: A Graphical Tool for Breeders, Geneticists, and Agronomists

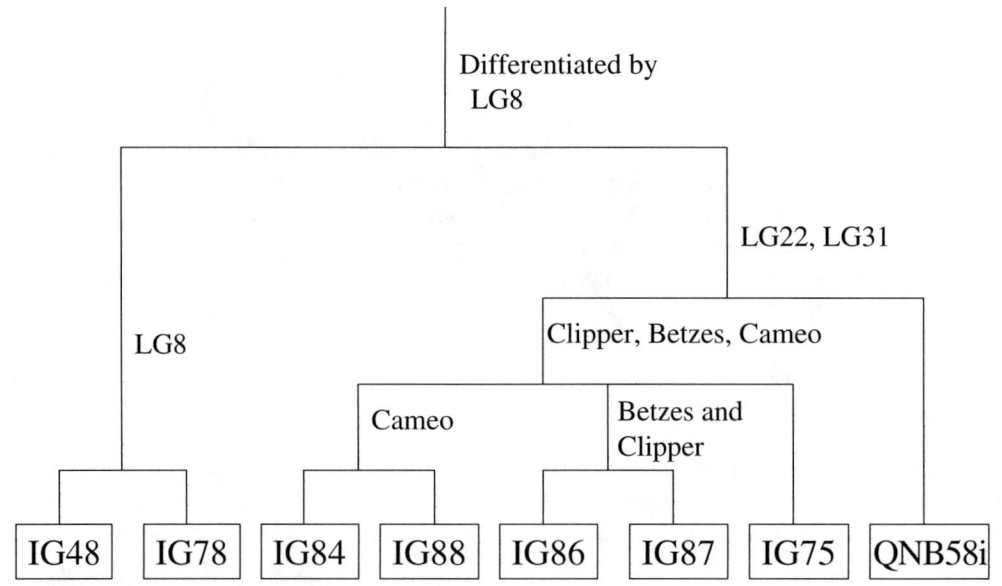

FIGURE 10.8 GGE biplot based on disease scores of 13 barley cultivars or LG inoculated with 5 net blotch IG from Western Australia. The IG are used as entries — in italics — and the barley LG are used as testers — uppercase.

FIGURE 10.9 Dendrogram showing clusters of IG based on their differential virulence to barley genotypes and LG.

TABLE 10.2
FHB Incidence of 17 Wheat Genotypes for 4 *F. culmorum* Strains in 3 Years

| Wheat Genotypes | Fusarium Strains ||||||||||||
| | 1986 |||| 1987 |||| 1988 ||||
	39	329	348	438	39	329	348	436	39	329	348	436
SVP1	2.0	1.5	3.0	1.5	7.3	0.8	0.5	2.1	5.3	2.7	2.0	2.7
SVP2	9.0	1.0	3.0	1.5	13.5	0.3	0.1	0.2	7.0	1.7	3.3	3.7
ARINA	8.0	2.5	5.0	4.0	12.0	0.3	0.1	2.8	6.3	1.0	2.0	3.3
SVP3	18.0	1.0	4.5	7.0	8.9	0.2	0.1	0.7	2.0	2.3	1.3	1.7
SVP4	6.0	3.0	1.0	1.5	11.1	0.1	0.4	2.7	7.0	4.7	5.0	8.3
SAIGA	4.5	7.5	4.5	9.0	15.5	1.1	0.3	1.7	9.3	9.7	6.0	11.7
SVP5	9.0	13.0	1.0	2.5	17.8	0.6	0.4	8.4	15.7	5.3	3.3	4.7
SVP6	23.5	4.0	9.5	3.5	16.3	0.7	0.4	1.4	5.0	6.3	7.3	13.7
SVP7	27.5	4.5	2.5	8.5	35.2	1.9	0.5	5.8	3.3	6.3	4.0	8.3
SVP8	11.0	3.5	3.0	1.5	54.0	1.2	1.2	19.3	4.0	2.7	1.3	6.3
SVP9	60.0	7.0	7.5	9.0	36.3	3.0	0.5	7.5	13.0	4.7	5.0	5.7
SVP10	47.0	18.0	14.5	22.5	44.1	5.4	3.5	11.6	6.7	12.5	3.3	4.0
SVP11	67.5	16.0	17.5	17.0	34.2	7.0	2.5	9.3	11.0	9.0	6.0	25.7
SVP12	25.5	5.0	5.0	10.5	69.3	5.0	1.7	13.2	30.0	22.3	20.0	20.3
NAUTICA	62.5	20.5	20.0	30.5	32.2	1.3	0.8	4.8	40.0	25.0	18.0	20.3
SVP13	32.5	5.0	9.0	42.5	57.3	5.2	4.3	30.5	37.7	14.7	38.3	31.0
SVP14	62.5	16.5	27.5	23.0	58.5	3.7	2.7	21.7	36.7	26.3	13.3	20.3

Source: Data from Snijders, C. H. A. and Van Eeuwijk, F. A., *Theor. Appl. Genet.*, 81:239–244, 1991.

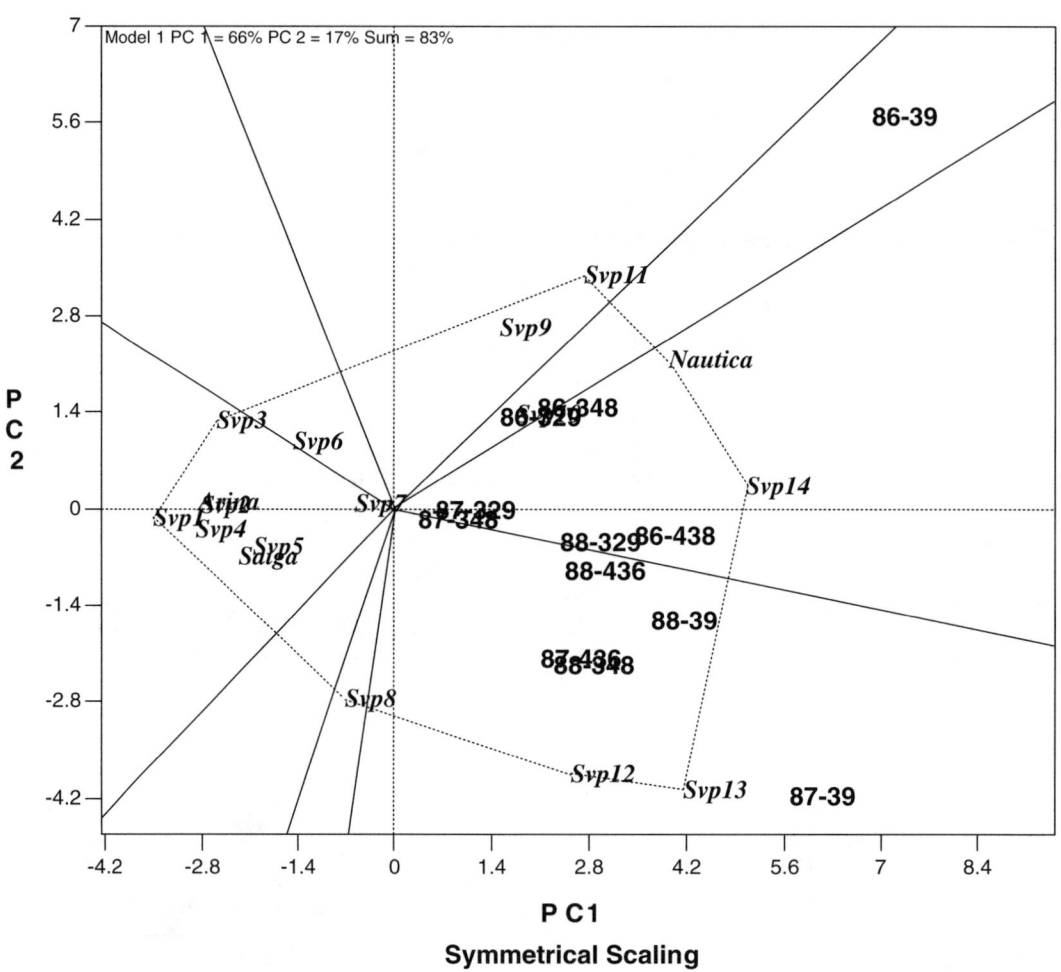

FIGURE 10.10 GGE biplot based on disease scores indicating susceptibility of 17 wheat genotypes inoculated with 4 *Fusarium culmorum* strains in 1986 to 1988. The wheat genotypes are used as entries — in italics — and the *Fusarium* strains are used as testers — uppercase.

Biplot Analysis of Host Genotype-by-Pathogen Strain Interactions

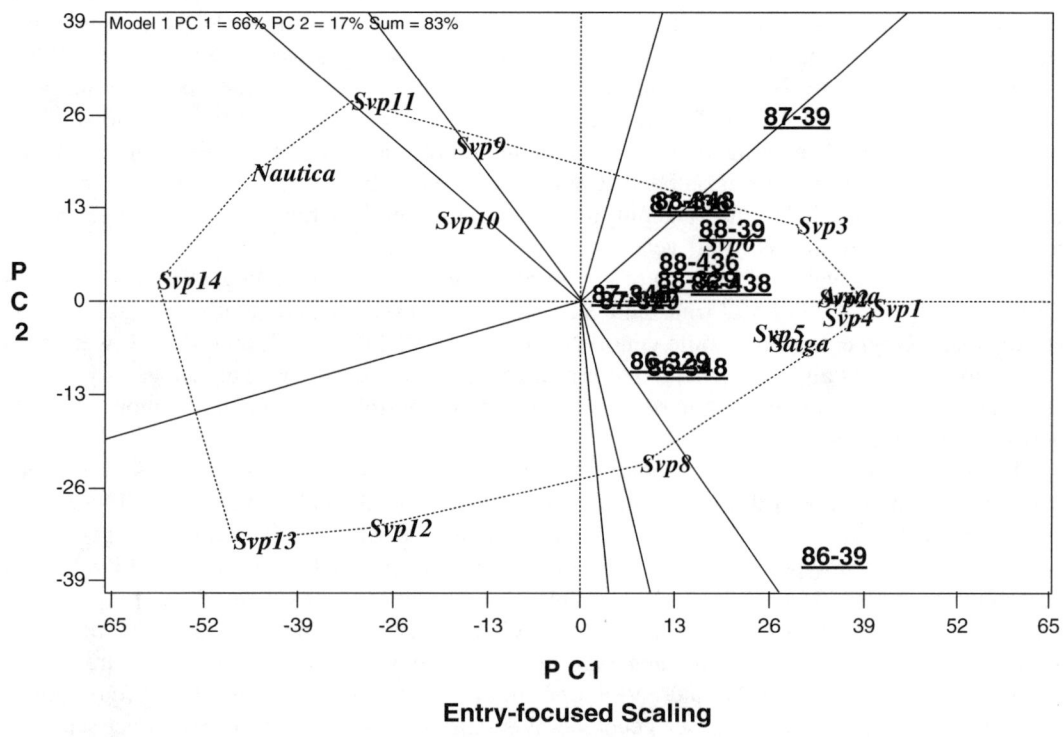

FIGURE 10.11 GGE biplot based on reserved disease scores indicating resistance of 17 wheat genotypes inoculated with 4 *Fusarium culmorum* strains in 1986 to 1988. The wheat genotypes are used as entries — in italics — and the *Fusarium* strains are used as testers — uppercase.

Thus, although there are large crossover genotype-by-strain-by-year interactions (Figure 10.10), these interactions have little impact on selection for FHB-resistant genotypes. Since resistant genotypes are resistant to all strains and in all years (Figure 10.11), they will be selected in all environments. Susceptible genotypes and resistant genotypes may not be easily differentiated in some strain–environments, however. Therefore, genotypes should be tested with highly virulent strains in a few environments and selected for mean resistance and stability across strains and environments. The average tester coordinate (ATC) view of the biplot facilitates this (Figure 10.12). As described in previous chapters, the ATC is drawn following these steps: 1) an average tester is defined by mean PC1 and PC2 scores across all testers, i.e., strain-year combinations; 2) the ATC abscissa is drawn, which is the line that passes through the biplot origin and the average tester; 3) the ATC ordinate is drawn, which passes through the biplot origin and is perpendicular to the ATC abscissa. The ATC abscissa represents the mean resistance, and the ATC ordinate represents the stability of genotypes across all strain–year combinations. When entry-focused scaling is used, the units of both PC1 and PC2, hence the units of both mean resistance and stability for the genotypes, are the same as the unit of the original data. This feature facilitates visualization of both mean and stability on the same scale. Assuming that mean resistance and stability are equally important, the genotypes can be compared for both aspects using concentric circles (Figure 10.12). The center of concentric circles is the ideal genotype, which is defined as the virtual genotype that is most resistant to all strains in all years.

Figure 10.12 reveals that a group of genotypes, which is represented by SVP1 and includes SVP2, SVP3, SVP4, SVP5, Arina, and Saiga, is both resistant and relatively stable across strain–years. On the contrary, some genotypes, notably SVP11, SVP12, and SVP13, were both susceptible and unstable. A general pattern appears to be that truly resistant genotypes are resistant in all environments, but performances of susceptible genotypes are more variable across environments.

Figures 10.10 to 10.12 present genotype-by-strain-by-year interaction, and the observed crossover interactions are mostly due to the interaction between strain 39 and the year factor. To visualize any genotype-by-strain interaction, the year factor should be removed by taking averages across years. Crossover genotype-by-strain interactions did exist (Figure 10.13): while SVP13 was most susceptible to strains 348 and 438, SVP14 was most susceptible to strains 39 and 329. This crossover interaction, however, does not affect selection for disease resistance. When the signs of the testers were reversed, no major crossover genotype-by-strain interaction was observed (Figure 10.14): genotypes SVP1, SVP2, SVP3, and SVP4 were the most resistant genotypes to all four strains. Thus, although there were crossover genotype-by-strain interactions, they do not affect selection for *Fusarium*-resistant genotypes.

For genotype-by-strain systems in which there is no clear resistance specificity, it is possible that some strains are more effective in selecting for resistant genotypes. Strains can be evaluated for their discriminating ability and representativeness of other strains. This can be easily achieved by examining the ATC view of a biplot using tester-focused scaling (Figure 10.15). The vector length of a tester, or the distance between the marker of a tester and the biplot origin, is a measure of its discriminating ability. The projection onto the ATC ordinate is a measure of representativeness and the shorter the projection of a tester, the more representative it is. The center of the concentric circles in Figure 10.15 represents the ideal tester, which is most discriminating and representative. Although none of the four strains was ideal, strain 39, which was collected from wheat grains, seems to be more ideal than others for selecting for FHB-resistant wheat genotypes (Figure 10.15).

Biplot Analysis of Host Genotype-by-Pathogen Strain Interactions

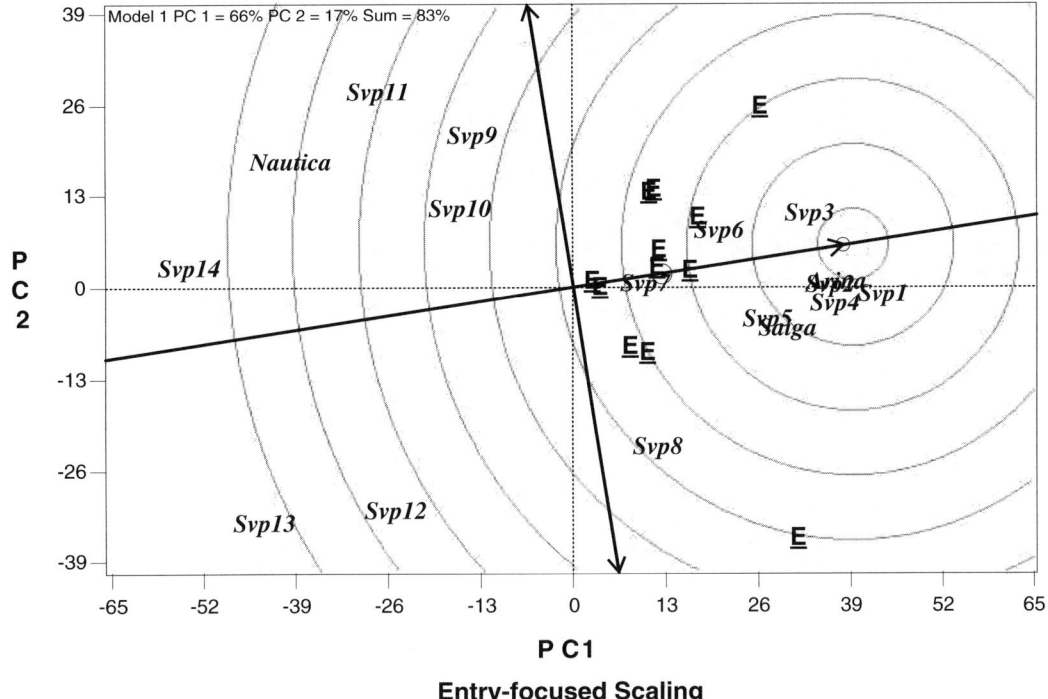

FIGURE 10.12 Mean resistance and stability of wheat genotypes across four *Fusarium culmorum* strains in 1986 to 1988. The center of the concentric circles represents the ideal wheat genotype, which is the most resistant to all Fusarium strains in all years.

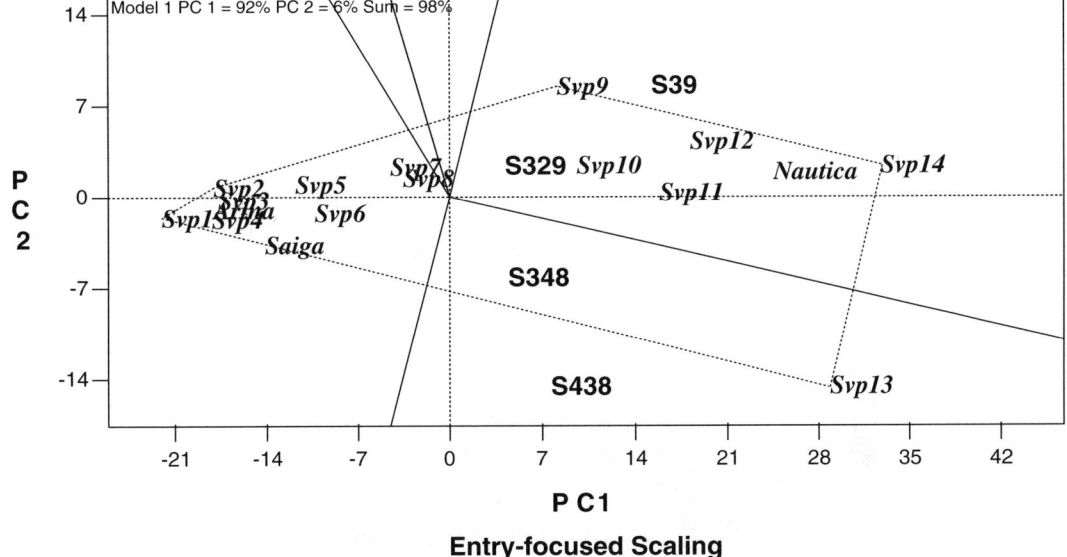

FIGURE 10.13 GGE biplot based on disease scores indicating susceptibility of 17 wheat genotypes inoculated with 4 *Fusarium culmorum* strains averaged across 1986 to 1988. The wheat genotypes are used as entries — in italics — and the *Fusarium* strains are used as testers — uppercase.

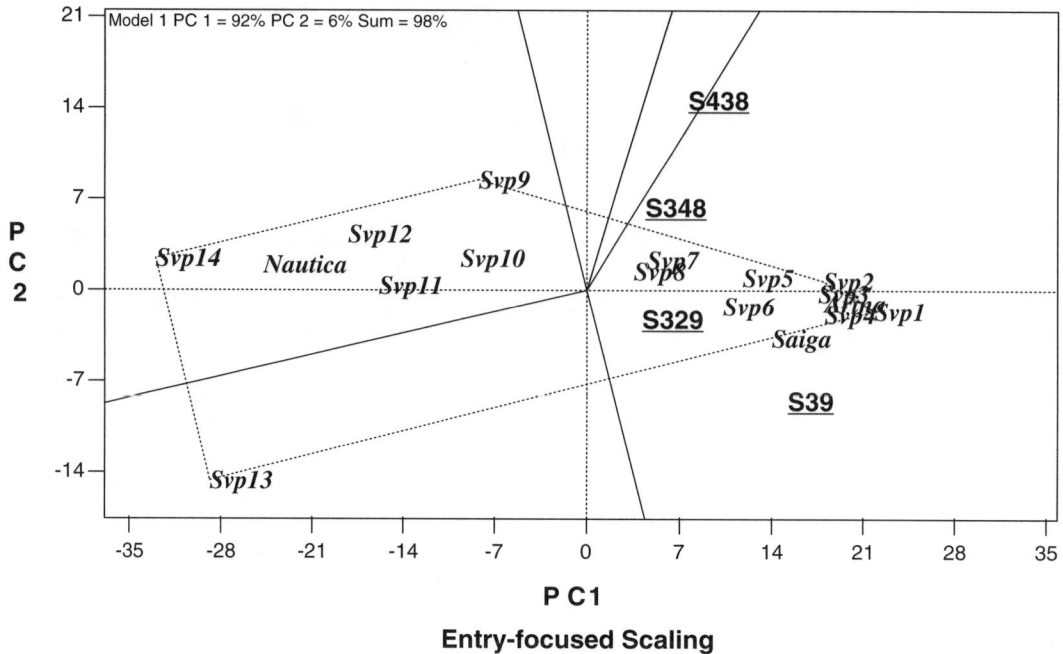

FIGURE 10.14 GGE biplot based on reserved disease scores indicating resistance of 17 wheat genotypes inoculated with 4 *Fusarium culmorum* strains averaged across 1986 to 1988. The wheat genotypes are used as entries — in italics — and the *Fusarium* strains are used as testers — uppercase.

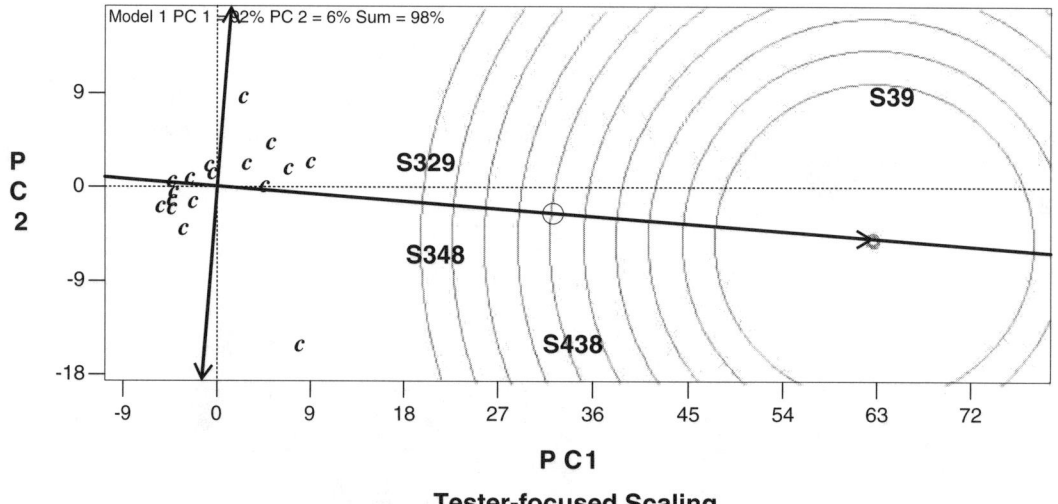

FIGURE 10.15 Ranking of the 4 *Fusarium culmorum* strains based on their discriminating ability and representativeness. The center of the concentric circles represents the ideal strain for selecting wheat FHB resistance.

11 Biplot Analysis to Detect Synergism between Genotypes of Different Species

SUMMARY

This chapter exemplifies biplot analysis of synergism between genotypes of different species using two datasets. The first dataset is about host-by-rhizobium interaction for nitrogen fixation, and the second is about wheat–maize interaction for wheat haploid embryo formation. Both represent a wide range of research interest in plant breeding and biological research.

In contrast to the host-by-pathogen relationship, in which pathogens have harmful effects on their hosts, symbiosis is an inter-species relationship in which two different species live together and benefit each other. A well-known example is the relationship between legume species and nitrogen-fixing rhizobial species. The first part of this chapter exemplifies biplot analysis of interactions between host genotype-by-nitrogen-fixing rhizobial strains using a dataset of four *Casuarina* host species inoculated with four *Frankia* strains (Mansour and Baker, 1994). *Casuarina* is a tropical plant used in agroforestry, and *Frankia* is a nitrogen-fixing rhizobial species.

The second part of this chapter discusses a wheat-by-maize hybridization dataset and demonstrates the application of biplots in identifying the best wheat and maize genotypes for wheat doubled-haploid production. In recent years, wheat breeding via doubled-haploids produced by pollinating wheat F_1 plants with maize pollen has become a common practice in some wheat breeding programs (O'Donoughue and Bennett, 1994; Laurie et al., 1990; Sadasivaiah et al., 1999). Cherkaoui et al. (2000) provided data on the interactions between ten wheat genotypes and eight maize genotypes for haploid wheat embryo formation (Table 11.2). Although maize–wheat interaction does not belong to the category of genotype-by-strain interaction, they share exactly the same data structure and nature of interaction.

11.1 GENOTYPE-BY-STRAIN INTERACTION FOR NITROGEN-FIXATION

The nitrogen content per plant of four *Casuarina* species, when inoculated with four *Frankia* strains, is presented in Table 11.1. The *Casuarina* main effect, *Frankia* main effect, and *Casuarina* × *Frankia* interaction explain, respectively, 21, 68, and 10% of the total variation. As was pointed out in Chapter 10, if the primary interest is in identifying the best *Casuarina* species for nitrogen fixation, the *Casuarina* species should be viewed as entries; if the primary interest is in identifying the best rhizobial strains for nitrogen fixation, the *Frankia* strains should be viewed as entries. Here we are interested in both; therefore, we have two different types of biplots. The GGEbiplot software function *Entry/Tester Switch Roles* makes visualization of the data in both ways very convenient.

TABLE 11.1
Nitrogen per Plant of Four *Casuarina* Species when Inoculated with Four *Frankia* Strains

Frankia Strains	*Casuarina* Species			
	C. cunninghaniana	*C. equisetifolia*	*C. glauca*	*C. hybrid*
Control	14.5	11.4	11.6	18.0
Cci3	95.1	35.2	59.8	121.6
Cef1–82	24.9	11.2	29.7	46.4
Cgi4	97.9	72.7	102.7	105.4
Jct287	95.5	31.7	60.6	100.5

Source: Data from Mansour, S. R. and Baker, D. D., *Soil Biol. Biochem.*, 26:655–658, 1994.

11.1.1 Biplot with *Frankia* Strains as Entries

The biplot treating *Frankia* strains as entries is presented in Figure 11.1. Since the biplot is based on tester-centered data, the main effects of the *Casuarina* species are removed, and the biplot contains only *Frankia* main effects and *Casuarina*-by-*Frankia* interactions. The biplot explained 99% of the total variation. First, the four *Frankia* strains fall into three groups; each strain is different except that Cci3 and Jct287 are similar. Strain Cef1–82 is the poorest; it is similar to the control. Next, the four *Casuarina* species fall into two groups: *C. equisetifolia* and *C. glauca*, and *C. cunninghaniana* and *C. hybrid* are two pairs of similar spieces that differ in response to the *Frankia* strains. Third, clear crossover host-by-strain interactions: strain Cgi4 was the best for species *C. equisetifolia* and *C. glauca*, whereas strain Cci3 was the best for species *C. cunninghaniana* and *C. hybrid* for nitrogen fixation. Strains Cgi4 and Cci3 should fix almost the same amount of nitrogen when species *C. cunninghaniana* is inoculated, since it is located right on the perpendicular line toward the polygon side that connects the two strains. These statements can be verified from the data (Table 11.1).

11.1.2 Biplot with *Casuarina* Species as Entries

When the *Casuarina* species are used as entries, there is no crossover interaction between *Casuarina* and *Frankia* (Figure 11.2). The species *C. hybrid* produced the most nitrogen, regardless of the *Frankia* strains; it also produced the most nitrogen without inoculation, i.e., control. Therefore, *C. hybrid* should be the ideal species for nitrogen fixation, since it produced the most nitrogen with all *Frankia* strains. One may question if *C. hybrid* is indeed more effective in nitrogen fixation, since it fixed more nitrogen even when not inoculated. Figure 11.2 indicates that its advantage over other species in fixing nitrogen when not inoculated is negligible, however, since the control treatment "control" is close to the origin of the biplot, implying that all species fixed similar amount of nitrogen in this treatment. This is consistent with the data (Table 11.1). Figure 11.2 also reveals that strains Cci3 and Jct287, and Cgi4 and Cef1–82 are pairs of similar strains that differ in discrimination of the *Casuarina* species, even though there is no crossover interaction.

Why are crossover interactions seen in Figure 11.1, where *Frankia* species are treated as entries, but not in Figure 11.2, where *Casuarina* species are treated as entries? This is because the *Casuarina* species had larger main effects. It explained 68% of the total variation, whereas the *Frankia* strains explained only 20% of the total variation. The *Casuarina*-by-*Frankia* interaction explained 10% of the total variation. This example illustrates one point that is common sense, i.e., presence vs. absence of crossover interactions depends on the relative magnitude of the entry main effects vs. the entry-by-tester interaction. The greater the main effects relative to GEI, the smaller the probability of the occurrence of crossover interactions, and more feasible the selection based on main effects alone and fewer testers.

Biplot Analysis to Detect Synergism between Genotypes of Different Species

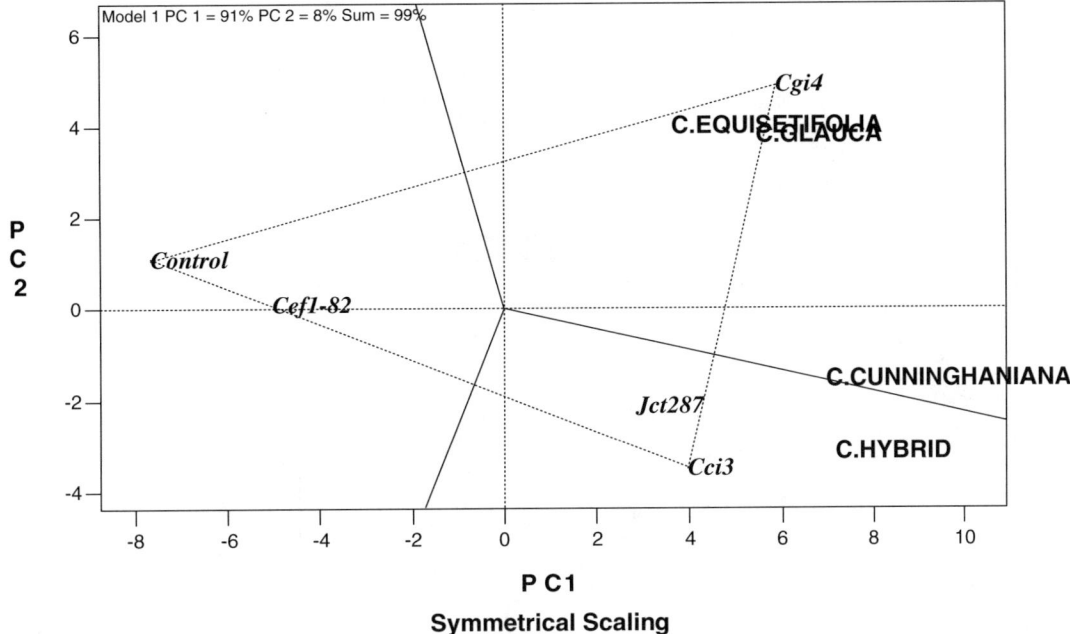

FIGURE 11.1 GGE biplot showing interactions between four *Casuarina* species (preceded by "C") and four *Frankia* strains for nitrogen fixation. The *Frankia* strains are treated as entries and the *Casuarina* species as testers. Based on data from Mansour, S. R. and Baker, D. D., *Soil Biol. Biochem.*, 26:655–658, 1994.

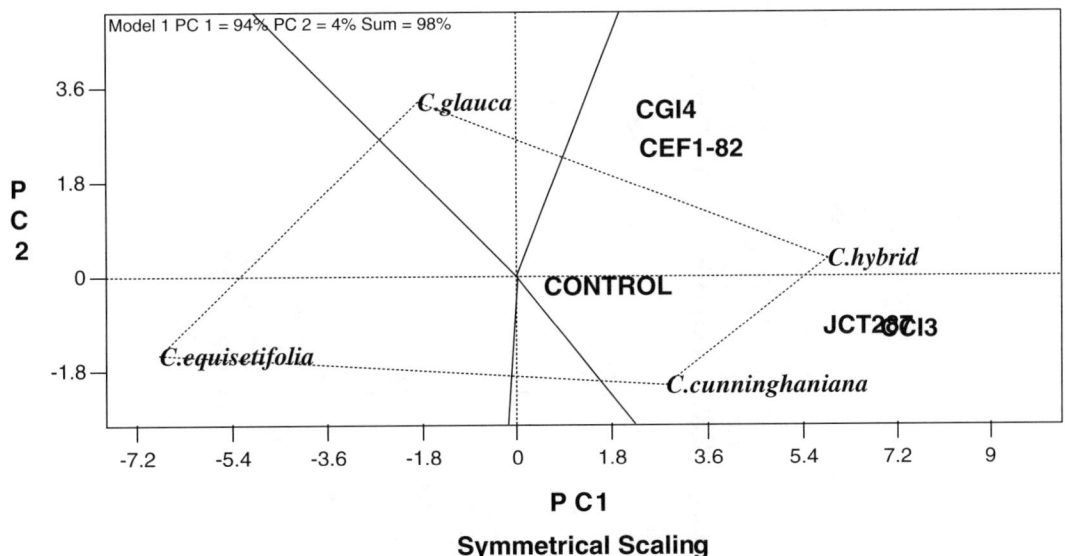

FIGURE 11.2 GGE biplot showing interactions between four *Casuarina* species (preceded by "C") and four *Frankia* strains for nitrogen fixation. The *Frankia* strains are treated as testers and the *Casuarina* species as entries. Based on data from Mansour, S. R. and Baker, D. D., *Soil Biol. Biochem.*, 26:655–658, 1994.

TABLE 11.2
Percentage of Embryo Formation for Ten Durum Wheat Genotypes When Pollinated with Eight Maize Genotypes

Maize Genotypes	Durum Wheat Genotypes							
	BELBACHIR	COCORIT	ISLY	KARIM	MARZAK	MASSA	ORABIA	SARIF	SEBOU	TENSIFT
KAMLA	20.0	9.0	12.6	9.3	9.8	22.8	13.3	9.5	9.5	12.1
GUICH	6.3	1.0	7.2	8.1	8.2	19.3	22.7	5.7	9.3	10.2
BERRECHID	1.3	1.6	5.8	2.2	5.1	7.8	7.5	3.6	7.5	6.0
ELBOURYA	10.5	4.4	5.5	6.0	7.4	12.6	11.8	4.8	9.8	13.6
VL90	10.0	0.0	11.5	2.8	6.1	4.9	6.2	0.0	1.0	0.0
MABCHOURA	2.9	0.0	0.0	1.1	2.5	8.2	1.7	0.0	2.9	6.8
V5	7.8	4.8	0.0	0.8	2.9	9.1	14.1	0.0	2.4	0.0
DOUKKALA	1.4	0.9	1.3	2.1	0.0	1.2	5.2	2.7	2.1	3.8

Source: Data from Cherkaoui, S. et al., *Plant Breeding*, 199:31–36, 2000.

11.2 WHEAT–MAIZE INTERACTION FOR WHEAT HAPLOID EMBRYO FORMATION

In the durum wheat haploid embryo-formation dataset (Table 11.2), wheat main effects, maize main effects, and wheat-by-maize interaction explained 24, 46, and 30% of the total variation, respectively. As done for the *Casuarina*-by-*Frankia* dataset, we will look at the data in two ways, using wheat genotypes and maize genotypes as entries in separate biplots.

11.2.1 BIPLOT WITH MAIZE GENOTYPES AS ENTRIES

With maize genotypes as entries, the GGE biplot revealed limited crossover interactions between maize genotypes and wheat genotypes for wheat haploid embryo formation (Figure 11.3). The maize genotype Kamla was the better mating partner with all durum wheat genotypes except Orabia, for which maize genotype Guich was the best partner. These statements can be easily verified from Table 11.2. The biplot explained 81% of the total variation in the tester-centered data (Figure 11.3).

An ideal maize genotype for wheat haploid embryo formation should be one that is most effective in promoting haploid embryo production for all wheat genotypes. The center of the concentric circles in Figure 11.4 represents such an ideal maize genotype. Although such an ideal genotype may not exist in reality, it can be used as a reference for comparing the real maize genotypes. Figure 11.4 indicates that maize genotype Kamla was closest to the ideal genotype; it is, therefore, the best among the tested maize genotypes for wheat haploid embryo production. The second and third best maize genotypes are Guich and Elbourya; they seem to be equally good in both mean and stability. Guich had a larger mean value than Elbourya, but Elbourya was more stable in inducing wheat haploid embryos. The other five maize genotypes all produced below-average wheat embryos.

11.2.2 BIPLOT WITH WHEAT GENOTYPES AS ENTRIES

If the main purpose is to select for wheat genotypes that are prone to haploid embryo induction upon pollination with maize, the wheat genotypes should be used as entries (Figure11.5). We see more crossover wheat–maize interactions in Figure 11.5, as compared with Figure 11.3. This is because the wheat main effect, which is 24% of total variation, is smaller than the maize main effect, which is 46%. Obviously, wheat genotypes Orabia and Massa were most amenable to embryo formation. Based on Figure 11.5, the maize pollinators fall into three sectors, and three best wheat genotypes are identified: 1) durum wheat genotype Orabia is best for haploid embryo production when pollinated with maize genotypes Doukkala, Mabchoura, Elbourya, Berrechid, and Guich; 2) durum wheat genotype Massa is best when pollinated with maize genotypes V5 and Kamla; and 3) durum wheat genotype Belbachir was best when pollinated with maize genotype VL90 (partially hidden). These statements are generally in agreement with the data (Table 11.2). Although the two durum-wheat genotypes, Orabia and Massa, were identified as being best for different maize pollinators, they are quite similar, as they are close to each other in the biplot (Figure 11.5).

On the basis of both mean performance and stability across different maize pollinators, Massa is identified as the best durum wheat genotype for haploid embryo production. The durum genotype Orabia was almost as good as Massa. All other durum wheat genotypes, except Belbachir, had below-average haploid embryo yields (Figure 11.6).

To summarize, the best two maize genotypes were Kamla and Guich, and the best two durum wheat genotypes were Orabia and Massa. The best two combinations were Kamla × Massa and Guich × Orabia, as they show positive interactions in both biplots (Figures 11.3 and 11.5).

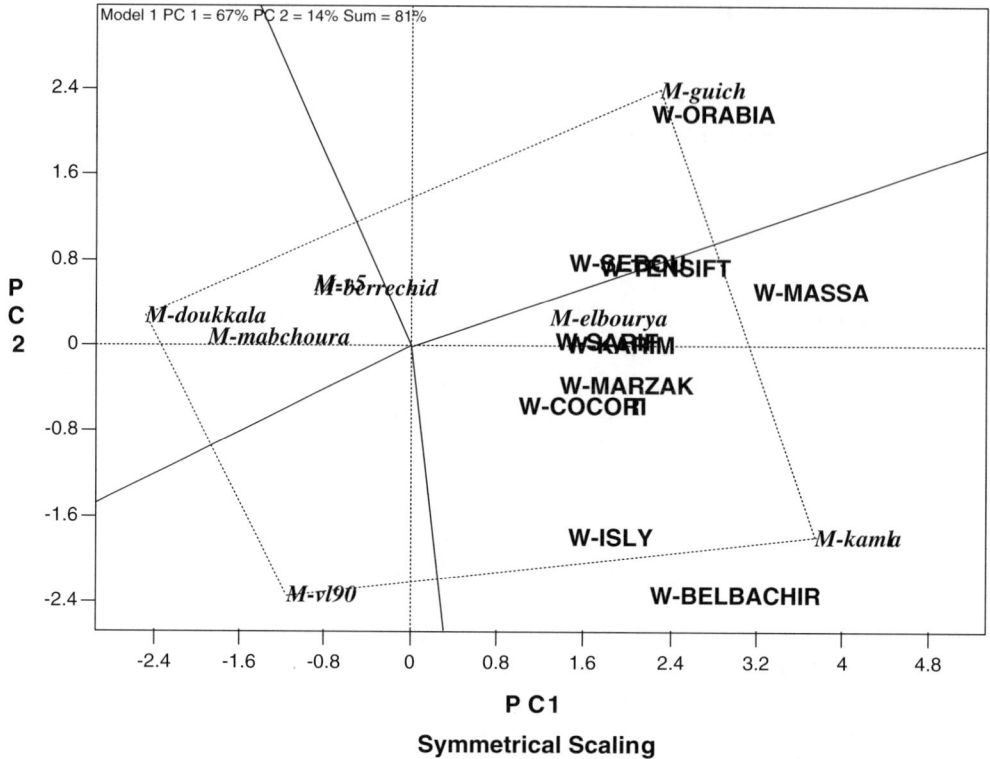

FIGURE 11.3 GGE biplot showing interactions between eight maize genotypes (preceded by "M-") and ten durum wheat genotypes (preceded by "W-") for durum wheat haploid embryo formation. Maize genotypes are treated as entries and durum wheat genotypes as testers. Based on data from Cherkaoui, S. et al., *Plant Breeding*, 199:31–36, 2000.

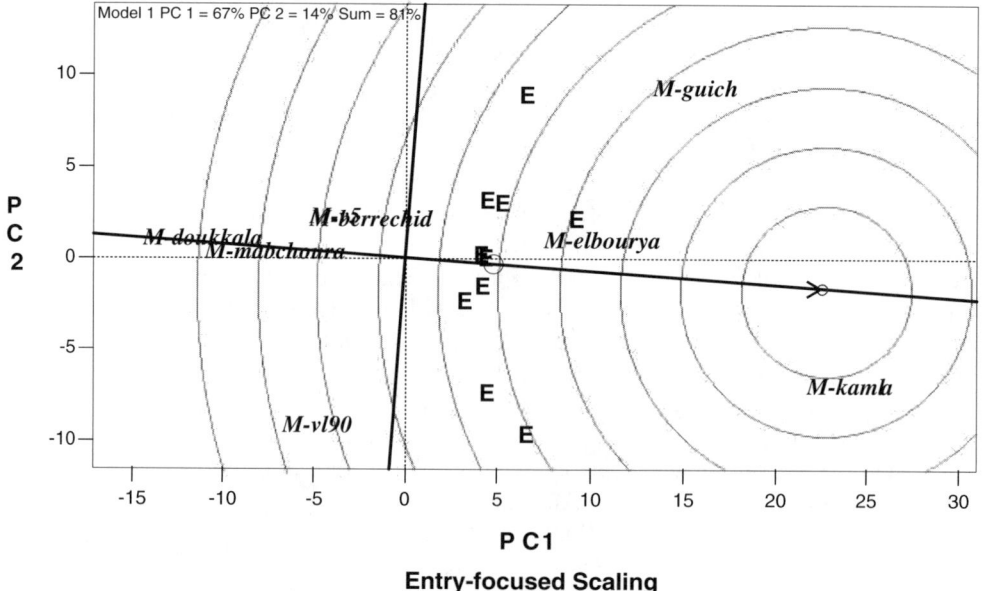

FIGURE 11.4 GGE biplot showing the rank order of the maize genotypes relative to their mean performance and stability for haploid embryo formation in durum wheat across different durum wheat genotypes. Based on data from Cherkaoui, S. et al., *Plant Breeding*, 199:31–36, 2000.

Biplot Analysis to Detect Synergism between Genotypes of Different Species 253

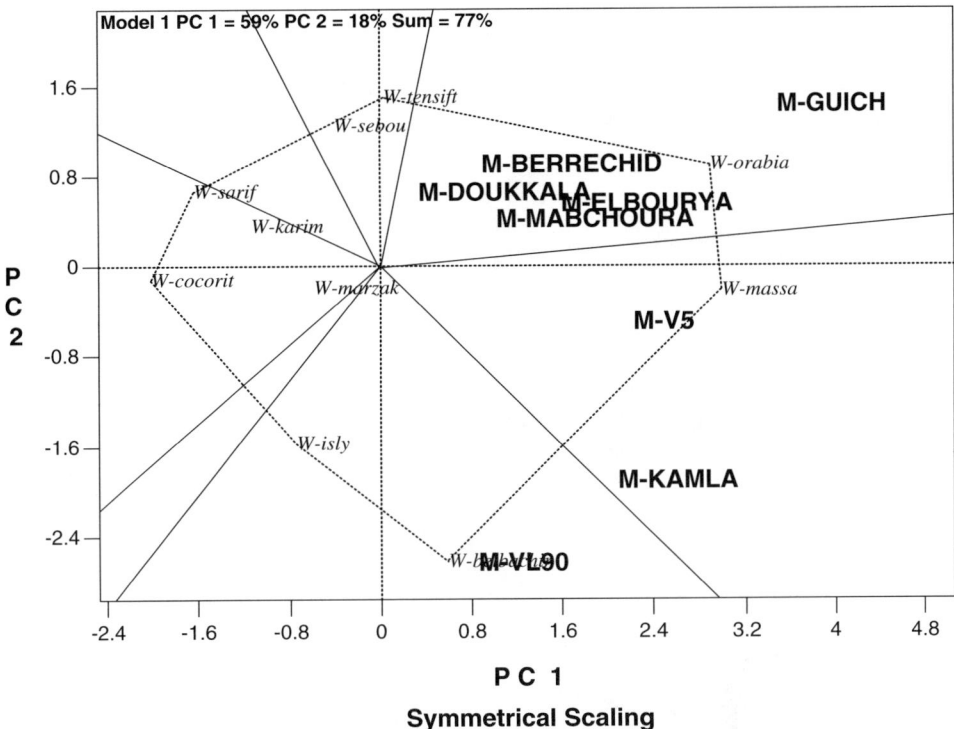

FIGURE 11.5 GGE biplot showing interactions between eight maize genotypes (preceded by "M-") and ten durum wheat genotypes (preceded by "W-") for durum wheat haploid embryo formation. Maize genotypes are treated as testers and durum wheat genotypes as entries. Based on data from Cherkaoui, S. et al., *Plant Breeding,* 199:31–36, 2000.

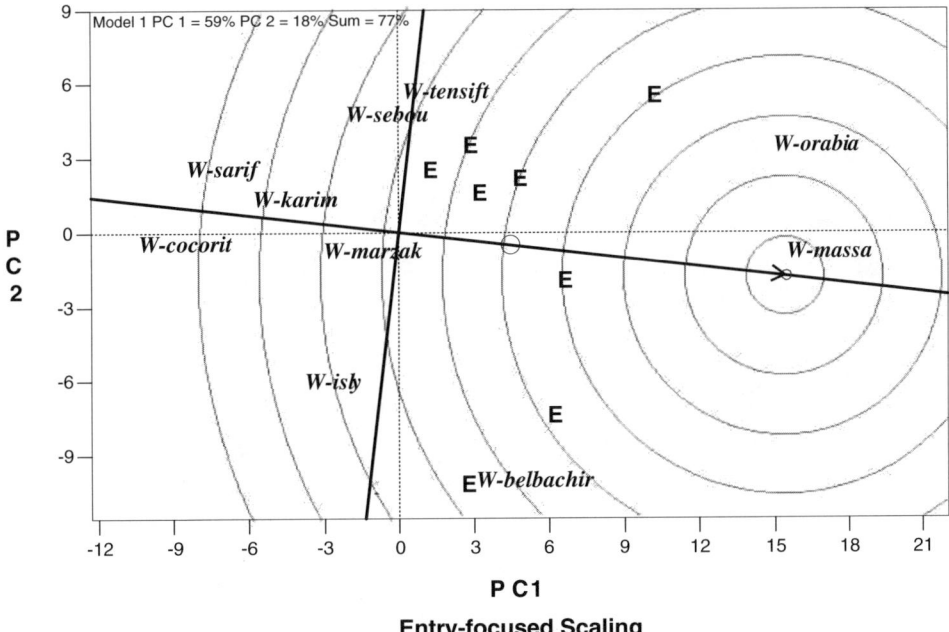

FIGURE 11.6 GGE biplot showing the rank order of the durum wheat genotypes relative to mean and stability for haploid embryo formation in wheat when pollinated with different maize genotypes. Based on data from Cherkaoui, S. et al., *Plant Breeding,* 199:31–36, 2000.

References

Alia, R., J. Moro, and J.-B. Denis. 1997. Performance of *Pinus pinaster* provenances in Spain: interpretation of genotype by environment interaction, *Can. J. For. Res.*, 27:1548–1559.

Allard, R.W. 1960. *Principles of plant breeding*, John Wiley & Sons, New York.

Allard, R.W. 1999. *Principles of plant breeding*, 2nd ed., John Wiley & Sons, New York.

Allard, R.W. and A.D. Bradshaw. 1964. Implications of genotype-environmental interactions in applied plant breeding, *Crop Sci.*, 4:503–508.

Annicchiarico, P. 1997. STABSAS: a SAS computer programme for stability analysis, *Ital. J. Agron.*, 1(1):7–9.

Annicchiarico, P. 1999. Variety × location interaction and its implications on breeding lucerne: A case study, p. 35–43. In: F. Veronesi and D. Rosellini (Eds.), *Proc. XIII Eucarpia Medicago spp. germplasm*. 13–16 Sep. 1999, Perugia, Italy.

Atlin, G.N. et al., 2000a. Selection response in subdivided target regions, *Crop Sci.*, 40:7–13.

Atlin, G.N, K.B. McRae, and X. Lu. 2000b. Genotype × region interaction for two-row barley yield in Canada, *Crop Sci.*, 40:1–6.

Baker, R.J. 1988a. Tests for crossover genotype-by-environment interactions, *Can. J. Plant Sci.*, 68:405–410.

Baker, R.J. 1988b. Differential response to environmental stress, p. 492–504. In: B.S. Weir et al., (Eds.), *Proc. 2nd Int. Conf. Quantitative Genetics*, Sinauer, Sunderland, MA.

Baker, R.J. 1990. Crossover genotype-environmental interaction in spring wheat, p. 42–51. In: M.S. Kang (Ed.), *Genotype-by-environment interaction and plant breeding*, Louisiana State University Agricultural Center, Baton Rouge, LA.

Balzarini, M., S.B. Milligan, and M.S. Kang. 2002. Best linear unbiased prediction: A mixed model approach in multi-environmental trials, p.353–364. In: M.S. Kang (Ed.), *Crop improvement: challenges in the twenty-first century*, Food Products Press, Binghamton, NY.

Baril, C.P. 1992. Factor regression for interpreting genotype-environment interaction in bread-wheat trials, *Theor. Appl. Genet.*, 83:1022–1026.

Basford, K.E., P.M. Kroonenberg, and M. Cooper. 1996. Three-mode analytical methods for crop improvement programs, p. 291–305. In: M. Cooper and G.L. Hammer (Eds.), *Plant adaptation and crop improvement*, CAB International, Wallingford, U.K.

Basford, K.E., P.M. Kroonenberg, and I.H. DeLacy. 1991. Three-way methods for multi-attribute genotype × environment data: an illustrated partial survey, *Field Crops Res.*, 27:131–157.

Beavis, W.D. and P. Keim. 1996. Identification of quantitative trait loci that are affected by environment, p. 123–149. In: M.S. Kang and H.G. Gauch, Jr. (Eds.), *Genotype-by-environment interaction*, CRC Press, Boca Raton, FL.

Becker, H.C. 1981. Correlations among some statistical measures of phenotypic stability, *Euphytica*, 30:835–840.

Becker, H.C. and J. Leon. 1988. Stability analysis in plant breeding, *Plant Breed.*, 101:1–23.

Blum, A. 1988. *Plant breeding for stress environments*, CRC Press, Boca Raton, FL.

Bradshaw, A.D. 1965. Evolutionary significance of phenotypic plasticity in plants, *Adv. Genet.*, 13:115–155.

Brancourt-Hulmel, M., J.-B. Denis, and C. Lecomte. 2000. Determining environmental covariates which explain genotype environment interaction in winter wheat through probe genotypes and biadditive factorial regression, *Theor. Appl. Genet.*, 100:285–298.

Bridges, W.C., Jr. 1989. Analysis of a plant breeding experiment with heterogeneous variances using mixed model equations, p. 145–151. In: *Applications of mixed models in agriculture and related disciplines*, So. Coop. Ser. Bull. No. 343, Louisiana Agricultural Experiment Station, Baton Rouge, LA.

Buerstmayr, H. et al., 1999. Combining ability of resistance to head blight caused by *Fusarium culmorum* (W.G. Smith) in the F1 of a seven-parent diallel of winter wheat (*Triticum aestivum* L.), *Euphytica*, 110:199–206.

Burns, G.W. 1969. *The science of genetics*, 2nd ed., Macmillan, New York.

Busey, P. 1983. Management of crop breeding, p. 31–54. In: D.R Wood, K.M. Rawal, and M.N. Wood (Eds.), *Crop breeding*, American Society of Agronomy, Crop Science Society of America, Madison, WI.

Butron, A. et al., 1998. Defense mechanisms of maize against pink stem borer, *Crop Sci.*, 138:1159–1163.

Carter, T.E., Jr. et al., 1983. Coefficients of variation, error variances, and resource allocation in soybean growth analysis experiments, *Agron. J.*, 75:691–696.

Chapman, S.C. et al., 1997. Genotype by environment effects and selection for drought tolerance in tropical maize. II. Three-mode pattern analysis, *Euphytica*, 95:11–20.

Charmet, G. et al., 1993. Genotype × environment interactions in a core collection of French perennial ryegrass populations, *Theor. Appl. Genet.*, 86:731–736.

Cherkaoui, S. et al., 2000. Durum wheat by maize crosses for haploid wheat production: influence of parental genotypes and various experimental factors, *Plant Breed.*, 199:31–36.

Comstock R.E. and R.H. Moll. 1963. Genotype-environment interactions, p.164–196. In: W.D. Hanson and H.F. Robinson (Eds.), *Statistical genetics and plant breeding*, National Academy of Science — National Research Council Publishing NAS-NRC, Washington, D.C.

Cooper, M. and G.L. Hammer (Eds.). 1996. *Plant adaptation and crop improvement*, CAB International, Wallingford, U.K., ICRISAT, Patancheru, India, and IRRI, Manila, Philippines.

Cooper, M., I.H. DeLacy, and R.L. Eisemann. 1993. Recent advances in the study of genotype × environment interactions and their application to plant breeding, p.116–131. In: B.C. Imrie and J.B. Hacker (Eds.), Focused plant improvement: towards responsible and sustainable agriculture, *Proc. 10th Australia plant breeding conference*. Vol. I. Organizing Committee, Australian Convention and Travel Service, Canberra, Australia.

Cooper, M. et al., 1997. Wheat breeding nurseries, target environments, and indirect selection for grain yield, *Crop Sci.*, 37:1168–1176.

Cornelius, P.L., J. Crossa, and M.S. Seyedsadr. 1996. Statistical tests and estimates of multiplicative models for GE interaction, p. 199–234. In: M.S. Kang and H.G. Gauch, Jr. (Eds.), *Genotype-by-environment interaction*, CRC Press, Boca Raton, FL.

Crispeels, M.J. (Ed.). 1994. Introduction to signal transduction in plants: A collection of updates, *Am. Soc. Plant Physiologists*, Rockville, MD.

Crossa, J. and P.L. Cornelius. 1997. Sites regression and shifted multiplicative model clustering of cultivar trial sites under heterogeneity of error variances, *Crop Sci.*, 37:405–415.

Crossa, J., P.L. Cornelius, and M.S. Seyedsadr. 1996. Using the shifted multiplicative model cluster methods for crossover GE interaction, p. 175–198. In: M.S. Kang and H.G. Gauch, Jr. (Eds.), *Genotype-by-environment interaction*, CRC Press, Boca Raton, FL.

Dashiell, K.E., O.J. Ariyo, and L. Bello. 1994. Genotype × environment interaction and simultaneous selection for high yield and stability in soybeans (*Glycine max* (L.) Merr.), *Ann. Appl. Biol.*, 124:133–139.

DeLacy, I.H. et al., 1996a. Analysis of multi-environment trials — a historical perspective, p. 39–124. In: M. Cooper and G.L. Hammer (Eds.), *Plant adaptation and crop improvement*, CAB International, Wallingford, U.K.

DeLacy, I.H., M. Cooper, and K.E. Basford. 1996b. Relationships among analytical methods used to study genotype-by-environment interactions and evaluation of their impact on response to selection, p. 51–84. In: M.S. Kang and H.G. Gauch, Jr. (Eds.), *Genotype-by-environment interaction*, CRC Press, Boca Raton, FL.

de la Vega, A.J. and S.C. Chapman. 2001. Genotype by environment interaction and indirect selection for yield in sunflower. II. Three-mode principal component analysis of oil and biomass yield across environments in Argentina, *Field Crops Res.*, 72:39–50.

Denis, J.-B. 1988. Two-way analysis using covariates, *Statistics*, 19:123–132.

Denis, J.-B., H.-P. Piepho, and F.A. Van Eeuwijk. 1996. *Mixed models for genotype by environment tables with an emphasis on heteroscedasticity*, Tech. Rep Dept. de Biometrie, Lab. de Biometrie, INRA, Versailles, France.

Devine, T.E. 1982. Genetic fitting of crops to problem soils, p. 143. In: M.N. Christensen and C.F. Lewis (Eds.), *Breeding plants for less favorable environments*, Wiley Interscience, New York.

Donmez, E. et al., 2001. Genetic gain in yield attributes of winter wheat in the Great Plains, *Crop Sci.*, 41:1412–1419.

Duvick, D.N. 1996. Plant breeding, an evolutionary concept, *Crop Sci.*, 36:539–548.

Dyke, G.V., P.W. Lane, and J.F. Jenkyn. 1995. Sensitivity (stability) analysis of multiple variety trials, with special reference to data expressed as proportions or percentages, *Expl. Agric.*, 31:75–87.

Eberhart, S.A. and W.A. Russell. 1966. Stability parameters for comparing varieties, *Crop Sci.*, 6:36–40.

References

Eberhart, S.A. and W.A. Russell. 1969. Yield stability for a 10-line diallel of single-cross and double-cross maize hybrids, *Crop Sci.,* 9:357–361.

Einhellig, F.A. 1996. Interactions involving allelopathy in cropping systems, *Agron. J.,* 88:886–893.

Eisemann, R.L., M. Cooper, and D.R. Woodruff. 1990. Beyond the analytical methodology, better interpretation and exploitation of GE interaction in plant breeding, p. 108–117. In: M.S. Kang (Ed.), *Genotype-by-environment interaction and plant breeding,* Louisiana State University Agric. Center, Baton Rouge, LA.

Eskridge, K.M. 1996. Analysis of multiple environment trials using the probability of outperforming a check, p. 273–307. In: M.S. Kang and H.G. Gauch, Jr. (Eds.), *Genotype-by-environment interaction,* CRC Press, Boca Raton, FL.

Evans, L.T. 1993. *Crop evolution, adaptation and yield,* Cambridge University Press, New York.

Fernandez, G.C.J. 1991. Analysis of genotype × environment interaction by stability estimates, *HortScience,* 26:947–950.

Finlay, K.W. and G.N. Wilkinson. 1963. The analysis of adaptation in a plant breeding programme, *Aust.J. Agr. Res.,* 14:742–754.

Flor, H.H. 1946. Genetics of pathogenicity in *Melampsora lini, J. Agr. Res.,* 73:335–357.

Flores, F. 1993. Interaccion genotipo-ambiente en *Vicia faba* L., Ph.D. thesis, University of Cordoba, Spain.

Flores, F., M.T. Moreno, and J.I. Cubero. 1998. A comparison of univariate and multivariate methods to analyze G × E interaction, *Field Crops Res.,* 56:271–286.

Fox, P.N. and A.A. Rosielle. 1982. Reducing the influence of environmental main-effects on pattern analysis of plant breeding environments, *Euphytica,* 31:645–656.

Fox, P.N. et al., 1990. Yield and adaptation of hexaploid spring triticale, *Euphytica,* 47:57–64.

Freeman, G.H. 1975. Analysis of interactions in incomplete two-way tables, *Appl. Stat.,* 24:46–55.

Gabriel, K.R. 1971. The biplot graphic display of matrices with application to principal component analysis, *Biometrika,* 58:453–467.

Gauch, H.G., Jr. 1988. Model selection and validation for yield trials with interaction, *Biometrics,* 44:705–715.

Gauch, H.G., Jr. and R.W. Zobel. 1988. Predictive and postdictive success of statistical analyses of yield trials, *Theor. Appl. Genet.,* 76:1–10.

Gauch, H.G., Jr. and R.W. Zobel. 1996. AMMI analysis of yield trials, p. 85–122. In: M.S. Kang and H.G. Gauch, Jr. (Eds.), *Genotype-by-environment interaction,* CRC Press, Boca Raton, FL.

Gauch H.G. and R.W. Zobel. 1997. Identifying mega-environments and targeting genotypes, *Crop Sci.,* 37:311–326.

Gimelfarb, A. 1994. Additive-multiplicative approximation of genotype-environment interaction, *Genetics,* 138:1339–1349.

Glaz, B., J.D. Miller, and M.S. Kang. 1985. Evaluation of cultivar-testing locations in sugarcane, *Theor. Appl. Genet.,* 71:22–25.

Gorman, D.P., M.S. Kang, and M.R. Milam. 1989. Contribution of weather variables to genotype × environment interaction in grain sorghum, *Plant Breed.,* 103:299–303.

Graham, R.D. 1982. Breeding for nutritional characteristics in cereals, *Adv. Plant Nutr.,* 1:57.

Gravois, K.A., K.A.K. Moldenhauer, and P.C. Rohman. 1990. GE interaction for rice yield and identification of stable, high-yielding genotypes, p. 181–188. In: M.S. Kang (Ed.), *Genotype-by-environment interaction and plant breeding,* Louisiana State University Agricultural Center, Baton Rouge, LA.

Griffing, B. 1956. Concept of general and specific combining ability in relation to diallel crossing systems, *Aust. J. Biol. Sci.,* 9:463–493.

Gupta, S. and R. Loughman. 2001. Current virulence of *Pyrenophora teres* on barley in Western Australia, *Plant Dis.,* 82:960–966.

Gutierrez, J.C., M. Lopez, and K.M. El-Zik. 1994. AMMI (additive main effects and multiplicative interactions analysis): A tool to determine adaptability of upland cotton genotypes in Spain, Cotton Improvement Conference, *Beltwide Cotton Conf. Proc.,* 2:688–689.

Haldane, J.B.S. 1946. The interaction of nature and nurture, *Ann. Eugenics,* 13:197–205.

Hall, A.E. 2001. *Crop responses to environment,* CRC Press, Boca Raton, FL.

Harsch, J. 2001. Perceptions of sustainability, p. 31–47. In: W.A. Payne, D.R. Keeney, and S.C. Rao (Eds.), *Sustainability of agricultural systems in transition,* ASA Special Publ. No. 64, ASA-CSSA-SSSA, Madison, WI.

Harville, D.A. 1977. Maximum-likelihood approaches to variance component estimation and to related problems, *J. Am. Stat. Assoc.*, 72:320–340.

Henderson, C.R. 1975. Best linear unbiased estimation and prediction under a selection model, *Biometrics*, 31:423–447.

Hill, M.O. 1974. Correspondence analysis: A neglected multivariate method, *Appl. Stat.*, 23:340–354.

Hühn, M. 1979. Beitrage zur erfassung der phanotypischen stabilitat, *EDV Med. Biol.*, 10:112–117.

Hühn, M. 1996. Nonparametric analysis of genotype × environment interactions by ranks, p. 235–271. In: M.S. Kang and H.G. Gauch, Jr. (Eds.), *Genotype-by-environment interaction*, CRC Press, Boca Raton, FL.

Hussein, M.A., A. Bjornstad, and A.H. Aastveit. 2000. SASG × ESTAB: A SAS program for computing genotype × environment stability statistics, *Agron. J.*, 92:454–459.

Jenns, A.E., K.J. Leonard, and R.H. Moll. 1982. Stability analyses for estimating relative durability of quantitative resistance, *Theor. Appl. Genet.*, 63:183–192.

Jensen, R.C. et al., 1995. Genotype-by-environment interaction in genetic mapping of multiple quantitative trait loci, *Theor. Appl. Genet.*, 91:33–37.

Johnson, R. and D.R. Knott. 1992. Specificity in gene-for-gene interactions between plants and pathogens, *Plant Pathol.*, 41:1–4.

Jones, H.G. 1992. *Plants and microclimate: a quantitative approach to environmental plant physiology*, 2nd ed., Cambridge University Press, Cambridge, U.K.

Kang, M.S. 1988. A rank-sum method for selecting high-yielding, stable corn genotypes, *Cereal Res. Commun.*, 16:113–115.

Kang, M.S. (Ed.). 1990. *Genotype-by-environment interaction and plant breeding*, Louisiana State University Agricultural Center, Baton Rouge, LA.

Kang, M.S. 1993. Simultaneous selection for yield and stability in crop performance trials: Consequences for growers, *Agron. J.*, 85:754–757.

Kang, M.S. 1998. Using genotype-by-environment interaction for crop cultivar development, *Adv. Agron.*, 62:199–252.

Kang, M.S. 2000. Genotype-by-environment interaction and performance stability in crop breeding, p. 1–42. In: F. Zavala Garcia and N.E. Treviño Hernández (Eds.), *Simposium interaccion genotipo × ambiente*, SOMEFI-CSSA-UG, Irapuato, Gto., Mexico.

Kang, M.S. (Ed.). 2002. *Quantitative genetics, genomics and plant breeding*, CABI Publishing, Wallingford, U.K.

Kang, M.S. and H.G. Gauch, Jr. (Eds.). 1996. *Genotype-by-environment interaction*, CRC Press, Boca Raton, FL.

Kang, M.S. and D.P. Gorman. 1989. Genotype × environment interaction in maize, *Agron. J.*, 81:662–664.

Kang, M.S. and R. Magari. 1995. STABLE: Basic program for calculating yield-stability statistics, *Agron. J.*, 87:276–277.

Kang, M.S. and R. Magari. 1996. New developments in selecting for phenotypic stability in crop breeding, p. 1–14. In: M.S. Kang and H.G. Gauch, Jr. (Eds.), *Genotype-by-environment interaction*, CRC Press, Boca Raton, FL.

Kang, M.S., J.D. Miller, and L.L. Darrah. 1987. A note on relationship between stability variance and ecovalence, *J. Hered.*, 78:107.

Kang, M.S. and H.N. Pham. 1991. Simultaneous selection for high yielding and stable crop genotypes, *Agron. J.*, 83:161–165.

Kang, M.S., B.G. Harville, and D.P. Gorman. 1989. Contribution of weather variables to genotype × environment interaction in soybean, *Field Crops Res.*, 21:297–300.

Kang, M.S. and J.D. Miller. 1984. Genotype × environment interactions for cane and sugar yield and their implications in sugarcane breeding, *Crop Sci.*, 24:435–440.

Kasha, K.J. and K.N. Kao. 1970. High frequency haploid production in barley (*Hordeum vulgare*, L.), *Nature*, 225:874–876.

Kasha, K.J. et al., 1995. The North American barley map on the cross HT and its comparison to the map on cross SM, p. 73–88. In: K. Tsunewaki (Ed.), *The plant genome and plastome: their structure and evolution*, Kodansha Scientific Ltd., Tokyo, Japan.

Kempton, R.A. 1984. The use of biplots in interpreting variety by environment interactions, *J. Agr. Sci.*, 103:123–135.

References

Khush, G.S. 1993. Breeding rice for sustainable agricultural systems, p. 189–199. In: D.R. Buxton et al. (Eds.), *International Crop Science*. I. Crop Science Society of America, Madison, WI.

Knott, D.R. 1990. Near-isogenic lines of wheat carrying genes for stem rust resistance, *Crop Sci.*, 30:901–905.

Knott, D.R. 2000. Inheritance of resistance to stem rust in 'Triumph 64' winter wheat, *Crop Sci.*, 40:1237–1241.

Korol, A.B., Y.I. Ronin, and V.M. Kirzhner. 1995. Interval mapping of quantitative trait loci employing correlated trait complexes, *Genetics*, 140:1137–1147.

Kroonenberg, P.M. 1983. *Three-mode principal component analysis: theory and applications*, DSWO Press, Leiden, The Netherlands.

Kroonenberg, P.M. 1995. *Introduction to biplots for G × E tables*, Dept. of Mathematics, Res. Rep. #51, University of Queensland.

Laurie, D.A., L.S. O'Donoughue, and M.D. Bennett. 1990. Wheat × maize and other wide sexual hybrids: their potential for genetic manipulation and crop improvement, p. 95–126. In: J.P. Gustafson (Ed.), *Gene manipulation in plant improvement*. II. Plenum Press, New York.

LeClerg, E.L. 1966. Significance of experimental design in plant breeding, p. 243–313. In: K.J. Frey (Ed.), *Plant breeding*. Iowa State University Press, Ames, IA.

Leigh, R.A. 1993. Perception and transduction of stress by plant cells, p. 223–237. In: L. Fowden, T. Mansfield, and J. Stoddart (Eds.), *Plant adaptation to environmental stress*, Chapman & Hall, New York.

Lin, C.S. 1982. Grouping genotypes by a cluster method directly related to genotype-environment interaction mean square, *Theor. Appl. Genet.*, 62:277–280.

Lin, C.S. and M.R. Binns. 1988a. A method of analyzing cultivar × location × year experiments: a new stability parameter, *Theor. Appl. Genet.*, 76:425–430.

Lin, C.S. and M.R. Binns. 1988b. A superiority measure of cultivar performance for cultivar × location data, *Can. J. Plant Sci.*, 68:193–198.

Lin, C.S. and M.R. Binns. 1994. Concepts and methods for analysis of regional trial data for cultivar and location selection, *Plant Breed. Rev.*, 12:271–297.

Lin, C.S., M.R. Binns, and L.P. Lefkovitch. 1986. Stability analysis: Where do we stand? *Crop Sci.*, 26:894–900.

Lopez, J. 1990. Estudio de la base genetica del contenido en taninos condensados en la semilla de las habas (*Vicia faba* L.), Ph.D. thesis, University of Cordoba, Spain.

Magari, R. and M.S. Kang. 1993. Genotype selection via a new yield-stability statistic in maize yield trials, *Euphytica*, 70:105–111.

Magari, R. and M.S. Kang. 1997. SAS-STABLE: Stability analyses of balanced and unbalanced data, *Agron. J.*, 89:929–932.

Magari, R., M.S. Kang, and Y. Zhang. 1996. Sample size for evaluating field ear moisture loss rate in maize, *Maydica*, 41:19–24.

Magari, R., M.S. Kang, and Y. Zhang. 1997. Genotype by environment interaction for ear moisture loss rate in corn, *Crop Sci.*, 37:774–779.

Mansour, S.R. and D.D. Baker. 1994. Selection trials for effective N_2-fixing *Casuarina-Frankia* combinations in Egypt, *Soil Biol. Biochem.*, 26:655–658.

Matheson, A.C. and P.P. Cotterill. 1990. Utility of genotype × environment interactions, *For. Ecol. Manage.*, 30:159–174.

Matheson, A.C. and C.A. Raymond. 1984. The impact of genotype × environment interactions on Australian *P. radiata* breeding programs, *Aust. For. Res.*, 14:11–25.

McCalla, A. 2001. World Bank opening remarks — an action plan to meet the challenges, p. 3–5. In: W.A. Payne, D.R. Keeney, and S.C. Rao (Eds.), *Sustainability of agricultural systems in transition*, ASA Spec. Publ. 64. ASA-CSSA-SSSA, Madison, WI.

McKeand, S.E. et al., 1990. Stability parameter estimates for stem volume for loblolly pine families growing in different regions in the southeastern U.S., *Forest Sci.*, 36:10–17.

Miller, D.A. 1996. Allelopathy in forage crop systems, *Agron. J.*, 88:854–859.

Mooers, C.A. 1921. The agronomic placement of varieties, *J. Am. Soc. Agron.*, 13:337–352.

Nassar, R. and M. Hühn. 1987. Studies on estimation of phenotypic stability: Test of significance for non-parametric measures of phenotypic stability, *Biometrics*, 43:45–53.

Nyquist, W.E. 1991. Estimates of heritability and prediction of selection response in plant populations, *Crit. Rev. Plant Sci.*, 10:235–322.

O'Donoughue, L.S. and M.D. Bennett. 1994. Durum wheat haploid production using maize wide-crossing, *Theor. Appl. Genet.*, 89:559–566.

Pandey, S. and C.O. Gardner. 1992. Recurrent selection for population, variety, and hybrid improvement in tropical maize, *Adv. Agron.*, 48:1–87.

Paterson, A.H. et al., 1991. Mendelian factors underlying quantitative traits in tomato: Comparison across species, generations, and environments, *Genetics,* 127:181–197.

Patterson, H.D. and R. Thompson. 1971. Recovery of inter-block information when block sizes are unequal, *Biometrika,* 58:545–554.

Pazdernik, D.L., L.L. Hardman, and J.H. Orf. 1997. Agronomic performance of soybean varieties grown in three maturity zones of Minnesota, *J. Prod. Agric.*, 10:425–430.

Piepho, H.-P. 1994. Missing observations in analysis of stability, *Heredity,* 72:141–145. (Correction 73 (1994): 58.)

Piepho, H.-P. 1998. Methods for comparing the yield stability of cropping systems — a review, *J. Agron. Crop Sci.*, 180:193–213.

Piepho, H.-P. 2000. A mixed model approach to mapping quantitative trait loci in barley on the basis of multiple environment data, *Genetics,* 156:2043–2050.

Pope, D.T. and H.M. Munger. 1953a. Heredity and nutrition in relation to magnesium deficiency chlorosis in celery, *Proc. Am. Soc. Hort. Sci.*, 61:472–480.

Pope, D.T. and H.M. Munger. 1953b. The inheritance of susceptibility to boron deficiency in celery, *Proc. Am. Soc. Hort. Sci.*, 61:481–486.

Rameau, C. and J.-B. Denis. 1992. Characterization of environments in long-term multi-site trials in asparagus, through yield of standard varieties and use of environmental covariates, *Plant Breed.*, 109:183–191.

Rao, M.S.S. et al., 2002. Genotype × environment interactions and yield stability of food-grade soybean genotypes, *Agron. J.*, 94:72–80.

Rédei, G.P. 1982. *Genetics,* Macmillan, New York.

Rédei, G.P. 1998. *Genetics manual: current theory, concepts, terms*, World Scientific, River Edge, NJ.

Robbertse, P.J. 1989. The role of genotype-environment interaction in adaptability, *So. African For. J.*, 150:18–19.

Ronin, Y.I., V.M. Kirzhner, and A.B. Korol. 1995. Linkage between loci of quantitative traits and marker loci. Multi-trait analysis with a single marker, *Theor. Appl. Genet.*, 90:776–786.

Rosielle, A.A. and J. Hamblin. 1981. Theoretical aspects of selection for yield in stress and nonstress environments, *Crop Sci.*, 21:943–946.

Ruttan, V.W. 1998. Meeting the food needs of the world, p. 98–104. In: V.W. Ruttan (Ed.), *International agricultural research: four papers*, Department of Applied Economics, University of Minnesota, Minneapolis, MN.

Sadasivaiah, R.S., B.R. Orshinsky, and G.C. Kozub. 1999. Production of wheat haploids using another culture and wheat × maize hybridization techniques, *Cereal Res. Commun.*, 27:33–39.

Saeed, M. and C.A. Francis. 1984. Association of weather variables with genotype × environment interaction in grain sorghum, *Crop Sci.*, 24:13–16.

Sari-Gorla, M. et al., 1997. Detection of QTL-environment interaction in maize by a least squares interval mapping method, *Heredity,* 78:146–157.

SAS Institute. 1996. *SAS/Stats user's guide*. SAS Institute, Cary, NC.

Scandalios, J.G. 1990. Response of plant antioxidant defense genes to environmental stress, p. 1–41. In: J.G. Scandalios and T.R.F. Wright (Eds.), *Advanced Genetics,* Academic Press, New York.

Searle, S.R. 1987. *Linear models for unbalanced data*, John Wiley & Sons, New York.

Shafii, B. and W.J. Price. 1998. Analysis of genotype-by-environment interaction using the Additive Main Effects and Multiplicative Interaction model and stability estimates, *J. Agric. Biol. Environ. Stat.*, 3:335–345.

Sherrard, J.H. et al., 1985. Use of physiological traits, especially those of nitrogen metabolism for selection of maize, p. 109–130. In: C.A. Neyra (Ed.), *Biochemical basis of plant breeding. Vol. II: Nitrogen metabolism*, CRC Press, Boca Raton, FL.

Shifriss, O. 1947. Developmental reversal of dominance in *Cucurbita pepo, Proc. Am. Soc. Hort. Sci.*, 50:330–346.

Shukla, G.K. 1972. Some statistical aspects of partitioning genotype-environmental components of variability, *Heredity,* 29:237–245.

Silvey, V. 1981. The contribution of new wheat, barley, and oat varieties to increasing yield in England and Wales 1947 to 1978, *J. Nat. Inst. Agric. Bot.*, 15:399–412.

Simmonds, N.W. 1981. Genotype (G), environment (E), and GE components of crop yields, *Expl. Agric.*, 17:355–362.

Singleton, R.W. 1967. *Elementary genetics*, Van Nostrand, New York.

Smith, H. 1990. Signal perception, differential expression within multigene families, and the molecular basis of phenotypic plasticity, *Plant Cell Environ.*, 13:585–594.

Smith, M.E., W.R. Coffman, and T.C. Barker. 1990. Environmental effects on selection under high and low input conditions, p. 261–272. In: M.S. Kang (Ed.), *Genotype-by-environment interaction and plant breeding*, Louisiana State University Agricultural Center, Baton Rouge, LA.

Snijders, C.H.A. and F.A. Van Eeuwijk. 1991. Genotype-by-strain interactions for resistant to Fusarium head blight caused by *Fusarium culmorum* in winter wheat, *Theor. Appl. Genet.*, 81:239–244.

Sokal, R.R. and C.D. Michener. 1958. A statistical method for evaluating systematic relationships, *Univ. Kansas Sci. Bull.*, 38:1409–1438.

Specht, J.E. and D.R. Laing. 1993. Selection for tolerance to abiotic stresses — discussion, p.381–382. In: D.R. Bruxton et al., (Eds.), *International crop science*. I. Crop Science Society of America., Madison, WI.

Sprague, G.F. and W.T. Federer. 1951. A comparison of variance components in corn yield trials. II. Error, year × variety, location × variety, and variety components, *Agron. J.*, 43:535–541.

Steiner, K.C., J.R. Barbour, and L.H. McCormick. 1984. Response of *Populus* hybrids to aluminum toxicity, *Forest Sci.*, 30:404–410.

Stuber, C.W. and J.R. LeDeaux. 2000. QTL × environment interaction in maize when mapping QTLs under several stress conditions, p. 101–112. In: F. Zavala Garcia and N.E. Treviño Hernández (Eds.), *Simposium interaccion genotipo × ambiente*, SOMEFI-CSSA-UG, Irapuato, Gto, Mexico.

Stuber, C.W., M. Polacco, and M.L. Senior. 1999. Synergy of empirical breeding, marker-assisted selection, and genomics to increase crop yield potential, *Crop Sci.*, 39:1571–1583.

Tai, G.C.C. 1971. Genotypic stability analysis and its application to potato regional trials, *Crop Sci.*, 11:184–190.

Tai, G.C.C. 1990. Path analysis of genotype-environment interactions, p. 273–286. In: M.S. Kang (Ed.), Genotype-by-Environment Interaction and Plant Breeding, Louisiana State University Agricultural Center, Baton Rouge, LA.

Tai, G.C.C. and W.K. Coleman. 1999. Genotype × environment interaction of potato chip colour, *Can. J. Plant Sci.*, 79:433–438.

Thomas, J.B., G.B. Schaalje, and M.N. Grant. 1993. Survival, height, and genotype by environment interaction in winter wheat, *Can. J. Plant. Sci.*, 73:417–427.

Tinker, N.A. 1996. Composite Interval Mapping. http://gnome.agrenv.mcgill.ca/tinker/pgiv/cim.htm.

Tinker, N.A. et al., 1996. Regions of the genome that affect agronomic performance in two-row barley, *Crop Sci.*, 36:1053–1062.

Tollenaar, M. 1989. Genetic improvement in grain yield of commercial maize hybrids grown in Ontario from 1959 to 1988, *Crop Sci.*, 29:1365–1371.

Troyer, A.F. 1996. Breeding for widely adapted popular maize hybrids, *Euphytica*, 92: 163–174.

Troyer, A.F. 1997. Breeding widely adapted, popular maize hybrids, p. 185–196. In: P.M.A. Tigerstedt (Ed.), *Adaption in plant breeding*, Kluwer Academic Publishers, The Netherlands.

Unsworth, M.H. and J. Fuhrer. 1993. Crop tolerance to atmospheric pollutants, p.363–370. In: D.R. Bruxton et al., (Eds.), *International crop science*. I. Crop Science Society of America, Madison, WI.

Van Eeuwijk, F.A., J.-B. Denis, and M.S. Kang. 1996. Incorporating additional information on genotypes and environments in models for two-way genotype by environment tables, p. 15–49. In: M.S. Kang and H.G. Gauch, Jr. (Eds.), *Genotype-by-environment interaction*, CRC Press, Boca Raton, FL.

Vanderplank, J.E. 1991. The two gene-for-gene hypotheses and a test to distinguish them, *Plant Pathol.*, 40:1–3.

Wallace, D.H. and W. Yan. 1998. *Plant breeding and whole-system plant physiology — breeding for maturity, adaptation, and yield*, CAB International, Wallingford, U.K.

Westcott, B. 1987. A method of assessing the yield stability of crop genotypes, *J. Agr. Sci.*, 108:267–274.

Wittwer, S.H. 1998. The changing global environment and world crop production, *J. Crop Prod.*, 1(1):291–299.

Woodend, J.J. and A.D.M. Glass. 1993. Genotype-environment interaction and correlation between vegetative and grain production measures of potassium use-efficiency in wheat (*T. aestivum* L.) grown under potassium stress, *Plant Soil,* 151:39–44.

Wricke, G. 1962. Über eine Methode zur Erfassung der ökologischen Streubreite, *Zeitschrift für Pflanzenzüchtung,* 47:92–96.

Yan, W. 1993. The interconnectedness among the traits of wheat and its implication in breeding for higher yield, *Cereal Crops,* 1993(1):43–45.

Yan, W. 1999. A study on the methodology of cultivar evaluation based on yield trial data — with special reference to winter wheat in Ontario, Ph.D. thesis, University of Guelph, Guleph, Ontario, Canada.

Yan, W. 2001. GGEBiplot — a Windows application for graphical analysis of multi-environment trial data and other types of two-way data, *Agron. J.,* 93:1111–1118.

Yan, W. and L.A. Hunt. 1998. Genotype by environment interaction and crop yield, *Plant Breed. Rev.,* 16:135–178.

Yan, W. and L.A. Hunt. 2001. Interpretation of genotype environment interaction for winter wheat yield in Ontario, *Crop Sci.,* 41:19–25.

Yan, W. and L.A. Hunt. 2002. Biplot analysis of diallel data, *Crop Sci.,* 42:21–30.

Yan, W. and I. Rajcan. 2002. Biplot analysis of sites and trait relations of soybean in Ontario, *Crop Sci.,* 42:11–20.

Yan, W. and D.H. Wallace. 1995. Breeding for negatively associated traits, *Plant Breed. Rev.,* 13:141–177.

Yan, W. et al., 2000. Cultivar evaluation and mega-environment investigation based on the GGE biplot, *Crop Sci.,* 40:597–605.

Yan, W. et al., 2001. Two types of GGE biplots for analyzing multi-environment trial data, *Crop Sci.,* 41:656–663.

Yates, F. and W.G. Cochran. 1938. The analysis of groups of experiments, *J. Agr. Sci.,* 28:556–580.

Zhang, Q. and S. Geng. 1986. A method of estimating varietal stability for long-term trials, *Theor. Appl. Genet.,* 71:810–814.

Zobel, R.W., M.J. Wright, and H.G. Gauch, Jr. 1988. Statistical analysis of a yield trial, *Agron. J.,* 80:388–393.

Index

A

Abiotic stress, 9, 17
Absolute stability, 88
Accessories menu, 108, 110, 116
Acid-soil adaptation, selection for, 89
Activated oxygen species, 9
Acute angle and correlation, 35
Adaptation of cultivar in different environments, 67
Adapted single crosses, identification of, 8
Adaptedness-related genes, 8
Additive main effects and multiplicative interaction (AMMI), 13, 93-95, 239
Adobe Writer, 110
AEC (average environment coordinate), 54-55, 85-87, 91
AEC view of GGE biplot, 54-55
AFLP, see Amplified fragment length polymorphisms
Agriculture and statistics, 228
Agronomic concept of stability, 11, 16, see also Dynamic concept; Type 2 stability
Alveograph, 135
AMMI (addtive main effects and multiplicative interaction), 13, 93-95, 239
AMMI analysis, 95
AMMI biplot vs. GGE biplot, 93-94
AMMI model, 93-94
Amplified fragment-length polymorphisms (AFLP), 17
Angle between vectors, 27
Arccosine of correlation coefficient, 34
Artificial environments, 89, 157
Artificial infestation, 218
ATC, see Average tester coordinate
ATC view, 86, 211, 221, 224
ATC view of biplot, 86, 211, 224
Auto find QTL function, 135, 144, 177, 182
Auto find QTL of – option, 113
Average environment, 91
   axis, 85
   coordinate, 51, 85
Average tester coordinate, 86, 244
Average tester coordination option, 112, 208, 221
Avoiding GEI, 16

B

Barley
   genome project, 160
   genomics dataset, 175
   genomics project, 161, 163-171
   linkage map, 172
   mapping dataset, 159
   net blotch, 230, 232-238, 240

   net blotch isolates, 234
   resistance to net blotch, 229
   yield, 202
Best linear unbiased estimator (BLUE), 15
Best linear unbiased predictor (BLUP), 15, 16
Best mating partners, 215, 221
Best tester, 221
Biadditive factorial biplot, 15
Biadditive factorial regression, 13, 15
   Biological concept of stability, 11, see also Static concept of stability; Type 1 stability
Biotic stress, 9, 17
Biplot
   advantages, 225
   analysis, 36, 39
   analysis of diallel, 207-208
   analysis of MET data, 63
   analysis of two-way data, 36
   disadvantages, 225
   division into sectors, 32-33
   for FHB data of wheat, 209
   of markers on
      barley chromosome 1, 165
      barley chromosome 2, 166
      barley chromosome 3, 167
      barley chromosome 4, 168
      barley chromosome 5, 169
      barley chromosome 6, 170
      barley chromosome 7, 171
   origin, 32-33
   selection, 107, 110
   size option, 114
   title option, 113
   with reversed disease scores, 234
Biplot, concept of, 23
Biplot, inner-product property of, 23, 26, 28-29
Biplot, theory of, 23
Biplots menu, 108, 115
BLUE, see Best linear unbiased estimator
BLUP (Best linear unbiased predictor), 16
Bread wheat quality, 135
Breeder's eye, 150, 157
Breeding for resistance to diseases, 229
Breeding for resistance to insects, 229
Breeding phase, 7-8
Breeding strategy selection, 229
Broad adaptation, 9, 12, 16

C

CA (correspondence analysis), 14
Carbon fixation, 9

Casuarina
  as entries, 248-249
  as testers, 249
  host species, 247
Casuarina-by-Frankia
  dataset, 251
  interaction, 248-249
Causes of GEI, strategies for studying, 96
Cereal rusts, 229
Change color option, 113
Change font option, 114
Clear contours/background noise menu, 113
Cluster analysis, 13-14, 16, 239
Coefficient of variance, 14
Color schemes, 113
Column vectors, 26, 34
Comma-delimited text file data, 104
Compare two entries option, 112, 127
Compare with
  a standard entry option, 144
  an entry/tester option, 112
  the ideal entry option, 112
  the ideal tester option, 113, 210
Comparing
  cultivars as packages of traits, 121
  new cultivars with check, 144-145
  testers with ideal tester, 224
  two cultivars, 69, 71-72, 74, 127
Comparison of
  environments with ideal environment, 93
  two cultivars, 67
Composite interval mapping, 159
Composite map-based interval mapping, 175
Connector line, 32
Conservation of resources, 16
Constant system capacity hypothesis, 156
Contribution of environmental factors to stability, 14
Contribution of environmental variables to GEI, 15
Contribution to yield by environment, 7
Contribution to yield by genotype, 7
Conventional diallel analysis, 225
Conventional statistics via GGEbiplot, 228
Convex hull, 73
Cookie spread, 144
Cookie width/thickness ratio, 144
Cookie-making quality, 142, 144-145, 148
Copper deficiency, single gene control of, 10
Correlation
  between G and P, reduction in, 8
  coefficient and vector angle, 89
  coefficients among environments, 90
  coefficients among wheat traits, 141, 149
Correlations among yield, associated traits and markers, 180
Correspondence analysis (CA), 14
Cosine of angle
  and correlation, 34
  between row vectors, 35
Cosine of vector angle vs. correlation, 92
Covariance and inner product, 47
Crop breeding, 7

Crop performance and environment, 5
Crossover interaction, 6-10, 41, 63, 77, 83, 230, 239, 244, 248, 251
Crossover rank change, 16
Culling
  based on a tester, 111
  environments, 92, see also Killer environments
  for oil and protein, 155
  for yield, 152-154
  for yield and oil, 153
  for yield and protein, 154
  inferior cultivars, 92
  intensity, 157
  of genotypes using single traits, 150
Cultivar as a biological system, 121
Cultivar development, 7
Cultivar evaluation, 39, 63-64
  using multiple traits, 121-122
  using single traits, 127
  using two traits, 135
Cultivar performance
  in an environment, 65-66
  in different environments, 68
Cultivar's vector, 67
Cultivar-by-environment interaction, 6
Cultivars as packages of traits, 142

D

Data file preparation, 104
Data format selection, 105
Data manipulation, 103-104, 107-109, 114
Dealing with GEI, 16, 88-89
Dendrogram of isolate groups, 240
Deoxynivalenol (DON), 239, see also Vomitoxin
Desirability index, 14
Determination coefficient for marker-trait association, 175
DH (doubled-haploid) lines, 6, 173-175, 192, 194, 198, 200, 202
Diallel
  analyses, 225
  analysis via GGEbiplot, 207
  cross, 19, 103-104, 207-208, 221, 229
  without parents option, 114, 225
Discard based on a tester option, 150
Discriminating ability, 221
  of Fusarium strains, 244, 246
  of test environments, 91
Disease resistance, 9
Disease screening in greenhouse, 89
DON, see Deoxynivalenol
Double crossovers, 162
Doubled-haploid (DH) lines, 160, 162
Doubled-haploid production, 119
Dough strength, 135
Drought adaptation, selection for, 89
Durable resistance, 18
Durum wheat, 250-253
Dynamic concept of stability, 11, see also Agronomic concept of stability

Index

E

E, see Environment main effect
Ear moisture loss rate in maize, 15, 19
Early multi-environment testing, 18
Eberhart and Russell stability analysis, 14
Ecovalence, 14
Ecovalence vs. stability variance, 12
Eigenvectors, 42
Embryo formation, 250
Entries and testers, 104
Entry eigenvectors, 231
Entry/tester switch roles option, 104, 114, 247
Entry-by-tester
    combination, 106
    data, 104, 231
    interaction, 207, 218, 248
    structure, 121
Entry-focused scaling, 43, 44, 115, 224, 243-246, 252-253
Environment, 3
    ability to discriminate among cultivars, 66
    and phenotype relationship, 5
    index, 4
    main effect, 39
    scores, 44, 47, 61, 94, 96
    vector angle vs. correlation, 89
    vector length vs. standard deviation, 91
    vectors, 47, 66
Environment, characterization of, 17
Environment, contribution of, 7
Environment, defined, 3
Environmental
    capacity, 158
    covariates, 96, 99
    factors, 9, 99
    index, 15
    PC (principal component)1 scores, 97-98
    PC2 scores, 98
    signals, plant response to, 9
    stress, plant's perception of, 9
    stress, transduction of, 9
    variance, 14
Environment-centered
    data, 44, 69-70, 75
    model, 64-65
    yield, 68
    yield of cultivars in an environment, 116
Environment-focused
    GGE biplot, 61-62
    scaling, 44-45, 47, 49, 90, 92-93
Environments for indirect selection, 89
Equality line, 69-70, 73, 75
Equal-space scaling, 44, 47-50, 52, 54, 116
Examine
    a tester option, 127
    an entry option, 144
    an entry/tester option, 111
Exit option, 111
Exotic germplasm, 17
Exploiting GEI, 16, 88
Extensibility, 135
Extensograph, 135

F

Factor regression (FANOVA), 13
Factor, see Gene
FANOVA (factor regression), 13
Farinograph, 135
FHB (Fusarium head blight), 209, 211-212, 215, 221, 230, 239, 241, 246
FHB resistance, 218, 244
File menu, 108, 110
Find QTL menu, 108, 109, 113
Find threshold for a QTL option, 113
Flood tolerance, 9
Flour ash, 135
Flour extraction, 127, 135
Flour protein, 142, 144
Format menu, 108, 109, 113
Frankia strains, 247-249
Full model, 41
Full name option, 111
Functions for
    a diallel-cross dataset, 119
    a GEI dataset, 117
    a genotype-by-strain dataset, 119
    a genotype-by-trait dataset, 118
    a QTL-mapping dataset, 118
Fusarium head blight (FHB), 127, 207
Fusarium strains, 239, 241, 243, 245-246

G

G, see Genotype main effect; Genotype
G (genotype) vs. PC1, 50
GCA (general combining ability) effects, 210, 215, 218, 225
GCA of parents, 208, 214, 224
GCA vs. ATC axis, 210
GEI (genotype-by-environment interaction), see also Gene-by-environment interaction
    analyses, 13
    and mega-environments, 7
    and QTL, 202
    and resource allocation, 19
    data, 229
    for potassium-use efficiency, 10
    heterogeneity, 15
    in crop breeding, implications of, 7
    model explained, 5
    two-way data, 37
GEI, as norm of reaction, 6
GEI, avoiding it, 83, 88-89
GEI, causes of, 9-10, 63-64, 96-97-99
GEI, contribution of, 7
GEI, correcting causes of, 17
GEI, detection of, 7
GEI, graphical representation, 6
GEI, implications of, 7-9
GEI, interpretation of, 7, 94
GEI, quantification of, 7
GEI, utilizing, 77

Gene, 3-5, 17, 94, 162, 175, 229
Gene expression, 3, 5, 119
Gene isolation, 175
Gene mapping using GGEbiplot, 162
Gene, coinage of term, 4
Gene-by-environment interaction, 94, 119
    strategies for exploiting beneficial potential, 17
    strategies for handling, 11
    strategies for minimizing undesirable aspects, 17
Gene-for-gene hypothesis, 229
General combining ability (GCA), 207, 221
General resistance, 18
Genes for adaptedness, 8
Genetic
    constitution of parents, 207, 215, 217-219, 221, 225
    erosion, 19
    factors, 158
    gain, loss of, 8
    markers, 159-160
Genomic regions with stable responses, 17
Genotype (G), 3-5
    eigenvector, 36-37
    evaluation using two traits, 150
    main effect (G), 39
    scores, 44, 47, 61, 94, 96
    selection and crossover interaction, 41
Genotype, characterization of, 17
Genotype, contribution of, 7
Genotype, defined, 3
Genotype-by-environment (GE), see also GEI; gene-by-environment interaction
    correlation, 11
    interaction (GEI), 3, 5
Genotype-by-environment-by trait interaction, 158
Genotype-by-genetic marker data, 103
Genotype-by-location interaction, 39-40, 77, 83, 97-98
Genotype-by-location-by-year interaction, 39, 77, 83
Genotype-by-marker data, 19, 104, 231
Genotype-by-planting date interaction, 19
Genotype-by-strain interaction, 230, 233, 239, 247
Genotype-by-strain-by-year interaction, 244
Genotype-by-trait
    biplot, 122, 124, 151-153, 155
    data, 19, 103-104, 121, 158, 160, 229, 231
Genotype-by-treatment interaction, 119
Genotype-by-year inteaction, 39
Genotype-focused GGE biplot, 61, 62
Genotype-focused scaling, 42-44, 47, 49, 59, 85-86, 88, 92, 145-147
Genotypes based on principal components (PC), 226-227
Genotypes of parents, 220
Genotypic
    covariates, 96, 99
    PC (principal component) scores, 97-98
    stability, 13-14
Geographical information system (GIS), 103
GGE (genotype + genotype-by-environment interaction)
    biplot, 11, 19, 37
    and regression model, 56
    construction via SAS, 39
    environment-focused scaling, 44
    of disease scores, 232
    software, 49, 103, see also GGEbiplot
    using Microsoft Excel, 62
    via conventional methods, 57
    vs. AMMI biplot, 93-94
    ways of visualizing, 64
GGE biplot, an alternative model for, 49
GGE biplot, comparisons via, 65
GGE biplot, construction of, 42
GGE biplot, introduction to, 39
GGE biplot, mathematical model for, 94
GGE model, 41-42, 98
GGE regression against G, 42
GGE variation and PCA model, 51
GGE variation and regression model, 51
GGE, concept of, 39
GGE, decomposition of, 94
GGEbiplot
    analysis of diallel data, 207
    applications to two-way data, 119
    does conventional statistics, 228
    evolution, 119
    for mapping QTL, 118
    for teaching quantitative genetics, 107
    software, 49, 62, 83, 94, 104-105, 208
        need for, 103
        organization of, 107
    system requirements, 108
GGEbiplot, a stand-alone program, 110
Gluten content vs. protein content, 142
Gluten strength, 135, 144
Grain
    falling number, 142
    moisture, 135, 142
    moisture vs. loaf volume, 142
    protein, 142
    sprouting, 135
    yield, 127, 150
Griffing's diallel methods, 207
Grouping linked markers, 160

H

Hairy peduncle gene in triticale, 10
Haploid embryo formation, 247, 251-253
Hard red spring wheat, 122-127, 131, 135, 137, 139, 142-143
Help menu, 108, 117
Hemisaprophytes, 229
Herbicide-by-insecticide interaction, 119
Heredity and environment, 3-4
Heritability, 7
    quantitative traits, 4
    stability statistics, 18
Heritability, impact of GEI on, 7
Heterosis, 17, 215
Heterotic groups, 207, 210, 218, 221, 225
Hide supplementary lines option, 111
High selection intensity, 156
Hordeum bulbosum, 160

Index

Horizontal resistance, 103, 119, 229
Host genotype-by-pathogen data, 104
Host genotypes as entries, 231
Host plant-by-pathogen interaction, 103, 229, 247
Host-by-rhizobium interaction, 247
Host-pathogen specificity, 229
Hühn's nonparametric stability statistics, 14
Hydrangeas, environmental effect on, 4

I

Ideal cultivar
    comparison of test cultivars with, 88
    criteria for, 85
Ideal environment, 91-93
Ideal maize genotype, 251
Ideal strain, 246
Ideal tester, 210, 224
Identifying
    best crosses, 207, 210, 215, 221
    parents for hybridization, 121
    superior parents, 207
    traits associated with a target trait, 121
Ignoring GEI, 16, 88-89
Image format, 107
Increased productivity sources, 17
Independent culling, 121, 150, 156
Index legend option, 111
Indirect selection, 89, 121
    for cookie-making quality, 144
    for loaf volume, 142
    identifying traits for, 135
    requirements for, 18
Inner product and covariance, 47
Inner-product property, 69, 94
Input PCA scores option, 110
Interaction principal component (IPC), 83, 94
Interactive QQE biplot, 120
Interconnectedness among traits, 182
Interrelationship among
    environments, 89
    parents, 215
    seed yield, protein, and oil, 151
    traits, 127, 130
Interval mapping, 159, 182
Investigating selection strategies, 121
IPC1 (interaction principal component 1), 93-95, 115
Iron deficiency, single gene control of, 10

J

Joint culling intensity, 157
Joint plot, 158
Joint selection intensity, 156-157
Joint two testers option, 112, 135

K

Kang's rank sum method for yield and stability, 13-14

Kang's yield-stability (YS_i) statistic, 14, 88
k-dimensional biplot, 37
Kernel weight, 135, 194, 198-202
    and associated markers, 197
    associates, 193
    of DH lines, 200
Killer environments, 92, see also Culling environments

L

Likelihood ratio, 159
Linear regression analysis, 12
Line-by-environment interaction, 202
Line-by-isolate interaction, 233
Loaf volume, 127, 135, 141-142
Location eigenvector, 36-37
LOD (log odds), 159
Log file menu, 117
Loss of membrane integrity, 9
Low selection intensity (culling), 156
Low-input environments, 18

M

Main effects, 94-95
Maize (Zea mays L.), 19
    as entries, 251-252
    as testers, 253
    diallel, 218, 220
    diallel subset, 224, 226
    hybrid trials, 15
    sun-red gene of, 4
Maize-by-wheat interaction, 253
Marker mwg502 association with barley yield, maturity, and quality, 191
Marker values of DH lines, 200
Marker-based selection, 159-160, 175, 206
Marker-by-environment interaction, 120
Marker-QTL association, threshold for, 175
Markers for barley yield, 176, 178-179
Markers for kernel weight, 199, 201
Marker-target trait association, 175
Maternal effects, 207
Matrix data format, 104, 106-107
Matrix multiplication, 23-25, 36
Maximizing system output, 156
Mean performance and stability, 83
Mean productivity, selection for, 18
Mean vs. IPC1 option, 115
Mega-environment, 16, 64, 83, 89, 91
Mega-environment criteria, 77
Mega-environment investigation, 63, 73, 77
Mega-environments and GEI, 7
MET (multi-environment trials) data analysis, 39, 64
Metabolic efficiency, 9
Micro-arrays, 119
Missing value estimation with GGE biplot, 198
Missing values, 106-107, 160, 192, 200, 202
Mitochondrial functions, 9

Mixed model approach, 15-16
Mixed-model method for QTL detection, 14
Mixing tolerance index, 135
MME, see Mixed model equations
Model 0, 0-1 data option, 114
Model 1, tester-centered option, 114
Model 2, within-tester STDEV standardized option, 114
Model 3, within-tester STDERR standardized option, 114
Model 4, double-centered option, 114
Model
    for biplot analysis of diallel, 207
    for genotype-by-strain interaction, 230
    selection, 107, 109
Models menu, 108, 114
Molecular markers, 17
Monogenic control of nutritional disorders, 9
Multi-environment trials (MET), 9, 39, 64
Multiple resistance to stress, 18
Multiple traits in cultivar evaluation, 121
Multiplicative model, complete, 93
Multi-year MET, 39

N

Negative selection, 92
Net blotch (Pyrenophora teres), 230, 232-234, 237-240
New job option, 110
Nitrogen fixation, 247, 249
Nitrogen production by Casuarina spp., 248
Nitrogen-fixing rhizobia, 247
Non-crossover interaction, 6, 218
Non-obligate parasites, 239
Nonparametric methods, 13
Nonproportional interaction, 218
Nonstress environments, 18
Norm of reaction, 3, 4, 6
Number of observations limit for GGEbiplot, 106
Numeric output option, 107, 117
Nutritional disorders, 9

O

Obligate parasites, 229
Observation data format, 104-107
Obtuse angle and correlation, 35
Oil content, 150
Origin of biplot method, 19
Overall culling intensity, 157

P

P, see Phenotype
Pairwise GEI of genotypes with check, 14
Parent selection for hybridization, 122
Parent-offspring regression, 13
Parents as testers, 213
Partial regression coefficients, 159
Partition factor, 42
Path coefficient analysis, 14

Pathogen strains as entries, 231
Pathogen-host interaction, 229
Pattern analysis, 13
PC, see Principal component
PCA (principal component analysis), 85, 110, 208
    a SAS program for, 59
    model, 51, 53, 55
    SAS output, 60
pdf file, 110
Performance trials, 39
Peroxidation of lipids, 9
Perpendicular line, 32-33
Phenotype (P), 3-5
Phenotypic stability variances, 15
Phenotypic standard deviation, 56
Piece-of-cake analogy, 150
Pink stem borer (PSB, Sesamia nonagrioides), 218, 221-222, 225
Plaisted and Peterson's _, 14
Plaisted's _i-bar, 14
Plant breeding
    and GEI, 7
    methodology, 121
Planting date-by-genotype interaction, 19
Plasticity, 3, see also Norm of reaction
Pleiotropic effects, 182
Polygon
    formation, 33
    vertex, 47
    view of genotype-by-trait biplot, 125-126, 146
    view of line-by-environment biplot for barley yield, 203
    view of maize diallel, 222-223
Positive selection, 92
Potassium-use efficiency, 10
Powdery mildew, 229
Primary scores for environment, 42
Primary scores for genotype, 42
Principal component (PC) 1, 42, 97, 160
Principal component 2, 42, 160
Principal component analysis (PCA)
    model, 50
    scaling, 47
Principal components (PC), 36, 42, 96-97, 115, 160-164, 177
Principal coordinate analysis, 13
Principal coordinates, 42, 44
PRINCOMP procedure of SAS, 59, 61
Print option, 111
Probability of outperforming a check, 13
Proportional interaction, see Non-crossover interaction
Protein content, 135, 150
PSB (pink stem borer, Sesamia nonagrioides) resistance, 221-222, 225
Pyramiding genes, 17, 230
Pyrenophora teres (net blotch), 230

Q

Q-by-E (QE) interaction, see QTL-by-environment interaction

Index

QQE (QTL + QE) biplot, 120
QTL (quantitative trait loci) analyses, 206
QTL and GEI, 202
QTL associated with marker 7mwg502, 190
QTL for barley
    agronomic traits, 188-189
    heading date, 181
    height, 184
    kernel weight, 182, 187
    lodging score, 182, 185
    maturity, 182-183
    test weight, 182, 186
    yield, 180, 203-206
QTL for
    kernel weight, 192, 196
    loaf volume, 140
    plant height, 182
QTL identification, 103, 120, 177, 192
    across environments, 203
    via GGEbiplot, 175
QTL mapping, 17, 118, 162, 175
    data, 229
    methods, 159
    model, 160
    strategies, 159
QTL stability, 204
QTL with negative effects, 206
QTL with positive effects, 206
QTL-by-environment interaction (Q-by-E), 14, 17-18, 103, 119, 159, 204
Qualitative traits, 4
Quality components, 121
Quality traits, 137, 143
Quantitative trait loci (QTL), 14
Quantitative traits, 4

R

Race-specific resistance, 119, 230
Ranking of maize genotypes for wheat haploids, 252
Reading of data, 107
Real angles vs. predicted angles, 36
Reciprocal crosses, 207
Redundant parents, 221
Redundant traits, 142
Regional performance trials, 107
Regression
    approach, 12, 14
    model, 39, 50-52, 54
    on environmental means model, 93
    ratio, 12-13
Regression-based GGE biplot, 83
Relative efficiency, 19
Reliability of performance, breeding for, 18
REML, see Restricted maximum likelihood
Repeatability of stability statistic, 18
Representativeness, 91, 221
Resistance and stability, 245
Resistance to
    abiotic stresses, 158

FHB (Fusarium head blight), 208, 246
Fusarium, 243, 246
net blotch (Pyrenophora teres), 236-237
pink stem borer (PSB, Sesamia nonagrioides), 221-222
pre-harvest sprouting, 142
stress, 17
Resource allocation, 19, 156
Responsive cultivars, 77
Responsiveness
    of cultivar to environment, 67, 77
    of strains, 244
    to environments, 73
Restricted maximum likelihood (REML), 15
Restriction fragment-length polymorphisms (RFLP), 17
Reversal of dominance in squash, 4
Reverse sign of all testers option, 114, 122, 233, 239
Reverse Y-axis option, 112
Reversed disease scores, 243
RFLP, see Restriction fragment length polymorphisms
Row vectors, 26, 35
Row vectors and correlation, 35
r-square vs. t-test, 175
Run any subset by... option, 114
Run balanced subset by... option, 114
Run subset by removing testers option, 233

S

SASG x ESTAB, 14
SAS (Statistical Analysis System) programs for stability statistics, 14
SAS_STABLE, 16
Save as, 110
SCA (specific combining ability), 213, 218, 221, 225
SCA of parents, 207-208, 210
Scaling menu, 108, 115
Scaling method selection, 107
Scaling methods, merits of, 47
Screening of possible QTL, 177
SDS, see Sedimentation
Secondary scores for environment, 42
Secondary scores for genotype, 42
Sedimentation (SDS), 135
Select data file to run option, 105
Selecting breeding strategies, 229
Selection efficiency, 156
Selection for
    dough properties, 142
    efficiency, 9
    loaf volume, 142
    mean productivity, 18
    tolerance to stress, 9, 18
Selection strategies, 150
Sensitivity analysis, 11
Sesamia nonagrioides, see Pink stem borer
Shifted multiplicative model (SHMM), 13
Show
    entry/tester option, 111
    tester vectors option, 221
Show/hide linear map option, 112

Show/hide logo option, 111
Show/hide polygon option, 112, 122, 144, 215
Show/hide tester vectors option, 111, 127, 210, 221
Simple
    interval mapping, 159, 177, 188-189
    map-based interval mapping, 175
    plot, 70, 75, 87
    plots option, 116
    regression, 175
Simultaneous selection for yield and stability, 14
Single letter entry option, 111
Single letter tester option, 111
Single nucleotide polymorphisms (SNP), 17
Single-environment trials vs. MET, 63
Single-year MET, 39
Singular value decomposition (SVD), 36
Singular value partitioning, 39, 42, 47, 94, 107, 231
Singular value scaling, 85
Singular value scaling selection menu, 110
SNP, see Single nucleotide polymorphisms
Soybean cultivars, 150, 151
Soybean trials, 39-40, 79-82
Spatial stability, 9
Spatial variability, 9
Specific adaptation, 16
Specific combining ability (SCA), 207-208, 210, 213, 218, 221, 225
Specific resistance, 18
Specificity, 229
Squash, reversal of dominance in, 4
Stability, 86, 244-245, 251-253
Stability analyses, 9-11
Stability index, measures of, 40
Stability statistic heritability, 18
Stability statistic repeatability, 18
Stability statistics, 10-11, 14, 16
Stability statistics classification (Type 1, Type 2, and Type 3), 12
Stability variance, 14, 16
    and ecovalence relationship, 12
    for unbalanced data, 15
Stability, benefits of, 12
Stability, breeding for, 18
Stability, concepts of, 11, 16
Stability, heritability of, 88
Stability, importance of, 88
Stability, measures of, 83
Stability-related statistics, 14
Stability-yield integration, 13
Stable cultivar selection, 92
STABLE, a BASIC program, 14
STABSAS, a SAS program, 14
Standard
    coordinates, 42, 44, 47
    deviation, 57, 91
    deviation-transformed data, 57
    mapping methods, 162
Start/Help/Exit option, 105
Static concept of stability, 11, see also Biological concept; Type 1 stability
Statistics and agriculture, 228

Stepwise regression, interactive, 120
Strategy for QTL identification, 175
Stratified rank analysis, 14
Sub-optimal levels of stress, 9
Subset panel, 115
Sun-red gene in maize, 4
Superiority measure, 13, 14
Super-optimal levels of stress, 9
Superoxides, 9
Susceptibility to
    boron deficiency, 9
    Fusarium, 242, 245
    magnesium deficiency, 9
    net blotch (Pyrenophora teres), 235, 237
SVD (singular value decomposition), 42, 47
    model, 39
    of environment-centered data, 65
    of standardized data, 122, see also Transformation of data
Symbiosis, 247
Symmetrical scaling, 44, 115
Synergism between species, 247
System capacity, 158
Systems understanding, 121, 150, 192

T

Table format, 105
Tai's _, 14
Technology contribution to yield, 7
Temporal stability, 9
Temporal variability, 9
Test environment evaluation, 64, 89
Test site-by-physical factor analysis, 103
Tester crosses, 214
Tester eigenvectors, 231
Tester vector view, 212, 221, 226
Tester/entry switch roles option, 231
Tester-by-entry interaction, 248
Tester-centered
    data, 231, 248, 251
    PCA, 208, see also Principal component analysis
Tester-focused scaling, 44-45, 163-171, 201-203
Testers and parents, 216
Test-location evaluation, 63
Three-way data visualization, 120
Tolerance to
    air pollutants, breeding for, 10
    biotic stresses, 158
    pink stem borer (PSB, Sesamia nonagrioides), 220
    stress, 17-18
Trait-by-environment factor interaction, 94, 97-98
Trait-focused scaling, 130-134, 136, 139, 151-155
Transformation of data, 39, 56
Transgenic germplasm, 17
Triticale, 10
Two-dimensional biplot, 37
Two-factor approach, 104
Two-rank matrix, 24

Index

Type 1 concept of stability, 12, see also Biological concept; static concept of stability
Type 2 concept of stability, 16, see also Agronomic concept; dynamic concept of stability
Type 3 stability defined, 12
Type 4 stability, 12, 14
Type II errors, 14

U

Unbalanced data, stability variance for, 15
Univariate GEI analyses, 13

V

Variance analysis, 93
Vector view of
 genotype-by-trait biplot, 130-131, 148
 GGE biplot, 89-90, 92, 139-140
 line-by-environment biplot, 202
Vertex cultivars, 73, 76, 78
Vertex of polygon, 47
Vertical resistance, 103, 119, 229
View menu, 108, 111
Viscosity, 144
Visual comparison of
 column/row elements, 29-30
 two cultivars, 128-129
Visual identification of largest column/row factor, 32
Visualization menu, 108, 111
Visualizing cultivar defects/merits, 121
Visualizing interrelationship among traits, 121
Visualizing various traits of a cultivar, 147
Vitreous kernels, 142
Vomitoxin, 239, see also Deoxynivalenol (DON)

W

Water-use efficiency, 9
Weather variables, contributions to GEI by, 15
Wheat
 as entries, 251, 253
 as testers, 252
 cultivar components, 156
 end-use quality, breeding for, 121
 FHB (Fusarium head blight), 211, 229-230, 239, 241
 genotype-by-maize genotype interaction, 119
 haploid embryo formation, 247, 251
 quality traits, 138
Wheat-by-maize
 hybridization, 119, 247, 250, 253
 interaction, 247, 251-252
Wheat-doubled haploid production, 247
Which-won-where
 issue, 94
 pattern, 47, 49-50, 57, 63-64, 73, 77, 83
Windows GGEbiplot software, 103
Winter wheat
 region of Ontario, 79
 resistance to FHB (Fusarium head blight), 207
 trials, 39-40, 43, 45-46, 48-55, 57-59, 62, 64-66, 84, 94, 96, 106
 yield, 97-99
Within-environment
 standard deviation-scaled data, 57
 standard deviation-standardized model, 56
 standard error-standardized model, 56
Within-tester standard deviation-standardized model, 160

Y

Yield components, 121
Yield-associated stability statistics, 13
Yield-stability integration, 13
Yield-stability-related statistics, 13-14, 88
YS_i (Kang's yield-stability) statistic, 88

Z

Zea mays L., see Maize